Limit Theorems for Random Fields with Singular Spectrum

Mathematics and Its Applications

Volume 465

Limit Theorems for Random Fields with Singular Spectrum

by

Nikolai Leonenko

Department of Mathematics,
University of Kiev,
Kiev, Ukraine

KLUWER ACADEMIC PUBLISHERS
DORDRECHT / BOSTON / LONDON

A C.I.P. Catalogue record for this book is available from the Library of Congress.

ISBN 0-7923-5635-7

Published by Kluwer Academic Publishers,
P.O. Box 17, 3300 AA Dordrecht, The Netherlands.

Sold and distributed in North, Central and South America
by Kluwer Academic Publishers,
101 Philip Drive, Norwell, MA 02061, U.S.A.

In all other countries, sold and distributed
by Kluwer Academic Publishers,
P.O. Box 322, 3300 AH Dordrecht, The Netherlands.

Printed on acid-free paper

Printed in the Netherlands.

Contents

Preface

This book is devoted to an investigation of the basic problems of the theory of random fields which are characterized by certain singular properties (e. g., unboundedness, or vanishing) of their spectral densities. These random fields are called, the random fields with singular spectrum, long-memory fields, random fields with long-range dependence, fields with slowly decaying correlations or strongly dependent random fields by various authors. This phenomenon has been observed empirically by many scientists long before suitable mathematical models were known. The methods and results differ significantly from the theory of weakly dependent random fields.

The first chapter presents basic concepts of the spectral theory of random fields, some examples of random processes and fields with singular spectrum, Tauberian and Abelian theorems for the covariance function of singular random fields.

In the second chapter limit theorems for non-linear functionals of random fields with singular spectrum are proved.

Chapter 3 summarizes some limit theorems for geometric functionals of random fields with long-range dependence.

Limit distributions of the solutions of Burgers equation with random data via parabolic and hyperbolic rescaling are presented in chapter 4.

And chapter 5 presents some problems of statistical analysis of random fields with singular spectrum.

I would like to thank the editor, Michiel Hazewinkel, for his support.

I am grateful to the following students and colleagues: I. Deriev, A. Olenko, K. Rybasov, L. Sakhno, M. Sharapov, A. Sikorskii, M. Šilac–Benšic.

I would also like to thank V.Anh, O. Barndorff-Nielsen,Yu. Belyaev, P. Biler, R. Bhattacharya, C. Boldrighini, R. Bradley, V. Buldigin, A. Bulinsky, Yu. Daletski,Yu. Davydov, E. Donskoi, M. Farge, R. Gill, L. Giraitis, M. Goria, D. Gurarie, A. Ivanov, M. Kelbert, Yu. Kondratiev, P. Kotelenez, Yu. Kozachenko, Li Zhanbing, M. Maejima, Yu. Mishura, A. Novikov, G. Peskir, V. Petrov, M. Priestley, E. Orsingher, N. Portenko, Yu. Prokhorov, Yu. Samoilenko, K. Schneider, O. Seleznev, A. Shiryaev, Yu. Suhov,V. Statulevičius, D. Surgailis, M. Taqqu, G. Terdik, E. Valkeila, A. Veretennikov, W. Woyczynski for many interesting discussions on random fields.

I consider it my pleasent duty to express here my deep gratitude to A. Skorokhod.

I learned theory of random fields from M.Yadrenko. I thank him now for all the help and encouragement he has provided.

I would like to thank my son Eugene and daughter Anna for their support. Finally, many thanks to my wife Olga for her understanding and help.

N. Leonenko
Kiev, Ukraine
April 1998.

Chapter 1

SECOND-ORDER ANALYSIS OF RANDOM FIELDS

1.1 Basic Concepts and Notation

This section presents basic concepts and statement that have important applications in this book.

1. Sets

Denote by R^n a real Euclidean space of dimension $n \geq 1$. Let $\langle x, y \rangle = x_1 y_1 + \ldots + x_n y_n$ be the scalar product of vectors $x = (x_1, \ldots, x_n)'$, $y = (y_{1,\ldots,} y_n)' \in R^n$. Then $\|x\| = \sqrt{\langle x, x \rangle}$ is the norm (length) of the vector x, and $\rho_{xy} = \|x - y\|$ is the distance between **x** and y in R^n. The symbol $dx = dx_1 \ldots dx_n$ denotes an element of Lebesgue measure in R^n. Denote by

$$R_+^n = \{x \in R^n : x_i \geq 0, \ i = 1, \ldots, n\}$$

the non-negative octant of R^n, by $Z^n, n \geq 1$ the n-dimensional integer lattice in R^n, and by $Z_+^n = Z^n \cap R_+^n$.

Let write

$$\Pi\,[a, b) = \{x \in R^n : a_i \leq x_i < b_i, \ i = 1, \ldots, n\}$$

be a parallelepiped in R^n. We shall write:

$$\Pi\,[0, b) = \Pi\,(b), \quad \Pi = \Pi\,[-\pi, \pi) = [-\pi, \pi)^n.$$

A ball and a sphere in R^n of radius r and with center at the origin are defined as the sets

$$v(r) = \{x \in R^n : \|x\| < r\}$$

and

$$s(r) = \{x \in R^n : \|x\| = r\},$$

respectively. If necessary we write $s_{n-1}(r)$ to specify the dimension of the sphere.

We denote by $|\Delta|$ the Lebesgue measure of a measurable set $\Delta \subset R^n$. For example,

$$|v(1)| = \frac{2\pi^{n/2}}{n\Gamma\left(\frac{n}{2}\right)} \tag{1.1.1}$$

is the volume of a unit ball in R^n. We denote by

$$|s(1)| = |s_{n-1}(1)| = \frac{2\pi^{n/2}}{\Gamma(n/2)} \tag{1.1.2}$$

is the area of surface of the sphere $s_{n-1}(1)$ when there is no danger of a misunderstanding.

Along with the Cartesian coordinate system in R^n we shall use the spherical coordinate system (r, u), where $r \geq 0$, $u = (\phi, \phi_1, \ldots, \phi_{n-2}) \in s_{n-1}(1)$, $0 \leq \phi < 2\pi$, $0 \leq \phi_j < \pi$, $j = 1, \ldots, n-2$, $n \geq 2$ which is related to the Cartesian system by the formulas: $x = (x_1, \ldots, x_n)' = (r, u)$, where

$$\begin{aligned} x_1 &= r \sin\phi_{n-2} \ldots \sin\phi_1 \sin\phi, \\ x_2 &= r \sin\phi_{n-2} \ldots \cos\phi, \ldots \\ x_{n-1} &= r \sin\phi_{n-2} \cos\phi_{n-3}, \\ x_n &= r \cos\phi_{n-2}. \end{aligned} \tag{1.1.3}$$

An element of the Lebesgue measure on the sphere $s_{n-1}(r)$ will be denoted by

$$d\sigma(x) = r^{n-1} \sin^{n-2}\phi_{n-2} \cdots \sin\phi_1 \, d\phi_1 \cdots d\phi_{n-2} \, d\phi \tag{1.1.4}$$

Denote by $\mathbf{C}$ a complex Euclidean space of on dimension one. Let $\bar{z}$ be the conjugate complex of $z \in \mathbf{C}$, then $|z|^2 = z\bar{z}$.

Denote by $\mathbf{L}_p(T), T \subseteq R^n$, a Banach space of a real or complex measurable function $f(x), x \in T$ such that

$$\|f\|_p = \left[\int_T |f(x)|^p \, dx \right]^{1/p} < \infty, \quad 1 \le p < \infty.$$

If $f(x) \in L_1(R^n)$, then the function $\tilde{f}(\|x\|) = \frac{1}{|s(1)|} \int_{s(1)} f(rx')dx'$ is called by radial part of function f, where $r = \|x\|$, $x' = x/r$ (for $x \neq 0$). The function $\tilde{f}$ is radial, that is $\tilde{f}$ depends on $r = \|x\|$ only. For $n > 1$ the function $\tilde{f}$ is radial if and only if $\tilde{f}(gx) = \tilde{f}(x)$ for all rotation g (see, below the section 1.4) and all $x \in R^n$. For radial function $\tilde{f}$ we write $\tilde{f}(x) = \tilde{f}(\|x\|)$ when is no danger of misunderstanding.

Let $\mathbf{L}_p(Z^n)$ be a Banach space of a function $f(x), x \in Z^n$, such that

$$\|f\|_p = \left[\sum_{x \in Z^n} |f(x)|^p \right]^{1/p} < \infty, \quad 1 \le p < \infty.$$

Let $\mathbf{L}_2(X, w(u)\, du)$ be a Hilbert space of measurable functions $f : X \to R^1$ satisfying the inequality

$$\int_X f^2(u) \; w\; (u) \; du < \infty$$

where $w : X \to R^1_+$ is a weight function such that $\int_X w(u)\; du = 1$.
We denote by

$$\Delta = \frac{\partial^2}{\partial x_1^2} + \cdots + \frac{\partial^2}{\partial x_n^2}$$

the Laplace operator, and by

$$\nabla = (\frac{\partial}{\partial x_1}, \cdots, \frac{\partial}{\partial x_n})',$$

the gradient.

2. Random elements

Let $(\Omega, \mathfrak{F}, P)$ be a complete probability space. A random element taking on values in a measurable space $(X, \mathfrak{B})$ is a mapping $\xi : \Omega \to X$ such that $\{\omega : \xi(x) \in A\} \in \mathfrak{F}$ for any $A \in \mathfrak{B}$. The measure $P_\xi(A) = P\{\xi \in A\}$

defined on the σ-algebra $\mathfrak{B}$ is called the distribution of ξ. If X is a metric space the σ-algebra of Borel sets is usually taken as $\mathfrak{B}$. For $X \subseteq R^m$ we denote the Borel σ-algebra by $\mathfrak{B}(X)$. A random element taking on values in $(R^m, \ \mathfrak{B}(R^m))$ is called a random variable if $m = 1$ and a random vector if $m \geq 2$. In either case we shall use the abbreviation 'r.v.'.

To define $P_\xi(A)$ for a r.v. $\xi = (\xi_1, \dots, \xi_m)' \in R^m$ it suffices to define the distribution function (d.f.)

$$F(x) = F_\xi(x_1, \dots, x_m) = P\{\xi_1 \leq x_1, \dots, \xi_m \leq x_m\}, \ x \in R^m.$$

A distribution $P_\xi(A)$ is said to be absolutely continuous if $P_\xi(A) = \int_A p(x)dx$, $A \in \mathfrak{B}(R^m)$. The function $p(x) = p(x_1, \dots, x_m)$, $x \in R^m$, is called the density function of the distribution of r.v. ξ. Clearly, $\int_{R^m} p(x)dx = 1$ and for almost all $x \in R^m$, $p(x) \geq 0$, $p(x) = (\partial^m / \partial x_1 \dots \partial x_m) \ F(x_1, \dots, x_m)$.

Let $g : R^m \to R^1$ be a Borel function. The mathematical expectation of the function $g(\xi)$ of the r.v. $\xi \in R^m$ is computed as the Lebesgue integral with respect to measure P, $Eg(\xi) = \int_\Omega g(\xi(\omega))P(d\omega)$ or as the m-fold Lebesgue -Stieltjes integral $Eg(\xi) = \int_{R^m} g(x)dF(x)$, provided this integral exists. In the case when the r.v.has a density function, $Eg(\xi) = \int_{R^m} g(x)p(x)dx$. For a r.v. $\xi = (\xi_1, \dots, \xi_m)' \in R^m$ by definition we set $E\xi = (E\xi_1, \dots, E\xi_m)'$. The symbols $cov(\xi, \eta) = E\xi\eta - E\xi E\eta$ and $\text{var}\,\xi = E\xi^2 - (E\xi)^2$ will be used to denote the covariance of the r.v.s ξ and η and the variance of the r.v. The matrix $B = (cov \ (\xi_i, \xi_j))_{i,j=1,\dots,m}$ is said to be the covariance matrix of r.v. $\xi = (\xi_1, \dots, \xi_m)' \in R^m$. For a r.v. ξ with $E\xi^2 < \infty$ Chebyshev's inequality holds: for any $\varepsilon > 0$, $P\{|\xi - E\xi| > \varepsilon\} \leq (\text{var} \ \xi)/\varepsilon^2$.

Let $\xi = \eta + i\zeta$ be a complex r.v., where $\eta = Re\,\zeta$ and $\zeta = Im\,\xi$ are real r.v.s . Then $\underline{E\xi = E\eta + iE\zeta}$, $|E\xi| \leq E\,|\zeta|$. The symbols $cov \ (\xi_1, \xi_2) = E[\xi_1 - E\xi_1] \ \overline{[\xi_2 - E\xi_2]}$ and $var \ \xi = cov \ (\xi, \xi)$ will be used to denote the covariance of the complex r.v.s . ξ_1, ξ_2, ξ.

Denote by $\mathbf{L}_2(\Omega)$ a Hilbert space of real or complex r.v. ξ with $E\xi = 0$ and scalar product $< \xi, \eta >= E\xi\bar{\eta}$. The characteristic function of real r.v. $\xi \in R^q$, $q \geq 1$ is defined as the function

$$f_\xi(t) = Ee^{i<t,\xi>} = \int_{R^q} e^{i<t,x>} dF_\xi(x), \quad t \in R^q.$$

The r.v. $\xi \in R^1$ is called Gaussian (or normal) if its characteristic function has the form

$$f_\xi(t) = Ee^{it\xi} = \exp\left\{ita - \frac{t^2\sigma^2}{2}\right\}, \quad t \in R^1, \quad a \in R^1, \quad \sigma^2 \geq 0$$

In the non-singular case ($\sigma^2 > o$) the Gaussian r.v. ξ has the density function

$$\phi_{a,\sigma^2}(u) = \frac{1}{\sqrt{2\pi}\sigma} \exp\left\{-\frac{(u-a)^2}{2\sigma^2}\right\}, \quad u \in R^1, \tag{1.1.5}$$

with $a = E\xi$, $\sigma^2 = \operatorname{var}\xi$. If $a = 0$, $E\xi^{2j} = (2j-1)!!\sigma^{2j}$, $E\xi^{2j+1} = 0$, $j = 1, 2, \ldots$. In what follows,a Gaussian r.v. $\xi \in R^1$ with density (1.1.5) will be denoted by $\mathcal{N}(a, \sigma^2)$ and the corresponding $d.f.$ by $\Phi_{a,\sigma^2}(x)$, $x \in R^1$. A r.v. $\xi \in R^q$, $q \geq 1$, is called Gaussian (or normal) if its characteristic function is of the form

$$f_\xi(t) = \exp\{i < t, a > -\frac{1}{2} < Bt, t >\}, \quad t \in R^q \tag{1.1.6}$$

where $a \in R^q$ and $B = (b_{ij})_{i,j=1,\ldots,q}$ is a non-negative definite symmetric matrix. If $B > 0$, the r.v. ξ has the density function

$$\phi_{a,B}(u) = (2\pi)^{-q/2}(\det\ B)^{-1/2} \exp\{-\frac{1}{2} < B^{-1}(u-a), u-a >\}, u \in R^q \tag{1.1.7}$$

with $a = (a_1, \ldots, a_q)'$, $a_j = E\xi_j, j = 1, \ldots, q$; $b_{ij} = \operatorname{cov}(\xi_i, \xi_j)$, $i, j = 1, \ldots, q$, are the elements of the covariance matrix of the r.v. ξ

In what follows, a Gaussian r.v. $\xi \in R^q$, $q > 1$ will be denoted by $\mathcal{N}_q(a, B)$. A system of r.v.s $\xi_\alpha \in R^1$, $\alpha \in \mathcal{U}$, where $\mathcal{U}$ is a set of indices is said to be Gaussian if for any integer $l \geq 1$ and any $\alpha_1, \ldots, \alpha_l \in \mathcal{U}$ the r.v. $(\xi_{\alpha_1}, \ldots, \xi_{\alpha_2})'$ is Gaussian (any linear combination $k_1\xi_{\alpha_1} + \cdots + k_l\,\xi_{\alpha_l}$, $k_i \in R^1$, $i = 1, \ldots, l$ is a Gaussian r.v.).

A family of r.v.s $\{\xi_T\}$, $T > 0$, converge in probability to the r.v. ξ if for any $\varepsilon > 0$, $P\,\{\|\xi_T - \xi\| > \varepsilon\} \to 0$ as $T \to \infty$. We denote this convergence as $T \to \infty$ by $\xi_T \xrightarrow{P} \xi$. A family $\{\xi_T\}$, $T > 0$, is said to converge almost surely to a r.v. ξ if there exists a measurable set $N \in \mathfrak{F}$ such that $P(N) = 0$ and $\xi_T(\omega) \to \xi(\omega)$ as $T \to \infty$ for every $\omega \bar{\in} N$. A family of r.v.s $\{\xi_T\}, T > 0$ is said to converge in mean square to the r.v. ξ ($\xi_T \overset{\text{m.s.}}{\to} \xi$) if $E\,\|\xi_T - \xi\|^2 \to 0$ as $T \to \infty$.

We denote this convergence as $\xi_T \overset{\text{a.s.}}{\to} \xi$ as $T \to \infty$. A family of r.v.s $\{\xi_T\}$, $T > 0$, is said to converge in distribution (law) to r.v. ξ with $d.f.$

F if $\lim_{T\to\infty} F_T(x) = F(x)$ at all continuity points x of $F(.)$. We denote this convergence by $\xi_T \xrightarrow{D} \xi$.

The following implications hold: $\xi_T \xrightarrow{\text{a.s.}} \xi \Longrightarrow \xi_T \xrightarrow{P} \xi \Longrightarrow \xi_T \xrightarrow{D} \xi$ and $\xi_T \xrightarrow{\text{m.s.}} \xi \Longrightarrow \xi_T \xrightarrow{P} \xi$.

A fundamental result that we shall have several occasions to owed is due to Slutsky (1925).

Lemma 1.1.1. (Slutsky's lemma). Let $\{\xi_T\}$ and $\{\eta_T\}$ be families of r.v.s such that $\xi_T \xrightarrow{D} \xi$ and $\eta_T \xrightarrow{P} c$,where $-\infty < c < \infty$.

Then

$$\xi_T + \eta_T \xrightarrow{D} \xi + c: \quad \xi_T\eta_T \xrightarrow{D} c\xi, \quad \xi_T/\eta_T \xrightarrow{D} \xi/c$$

if $c \neq 0$.

It is easy to see that first statement of Slutsky's Lemma can be extended easily to a family of k-dimensional r.v.s. The third statement is true for k-dimensional r.v.s ξ_T and a one-dimensional r.v. η_T.

We shall now formulate the Borel–Cantelli Lemma. For more details and proofs,see Gikhman and Skorokhod (1974).

Lemma 1.1.2. (Borel–Cantelli Lemma). Let $\{A_n,\ n \geq 1\}$ be a sequence of events with $\sum_{n=1}^{\infty} P(A_n) < \infty$.

Then $P(A_n$ occurs infinity often$) = 0$. Conversely, if the events $\{A_n, n \geq 1\}$ are independent and $\sum_{n=1}^{\infty} P(A_n) \neq \infty$, then $P(A_{n.}$ occurs infinity often$) = 1$.

One can study the convergence in distribution for multivariate vectors via the univariate case in the light of the following proposition.

Lemma 1.1.3. (Cramer–Wold). A family of k-dimensional r.v.s $\xi_T = (\xi_{T_1}, \ldots \xi_{T_k})'$ converge in distribution to a k-dimensional r.v. $\xi = (\xi_1, \ldots, \xi_k)'$ if and only if, for every $\lambda = (\lambda_1, \ldots, \lambda_k)' \in R^k$,

$$< \lambda, \xi_T >= \sum_{i=1}^{k} \lambda_i \xi_{Ti} \xrightarrow{D} \sum_{i=1}^{k} \lambda_i \xi_i =< \lambda, \xi >$$

as $T \to \infty$.

The r.v.s $\xi_1, \ldots, \xi_m$ are said to be independent if $P\{\xi_1 \in B_1, \ldots, \xi_m \in B_m\} = P\{\xi_1 \in B_1\} \cdots P\{\xi_m \in B_m\}$ for any $B_i \in \mathfrak{B}(R^k),\ i = 1, \ldots, m$.

A sequence of r.v.s $\xi_i, i = 1, 2, \dots$, is said to be a sequence of independent r.v.s (i.i.d. r.v.) if the r.v.s $\xi_1, \dots, \xi_m$ are independent for any $m \geq 1$.

The symbol $\chi(A)$ will denote the indicator function of event A, that is,

$$\chi(A) = \begin{cases} 1, & \omega \in A, \\ 0, & \omega \notin A. \end{cases}$$

Let δ^r_m be the Kronecker symbol, that is

$$\delta^r_m = \begin{cases} 1, & r = m, \\ 0, & r \neq m. \end{cases}$$

3. Random fields

Let $T \subseteq R^n$ be a set. A random field is defined as a function $\xi(\omega, x) : \Omega \times T \to R^m$ such that $\xi(\omega, x)$ is a r.v. for each $x \in T$. A random field will also be denoted as $\xi(x)$, $x \in R^n$. If $n = 1$ $\xi(x)$ is a random process. If $T \subseteq Z^n$ $\xi(x)$ is a random field with discrete parameter; if $T \subseteq R^n$ $\xi(x)$ is a random field with continuous parameter. For $n > 1$ and $m = 1, \xi(x), x \in T$ is a scalar random field; for $m > 1$ it is a vector random field.

In particular, if $\xi(x)$, $x \in T$, is a Gaussian system of r.v.s, the (scalar) field $\xi(x)$, $x \in T$, is said to be Gaussian.

Finite-dimensional distributions of the random field $\xi(x) \in R^p, x \in R^n$, are defined as a set of functions

$$F_{x^{(1)},\dots,x^{(r)}}(B_1, \dots, B_r) = P\{\xi(x^{(1)}) \in B_1, \dots, \ \xi(x^{(r)}) \in B_r\}, \qquad (1.1.8)$$

where $B_i \in \mathfrak{B}(R^p)$, $i = 1, \dots, r$; $r = 1, 2, \dots$.

We shall write $\xi(x) \stackrel{D}{=} \eta(x)$, $x \in T$, if finite-dimensional distributions of a random fields $\xi(x)$, $x \in T$, and $\eta(x)$, $x \in T$, coincide ($\xi(x)$ and $\eta(x)$ are stochastically equivalent).

For fixed ω the function $\xi(x)$, $x \in T$, is called a realization of the random field or a sample function.

The finite-dimensional distributions (1.1.8) possess the following properties:

1) functions $F_{x^{(1)},\dots,x^{(r)}}(B_1, \dots, B_r)$ are distributions of r.v.s with fixed $x^{(1)}, \dots, x^{(r)}$.

2) $F_{x^{(1)},\dots,x^{(r)}}(B_1, \dots, B_r) = F_{x^{(i_1)},\dots,x^{(i_r)}}(B_{i_1}, \dots, B_{i_r})$ for any permutation $i_1, \dots, i_r$ of indices $i_1, \dots, i_r$;

3) If $\xi(x) \in R^p$, then

$$F_{x^{(1)},\dots,x^{(r)}}(B_1,\dots,B_{r-1},R^p) = F_{x^{(1)},\dots,x^{(r-1)}}(B_1,\dots,B_{r-1}).$$

Theorem 1.1.1. (Kolmogorov (1936)).

Let functions $F_{x^{(1)},\dots,x^{(r)}}(B_1,\dots,B_r)$ are defined for $x^{(1)},\dots,x^{(r)} \in T$, $B_1,\dots,B_r \in \mathfrak{B}(R^p)$. Then $F_{x^{(1)},\dots,x^{(r)}}(B_1,\dots,B_r)$ are the finite-dimensional distributions of some random field if and only if the properties 1)-3) are fulfilled.

For the proof of theorem 1.1.1. see Kolmogorov (1936) or Gikhman and Skorokhod (1974).

A vector random field $\xi(x) = (\xi_1(x),\dots,\xi_p(x))'$, $x \in T$ is called Gaussian if for any $x^{(1)},\dots,x^{(r)} \in T$ the joint distribution of r.v.s $(\xi_1(x^{(1)}),\dots, \xi_p(x^{(1)}),\dots, \xi_1(x^{(r)}),\dots, \xi_p(x^{(r)}))'$ are Gaussian.

Finite-dimensional distributions of a Gaussian field $\xi(x)$, $x \in T$ are defined its mean valued $a(x) = (a_1(x),\dots,a_p(x))' = E\xi(x)$ and matrix-valued covariance function

$$\begin{aligned} B(x,y) &= E[\xi(x) - a(x)]\ [\xi(y) - a(y)]' \\ &= (E\xi_i(x)\ \xi_j(y) - a_i(x)\ a_j(y))_{i,j=1,\dots,p}. \end{aligned}$$

A real random field $\xi(x)$, $x \in T$, having a finite second moment is said to be mean-square continuous at point x_0 if $E\,|\xi(x) - \xi(x_0)|^2 \to 0$ as $\rho_{xx_0} \to 0$. If this relation holds for any $x_0 \in T$, the field $\xi(x)$, $x \in T$, is called mean square continuous on T. In order for a field $\xi(x)$, $x \in T$, having $E\xi^2(x) < \infty$ to be mean square continuous it is necessary and sufficient that the function *cov* $(\xi(x), \xi(y))$ be continuous along the diagonal $\{(x,y) \in T \times T : x = y\}$. Clearly, a mean square continuous field $\xi(x), x \in T$, is stochastically continuous, that is, for any $x_0 \in T$, $P\{|\xi(x) - \xi(x_0)| > \varepsilon\} \to 0$ for any $\varepsilon > 0$ as $\rho_{xx_0} \to 0$.

Let $T \in \mathfrak{B}(R^n)$. A random field $\xi(\omega,x) : \Omega \times T \to R^m$ is said to be measurable if for any $A \in \mathfrak{B}(R^m)$, $\{(\omega,x) : \xi(\omega,x) \in A\} \in \mathfrak{F} \times \mathfrak{B}(T)$. If $\xi(x)$, $x \in T$, is stochastically continuous on T, a measurable random field $\tilde{\xi}(x) \stackrel{D}{=} \xi(x)$, $x \in T$,exists.

The next theorem will be used throughout this book without special reference.

Theorem 1.1.2. (Fubini–Tonelli).

Let $\xi(\omega, x) : \Omega \times R^n \to R^1$ be a measurable random field. The following assertions are valid.

1. Sample function $\xi(x)$ are a.s. $\mathfrak{B}(R^m)$–measurable function of $x \in R^n$.

2. If $E\xi(x)$ exists for all $x \in R^n$, then $m(x) = E\xi(x)$ is a $\mathfrak{B}(R^n)$–measurable function of $x \in R^n$.

3. If $\Delta \in \mathfrak{B}(R^m)$ then the field $\xi(x)$ is integrable over $\Omega \times \Delta$ if at least one of the integrals $\int_\Delta \{E|\xi(x)|\}\, dx$ and $E\{\int_\Delta |\xi(x)|\, dx\}$ is finite; in this case the two-fold integral $(\int_\Delta E\xi(x)dx)$ is identical to the integral $\int_\Delta \{E\xi(x)\}\, dx$ and $E\{\int_\Delta \xi(x)\, dx\}$.

For the proof see, for example, Gikhman and Skorokhod (1974).

If the function $g_t(\cdot)$ is integrable over Δ and the measurable scalar field $\xi(x)$, $x \in R^n$, possesses the property $\sup\{E|\xi(x)|,\ x \in R^n\} < \infty$, then by Theorem 1.1.2., the integrals

$$s_t = \int_\Delta g_t(x)\xi(x)\, dx \tag{1.1.9}$$

are a.s. finite any $A \in \mathfrak{B}(R^n)$. In the same manner integrals of type (1.1.9) may be defined for vector field or vector function $g_t(x)$. The integral $\int_{s(r)} g_t(x)\xi(x)d\sigma(x)$ may be defined by similar way.

The field $\xi(\omega, x) : \Omega \times R^n \to R^1$ is said to be separable with respect to the set $I \subset T$ if I is countable and dense in T and there exists a set $N \in \mathfrak{F}, P(N) = 1$, such that for any ball $v(r) \subset R^n$

$$\{\omega : \sup_{x \in I\, \cap\, v(r)} \xi(x) = \sup_{x \in T\, \cap\, v(r)} \xi(x)\} \supset N,$$

$$\{\omega : \inf_{x \in I\, \cap\, v(r)} \xi(x) = \inf_{x \in T\, \cap\, v(r)} \xi(x)\} \supset N$$

For any random field $\xi(x)$, $x \in T$, there exists a separable field $\tilde{\xi}(x) \stackrel{D}{=} \xi(x)$, $x \in T$. Henceforth only measurable separable fields will be considered.

If F is some measure on a measurable space $(\Lambda, \mathfrak{B}(\Lambda))$, complex-valued random measure $Z(\Delta) = Z(\omega, \Delta) : \Omega \times \mathfrak{B}(\Lambda) \to \mathbf{C}$ is called of orthogonal random measure with the control measure F, if

1) $Z(\Delta) = \overline{Z(-\Delta)}$, where $-\Delta = \{\lambda : -\lambda \in \Delta\}$;

2) $Z(\bigcup_{j=1}^{r} \Delta_j) = \sum_{j=1}^{r} Z(\Delta_j)$ a.s. for any disjoint $\Delta_1, \ldots, \Delta_r \in \mathfrak{B}(\Lambda)$;

3) $EZ(\Delta) = 0$, $E|Z(\Delta)|^2 < \infty$, $EZ(\Delta_1)\overline{Z(\Delta_2)} = F(\Delta_1 \cap \Delta_2)$, $\Delta_1, \Delta_2 \in \Lambda$.

Theorem 1.1.3. (Karhunen (1947)).

Let $\xi(x), x \in T \subseteq R^n$ be a real random field with $E\xi(x) = 0, \quad E\,|\xi(x)|^2 < \infty$.If the covariance can be expressed as

$$B(x,y) = E\,\xi(x)\,\xi(y) = \int\limits_{\Lambda} f(x,\lambda)\,\overline{f(y,\lambda)}\,F(d\lambda), \quad f(x,\cdot) \in \mathbf{L}_2(\Lambda),\ x,y \in T \tag{1.1.10}$$

then there exists an orthogonal random measure Z on Λ with the control measure F such that

$$\xi(x) = \int\limits_{\Lambda} f(x,\lambda)\ Z(d\lambda), \tag{1.1.11}$$

where $E\,|Z(\Delta)|^2 = F(\Delta), \quad \Delta \in \mathfrak{B}(\Lambda)$, and the stochastic integral in (1.1.11) is viewed as on $\mathbf{L}_2(\Lambda)$ integral with the control measure F (see, Grenander (1981),p.57).

For proof of Theorem 1.1.3. see Karhunen (1947), Gikhman and Skorokhod (1972), pp.293-294 or Grenander (1981), pp.58-62.

We shall need one classical result on integral kernels (see Grenander(1981), p.p. 62-64).

Theorem 1.1.4. (Merser (1909)).

Consider a symmetric nonnegative definite and continuous function $B(s,t)$ on $[a,b]^2 \subset R^2$ and the integral equation

$$\lambda\phi(s) = \int\limits_a^b B(s,t)\ \phi(t)\ dt$$

The eigenvalues $\lambda_1, \lambda_2, \ldots$ and associated normalized eigenfunctions $\phi_1, \phi_2, \ldots$ satisfy

$$B(s,t) = \sum_{\nu=1}^{\infty} \lambda_\nu\ \phi_\nu(s)\ \phi_\nu(t)$$

with absolute and uniform convergence.

4. Constants

We note that the symbols k, k_i , etc. denote positive constants whose particular values are usually non-essential. The numbering of such constants holds only within the relevant subsection. Symbols $c, c_i(n,\alpha)$, etc., will be used to denote constants essential for the exposition. The numbering of such constants holds within the book. Assumptions are numbered A,A',A",B,C, etc. They are valid (without special mention) throughout the relevant chapter.

1.2 Elements of Spectral Theory of Random Fields

We present an overview of basic concepts of the spectral theory of random fields.

1. Discrete parameter homogeneous random fields

We generalize the concept of stationary random process by introducing the concept of homogeneous random field on n-dimensional lattice Z^n.

A real random field $\xi(x),\ x \in Z^n$ satisfying $E\xi^2(x) < \infty$ is called homogeneous (in the wide sense) if its mathematical expectation $m(x) = E\xi(x) = k = \text{const}$ and the covariance function $B(x,y) = cov\ (\xi(x)\xi(y)) = B(x-y)$ depends only on the difference $x - y$.

It means that its mathematical expectation $m(x)$ and covariance function $B(x,y)$ are invariant with respect to the Abelian group $G = (\mathbf{Z}^d, +)$ of shifts τ in Z^n, that is $m(x) = m(x+\tau),\quad B(x,y) = B(x+\tau,\ y+\tau)$ for any $x, y, \tau \in Z^n$. The covariance function $B(x-y)$, $x-y \in Z^n$, of a homogeneous random field is a non-negative definite kernel on $Z^n \times Z^n$, that is, for any $x^{(j)} \in R^n$, $z_j \in \mathbf{C}, j = 1, \ldots, r$,

$$\sum_{i,j=1}^{r} B(x^{(i)} - x^{(j)}) z_i \bar{z}_j \geq 0$$

Gerglotz' Theorem gives the following spectral representation of the covariance function:

$$B(x) = \int_{\Pi} e^{i<x,\lambda>} F(d\lambda) = \int_{\Pi} \cos < x, \lambda > F(d\lambda), \quad \Pi = [-\pi, \pi)^n, \quad (1.2.1)$$

where F is a finite measure on a measurable space $(\Pi, \mathfrak{B}(\Pi))$, which is called the spectral measure of the field $\xi(x)$, $x \in Z^n$. The function $F(\lambda) = F(\Pi[-\pi, \lambda))$, $\lambda = (\lambda_1, \ldots, \lambda_n)' \in \Pi$ is called the spectral function of the field.

If F is absolutely continuous with respect to Lebesgue measure on Π, the density function f of F is called the spectral density of the field. The function

f is even for real field. The density function $f \in \mathbf{L}_1(\Pi)$. The spectral density f allows the Fourier representation:

$$f(\lambda) = (2\pi)^{-n} \sum_{x \in Z^n} B(x)\, e^{-i<\lambda,x>}, \quad \lambda \in \Pi. \tag{1.2.2}$$

If $B \in \mathbf{L}_1(Z^n)$ the convergence in this expansion is uniform and f is continuous. The spectral density $f \in \mathbf{L}_2(\Pi)$ if and only if $B \in \mathbf{L}_2(Z^n)$.

Example 1.2.1. If the covariance function $B(x)$, $x \in Z^n$, appearing in representation (1.2.1) is of the form

$$B(x) = k \prod_{j=1}^{n} a_j^{|x_j|}, k > 0, \quad |a_j| < 1, \ j = 1, \ldots, n, \quad x = (x_1, \ldots, x_n)' \in Z^n,$$

then the spectral density

$$f(\lambda) = k(2\pi)^{-n} \prod_{j=1}^{n} (1 - a_j^2) \left| e^{i\lambda_j} - a_j \right|^{-2}, \quad \lambda = (\lambda_1, \ldots, \lambda_n)' \in \Pi.$$

By Karhunen's Theorem 1.2.1., the spectral representation (1.2.1) implies the spectral decomposition of the homogeneous random field $\xi(x)$, $x \in Z^n$, with $E\xi(x) = 0$, itself, that is, there exists a complex-valued orthogonal random measure $Z(\Delta)$, $\Delta \in \mathfrak{B}(\Pi)$, with control measure F, such that for every $x \in Z^n$ (P -a.s.)

$$\xi(x) = \int_{\Pi} e^{i<\lambda,x>} Z(d\lambda), \tag{1.2.3}$$

where $E\,|Z(\Delta)|^2 = F(\Delta)$.

2. Continuous-parameter homogeneous random fields

We generalize the concept of stationary random process with continuous parameter by introducing the concept of homogeneous random field on n-dimensional Euclidean space R^n.

A real random field $\xi(x), x \in R^n$ satisfying $E\xi^2(x) < \infty$ is called homogeneous (in the wide sense) if its mathematical expectation $m(x) = E\xi(x)$ and covariance function $B(x, y) = cov\ (\xi(x), \xi(y))$ are invariant with respect to the Abelian group $G = (R^n, +)$ of shifts in R^n, that is $m(x) = m(x + \tau)$, $B(x, y) = B(x + \tau,\ y + \tau)$ for any $x, y, \tau \in R^n$.

If means that $E\xi(x) = k = \text{const}$, and the covariance function $B(x,y) = B(x-y)$ depends only on the difference $x - y$. Without loss of generality it will usually be assumed that $k = 0$.

The covariance function $B(x-y)$ of a homogeneous random field is a non-negative definite kernel on $R^n \times R^n$, that is, for any $r \geq 1$, $x^{(j)} \in R^n$, $z_j \in \mathbf{C}$, $j = 1, \ldots r$,

$$\sum_{i,j=1}^{r} B(x^{(i)} - x^{(j)})\ z_i \bar{z}_j \geq 0.$$

If the function $B(x)$ is continuous at the point $x = 0$ the field is mean-square continuous at each point $x \in R^n$ and vice versa. If the function $B(x)$ is continuous at the point $x = 0$ then the $B(x)$ is continuous for all $x \in R^n$.

Let K be a class of functions which can be serve as correlation functions of homogeneous random fields. Now if $B_1(x), B_2(x) \in K$ and k_1, k_2 are constants, then $k_1 B_1(x) + k_2 B_2(x) \in K$, $\quad B_1(x) B_2(x) \in K$.

The Bochner-Khinchin theorem implies that a function $B(x), x \in R^n$ is the covariance function of a mean-square continuous random field $\xi(x), x \in R^n$, if and only if there exists a finite measure $F(\cdot)$ on $(R^n, \mathfrak{B}(R^n))$ such that

$$B(x) = \int_{R^n} e^{i<\lambda,x>} F(d\lambda) = \int_{R^n} \cos < \lambda, x > F(d\lambda) \tag{1.2.4}$$

with $F(R^n) = B(0) < \infty$.

Representation (1.2.4) is called the spectral decomposition of the correlation function, F is called the spectral measure of the field $\xi(x)$, $x \in R^n$.

The function $F(\lambda) = F(\Pi(-\infty, \lambda))$ is called the spectral function of the field $F(\infty, \ldots, \infty) = B(0), F(-\infty, \ldots, -\infty) = 0$.

A spectral measure F admits the Lebesgue decomposition into absolutely continuous, discrete and singular parts. If the last two components are missing, the spectral measure is absolutely continuous:

$$F(\Delta) = \int_{\Delta} f(\lambda)\ d\lambda, \quad \Delta \in \mathfrak{B}(R^n).$$

The function $f(\lambda)$, $\lambda \in R^n$, which is integrable over R^n, is called the spectral density.

If the spectral density exists, then the spectral decomposition (1.2.4) may be written as

$$B(x) = \int_{R^n} e^{i<\lambda,x>} f(\lambda)\ d\lambda. \tag{1.2.5}$$

If $B(x) \in \mathbf{L}_1(R^n)$, then, clearly, $f(\lambda)$ exists, and

$$f(\lambda) = \int_{R^n} B(x) e^{-i<\lambda,x>} dx, \quad \lambda \in R^n.$$

Example 1.2.2. Let the covariance function be of the form

$$B(x) = k \exp\{-\sum_{j=1}^{n} a_j |x_j|\}, \ x \in R^n, \ k > 0, \ a_j > 0, \ j = 1, \ldots, n.$$

Then the representation (1.2.5) holds with

$$f(\lambda) = k\pi^{-n} \prod_{j=1}^{n} \frac{a_j}{\lambda_j^2 + a_j^2}, \quad \lambda \in R^n.$$

By to Karhunen's theorem 1.1.3., the representation (1.2.4) implies that the spectral decomposition of the field $\xi(x), x \in R^n$, itself, that is, there exists a complex-valued orthogonal random measure $Z(\Delta), \ \Delta \in \mathfrak{B}(R^n)$ such that for every $x \in R^n$ (P-a.s.)

$$\xi(x) = \int_{R^n} e^{i<\lambda,x>} Z(d\lambda) \tag{1.2.6}$$

where $E|Z(\Delta)|^2 = F(\Delta)$.

3. Discretization problem

Consider a continuous parameter homogeneous random field $\xi(x), x \in R^n$, with continuous covariance function $B_c(x), x \in R^n$ and spectral density $f_c(\lambda)$, $\lambda \in R^n$. We will consider two types of discretization in $x \in R^n$.

I. *Instantenous sampling.*

For a fixed step size $h > 0$,define

$$\xi_{d_1}(x) \equiv \xi_c(hx), \quad x \in Z^n. \tag{1.2.7}$$

It is clear that $\xi_{d_1}(x), \ x \in Z^n$ is homogeneous with covariances

$$\begin{aligned} B_{d_1}(\nu) &= cov\ (\xi_{d_1}(x), \ \xi_{d_1}(x+\nu)) \\ &= B_c(\nu h) \\ &= \int_{R^n} e^{ih<\nu,\lambda>} f_c(\lambda) \ d\lambda \\ &= \int_{\Pi} e^{i<\nu,\lambda>} f_{d_1}(\lambda) \ d\lambda, \qquad \nu \in Z^n, \end{aligned}$$

where we have introduced

$$f_{d_1}(\lambda) = h^{-n} \sum_{k \in Z^n} f_c\left(\frac{\lambda + 2k\pi}{h}\right), \qquad \lambda \in \Pi, \tag{1.2.8}$$

$\lambda, k, \nu \in Z^n$, $\lambda + 2k\pi = (\lambda_1 + 2k_1\pi, \ldots, \lambda_n + 2k_n\pi)'$.

The series in (1.2.8.) converge a.s. since $f_c(\lambda) \in \mathbf{L}_1(R^n)$ so that $f_{d_1}(\lambda)$, $\lambda \in \Pi$, is well defined and in $\mathbf{L}_1(\Pi)$. Hence if we sample the ξ_c-field equidistant with the step h, the new discrete parameter homogeneous field $\xi_{d_1}(x)$, $x \in Z^n$, has a spectral density $f_{d_1}(\lambda)$, $\lambda \in \Pi$, given by (1.2.8).

II. *Locally average sampling.*

For a fixed step-size $h > 0$ define

$$\xi_{d_2}(x) = h^{-n} \int_{A_h} \xi_c(y)\, dy, \quad x \in Z^n \tag{1.2.9}$$

where

$$A_h = \{y \in R^n : (x_i - \frac{1}{2})h \le y_i < (x_i + \frac{1}{2})h, \quad i = 1, \ldots, n\}.$$

The new discrete parameter homogeneous random field $\xi_{d_2}(x)$, $x \in Z^n$, has covariance given by

$$\begin{aligned} B_{d_2}(\nu) &= cov\ (\xi_{d_2}(x),\ \xi_{d_2}(x+\nu)) \\ &= h^{-2n} \iint B_c(u - v)\, du\, dv, \qquad x, \nu \in Z^n \end{aligned} \tag{1.2.10}$$

integrated over

$$(x_i - \frac{1}{2})h \le u_i < (x_i + \frac{1}{2})h,\ i = 1, \ldots, n,$$

and

$$(x_i + \nu_i - \frac{1}{2})h \le v_i < (x_i + \nu_i + \frac{1}{2})h,\ i = 1, \ldots, n.$$

Expressing the covariance kernel B_c in (1.2.10) by its spectral representation (1.2.4), and combining with (1.2.8), we see after exchanging the order of integration that (1.2.10) is equal to

$$B_{d_2}(\nu) = \int_{\Pi} e^{i<\lambda,\nu>} f_{d_2}(\lambda)\, d\lambda, \quad \nu \in Z^n,$$

where

$$f_{d_2}(\lambda) = \left(\frac{4}{h}\right)^n \left\{\prod_{j=1}^{n} \sin^2 \frac{\lambda_j}{2}\right\} \sum_{k \in Z^n} f_c\left(\frac{\lambda + 2k\pi}{h}\right) \prod_{i=1}^{n} (\lambda_i + 2\pi k_i)^{-2}, \quad \lambda \in \Pi, \tag{1.2.11}$$

Hence the spectral density of a discrete parameter homogeneous random field $\xi_{d_2}(x)$, $x \in Z^n$, has the form (1.2.11).

Remark 1.2.1.The interesting problem of behavior of the spectral density (1.2.8) as $h \to 0$ is discussed in a paper of Stein (1995).

4. Homogeneous and isotropic random field

Rotations of the Euclidean space R^n are defined as linear transformations g of this space that do not change its orientation and preserve the distance of the points from the origin: $\|gx\| = \|x\|$. The rotation of R^n generate the group $SO(n)$.

The motion of R^n are defined as non-homogeneous linear transformations which preserve the distances between points of this space and its orientations. It is well known that any motion in R^n can be written in the form $x \to gx + \tau$, $g \in SO(n)$, where $\tau \in R^n$ may be treated as an element of the group $G = (R^n, +)$ of shifts in R^n. The motion of a Euclidean space generate the group $M(n)$.

A real random field $\xi(x)$, $x \in R^n$, satisfying $E\xi^2(x) < \infty$ is called homogeneous and isotropic (in the wide sense) if its mathematical expectation $m(x) = E\xi(x) = k = \text{const}$, and the covariance function $B(x,y) = cov(\xi(x), \xi(y)) = B(\|x - y\|)$ depends only on the Euclidean distance $\rho_{xy} = \|x - y\|$ between x and y. It means that its mathematical expectation $m(x)$ and covariance function $B(x,y)$ are invariant with respect to shifts, rotations and reflections in R^n or invariant into a group of motion

$$M(n) : m(x) = m(x + \tau), \quad B(x,y) = B(x + \tau, y + \tau), \quad x, y, \tau \in R^n;$$

$$m(x) = m(gx), \quad B(x,y) = B(gx, gy), \quad x, y \in R^n, \quad g \in SO(n). \tag{1.2.12}$$

The spectral measure F (see, (1.2.4)) of homogeneous isotropic random field is invariance with respect to the group $SO(n)$: $F(\Delta) = F(g\Delta)$ for every $g \in SO(n)$, $\Delta \in \mathfrak{B}(R^n)$.

For $\nu > -\frac{1}{2}$ we introduce by

$$J_\nu(z) = \sum_{m=0}^{\infty}(-1)^m(\frac{z}{2})^{2m+\nu}[m!\,\Gamma(m+\nu+1]^{-1}, \quad z > 0 \qquad (1.2.13)$$

the Bessel function of the first kind of order ν and by

$$\begin{aligned} Y_1(z) &= \cos z, \\ Y_n(z) &= 2^{(n-2)/2}\Gamma(\frac{n}{2})\; J_{(n-2)/2}(z)\; z^{(2-n)/2}, \quad z \geq 0, n \geq 2 \end{aligned} \qquad (1.2.13a)$$

the spherical Bessel function.

Remark 1.2.2.The spherical Bessel function in n-dimension may be defined also as

$$\begin{aligned} Y_1(z) &= \cos z, \\ Y_n(z) &= \frac{\int_0^\pi e^{iz\cos\theta}\sin^{n-2}\theta d\theta}{\int_0^\pi \sin^{n-2}\theta\; d\theta}, \quad n \geq 2,\; z \in R^1. \end{aligned}$$

The function $Y = Y_n(z)$ satisfies

$$\begin{aligned} Y_n(-z) &= Y_n(z); \quad Y_n(0) = 1; \\ Y_n'' + \frac{n-1}{z}Y_n' + Y_n &= 0, \quad z \neq 0; \\ Y_n' &= \frac{-z}{n}Y_{n+2}; \\ Y_n(z) &= 1 - \frac{z^2}{2n} + \cdots, \quad z \to 0; \\ Y_n(z) &= \frac{2^{n/2}\Gamma(n/2)}{\sqrt{2\pi}z^{(n-1)/2}}[\cos(z - \frac{(n-1)\pi}{4}) + O(\frac{1}{z})], \quad z \uparrow \infty. \end{aligned}$$

For more details and proof, see Gikhman and Skorokhod (1974), Pinsky (1993).

The Schoenberg theorem implies that a function $B(\rho)$, $\rho = \|x\| \geq 0$ is the covariance function of a mean-square continuous homogeneous isotropic random field $\xi(x)$, $x \in R^n$, if and only if there exists a finite measure G on

$(R^1_+, \mathfrak{B}(R^1_+))$ such that

$$B(\rho) = \int_0^\infty Y_n(u\rho)\ G(du), \tag{1.2.14}$$

with

$$G(R^1_+) = F(R^n) = B(0) < \infty.$$

The bounded non-decreasing function

$$G(u) = \int_{\{\lambda:\|\lambda\|<u\}} F(d\lambda), \qquad u \geq 0 \tag{1.2.14a}$$

is called the spectral function of the homogeneous isotropic random field. Clearly

$$\int_0^\infty dG(u) = F(R^n) < \infty$$

and the inversion formula

$$G(u) = 2^{(2-n)/2}\Gamma^{-1}(n/2) \int_0^\infty J_{n/2}(ur)\ (ur)^{n/2} \frac{B(r)}{r}\ dr \tag{1.2.14b}$$

holds. Thus, for $n = 2$ formula (1.2.14) becomes

$$B(\rho) = \int_0^\infty J_0(u\rho)\ G(du), \tag{1.2.15}$$

and for $n = 3$

$$B(\rho) = \int_0^\infty \frac{\sin\ u\rho}{u\rho}\ G(du). \tag{1.2.16}$$

Formula (1.2.14) follows from (1.2.4) if we proceed to spherical coordinates (1.1.3) and use the relation

$$\frac{1}{|s(u)|} \int_{s(u)} e^{i<\lambda,x>} d\sigma(\lambda) = Y_n(ux),$$

where $\sigma(\cdot)$ is the Lebesgue measure on the sphere $s(u) = s_{n-1}(u)$ (cf. (1.1.4)).

Remark 1.2.3. If $B(0) = 1$ then $G(\cdot)$ is a probabilistic measure on R^1_+. For $n > 1$ consider the symmetric measure on $[-1, 1]$

$$G_{(n-2)/2}(du) = (1-u^2)^{(n-3)/2}[\sqrt{\pi}\,\Gamma((n-1)/2)]^{-1}\Gamma(n/2)\,du, \quad u \in [-1,1].$$

Using the Poisson integral

$$J_\nu(z)\,\Gamma(\nu+\frac{1}{2}) = \pi^{-1/2}(z/2)^\nu \int_{-1}^{1} e^{itz}(1-t^2)^{\nu-1/2}dt, \quad \text{Re}\,\nu > -\frac{1}{2},$$

we find that the characteristic function of the distribution $G_{(n-2)/2}$ equals $Y_n(t)$; thus formula (1.2.14) may be written as

$$B(t) = E\,Y_n(t\eta) = E\,\exp\{it\eta\zeta\},$$

where η and ζ are independent r.v.s with distributions G and $G_{(n-2)/2}$ respectively. On the other hand, (1.2.14) may be treated as a Hankel type of transform

$$H_\nu(G,t) = B(t) = \int_0^\infty Y_n(ut)\,G(du)$$

with the kernel $Y_\nu(z) = \Gamma(\nu+1)\,Y_\nu(t)\,(t/2)^{-\nu}$ defined for $\nu > -\frac{1}{2}$(or $n > 1$). However, as $\nu \to -\frac{1}{2}+0$, $\lim Y_\nu = \cos t$ while the function $\cos t$ is the characteristic function of the probability measure $M = (e_1 + e_{-1})/2$ concentrated at the points 1 and -1 and assigning masses $\frac{1}{2}$ to each of them. Therefore, $\cos t$ may be denoted by $Y_{-1/2}(t)$ and the measure M by $G_{-1/2}(\cdot)$; thus one can assume by continuity that $\nu \geq -\frac{1}{2}$, this includes the case of dimension $n = 1$.

Let $n \geq 2$ and (r, u), $u \in s_{n-1}(1)$, $r \geq 0$, be spherical coordinates of the point $x \in R^n$ (cf.(1.1.3)). It is known (see Bateman and Erdelyi (1953) or Muller (1966),p.3) that there exists

$$h = h(m,n) = (2m+n-2)\frac{(m+n-3)!}{(n-2)!\,m!} \tag{1.2.17}$$

linearly independent orthonormalized real spherical harmonics $S^l_m(u)$, $u \in s_{n-1}(1)$, $l = 1, 2, \ldots, h(m,n)$, $S^{\,l}_{\,0}(u) = 1$ of degree $m = 0, 1, 2, \ldots$

Remark 1.2.4. Denote by $\Delta = \sum_{j=1}^{n} (\partial/\partial x_j)^2$ the Laplace operator in R^n, $n \geq 2$, and $x = (x_1, \ldots, x_n)' = ru$, $r = \|x\| \geq 0$, $u \in s_{n-1}(1)$,

$u_1^2 + \cdots + u_n^2 = 1$. Let $Q_m(x)$ be a homogeneous polynomial of degree m which satisfies $\Delta Q_m(x) = 0$ and $Q_m(ru) = r^m Q_m(u)$. Then $S_m(u) = Q_m(u) = r^{-m}\, Q_m(ru)$ is called a (regular) spherical harmonic of degree m. We obtain immediately

$$\int_{s(1)} S_m(u)\; S_k(u)\; d\sigma(u) = 0, \quad m \neq k,$$

where $\sigma(\cdot)$ is Lebesgue measure on the sphere $s(1) = s_{n-1}(1)$, defined in (1.1.4). The functions $S_m(u) = S_m\left(\frac{x}{\|x\|}\right)$ are continuous functions on the unit sphere and can be expressed as trigonometric polynomial of $\phi_1, \ldots, \phi_{n-1}, \phi$ (cf. (1.1.3)). Every spherical harmonicS_m of degree m, can be expressed as a linear combination the lineally independent spherical harmonics S_m^l, that is

$$S_m(u) = \sum_{l=1}^{h(m,n)} S_m^l(u),$$

where $h(m,n)$ is given by (1.2.17).

Remark 1.2.5. It is known that the spherical harmonics $\{S_m^l(u)\}$, $l = 1, \ldots, h(m,n)$, $m = 0, 1, 2, \ldots$, form a complete orthogonal system in the Hilbert space $\mathbf{L}_2(s_{n-1}(1), \sigma(du))$ of function $f : s(1) \to R^1$ satisfying the inequality

$$\int_{s(1)} f^2(x)\; \sigma(dx) < \infty.$$

We shall use the additional theorem for the spherical Bessel function defined by (1.2.13)

$$Y_n(\lambda\rho) = c_1^2(n) \sum_{m=0}^{\infty} \sum_{l=1}^{h(m,n)} S_m^l(u)\; S_m^l(v) \frac{J_{m+\frac{n-2}{2}}(\lambda r_1)}{(\lambda r_1)^{\frac{n-2}{2}}} \frac{J_{m+\frac{n-2}{2}}(\lambda r_2)}{(\lambda r_2)^{\frac{n-2}{2}}} \qquad (1.2.18)$$

where

$$\begin{aligned} x &= (r_1, u), \quad r_1 \geq 0, \quad u = x/\|x\| \in s(1), \\ y &= (r_2, v), \quad r_2 \geq 0, \quad v = y/\|y\| \in s(1), \\ \rho &= \|x - y\| = \sqrt{r_1^2 + r_2^2 - 2r_1 r_2 \cos\gamma}, \quad \cos\gamma = \frac{<x, y>}{\|x\| \cdot \|y\|}, \quad \lambda \geq 0, \end{aligned}$$

and

$$c_1^2(n) = 2^{n-1}\Gamma(n/2)\ \pi^{n/2} \tag{1.2.19}$$

For the proof of (1.2.18) see Bateman and Erdelyi (1953) or Muller (1966).

Remark 1.2.6. For $n = 2$, $c_1(n) = 2\pi$, $h(m,2) = 2$ and there exist for each $m = 0, 1, 2, \ldots$ two real spherical harmonics

$$S_m^1(\phi) = \frac{\cos m\phi}{\sqrt{2\pi}}, \quad S_m^2(\phi) = \frac{\sin m\phi}{\sqrt{2\pi}}.$$

The additional theorem (1.2.18) can be expressed in terms of complex orthogonal spherical harmonics

$$\tilde{S}_m(\phi) = \frac{\exp\{im\varphi\}}{\sqrt{2\pi}}, \quad 0 \le \varphi < 2\pi, \quad m \in Z^1,$$

as follows

$$J_0(\lambda\rho) = \sum_{m=-\infty}^{\infty} J_m(\lambda r_1)\ J_m(\lambda r_2)\ e^{im(\phi_1-\phi_2)} \tag{1.2.20}$$

where

$$\begin{aligned} \lambda &\ge 0, \quad \rho = \|x-y\|, \quad x = (r_1,\phi_1), \quad y = (r_2,\varphi_2), \\ r_j &\ge 0, \quad 0 \le \phi_j < 2\pi, \quad j = 1, 2, \end{aligned}$$

and

$$J_m(z) = \frac{1}{2\pi}\int_0^{2\pi} \exp\{iz\sin\theta - im\theta\}\ d\theta, \quad m \ge 0$$

is a Bessel function of the first kind of integer order, $J_{-m}(z) = (-1)^m J_m(z)$.

In view of Karhunen's theorem 1.1.3. and (1.2.14), (1.2.18) we arrive at the following spectral decomposition for the mean-square continuous homogeneous isotropic random field $\xi(x)$, $x = (r,u) \in R^n$, $u \in s(1)$, $r \ge 0$, with zero mean:

$$\begin{aligned} \xi(x) &= \xi(r,u) \\ &= c_1(n) \sum_{m=0}^{\infty} \sum_{l=1}^{h(m,n)} S_m^l(u) \int_0^{\infty} J_{m+\frac{n-2}{2}}(\mu r) \\ &\quad \times (\mu r)^{(2-n)/2} Z_m^l(d\mu), \end{aligned} \tag{1.2.21}$$

where $c_1(n)$ is defined by (1.2.19) and $Z^l_m(\cdot)$ is a family of real-valued random measures on $(R^1_+, \mathfrak{B}(R^1_+))$ such that the relations

$$\begin{aligned} EZ^l_m(\Delta) &= 0, \\ EZ^l_m(\Delta_1)\; Z^q_p(\Delta_2) &= \delta^p_m\; \delta^q_l\; G(\Delta_1 \cap \Delta_2), \quad \Delta_1, \Delta_2 \in \mathfrak{B}(R^1_+)) \end{aligned}$$

hold.

The series in formula (1.2.21) converges in the mean square.

For $n = 2$ the spectral representation (1.2.21) can be written as follows

$$\xi(r, \phi) = \sum_{m=0}^{\infty} \cos m\phi \int_0^{\infty} J_m(\mu r)\; Z^1_m(d\mu) + \sum_{m=1}^{\infty} \sin m\phi \int_0^{\infty} J_m(\mu t)\; Z^2_m(d\mu)$$

where

$$\begin{aligned} E\; Z^j_m(\Delta_1)\; Z^q_p(\Delta_2) &= \delta^p_m\; \delta^q_j\; d_m\; G(\Delta_1 \cap \Delta_2), \quad j, q = 1, 2, \\ d_m &= \begin{cases} 1, & m = 0, \\ 2, & m \neq 0. \end{cases} \end{aligned}$$

For $n = 2$ the spectral representation (1.2.21) can be written also in terms of complex spherical harmonics (see remark 1.2.5). In view of Karhunen's theorem 1.1.3. and (1.2.15), (1.2.20) we obtain the following spectral representation of the homogeneous isotropic random field

$$\xi(\rho, \phi) = \sum_{m=-\infty}^{\infty} e^{im\phi} \int_0^{\infty} J_m(\mu\rho)\; Z_m(d\mu),$$

where $Z_m(\cdot)$ are complex random measures such that

$$E\; Z_m(\Delta_1)\, \overline{Z_m(\Delta_2)} = \delta^{m'}_m\; G(\Delta_1 \cap \Delta_2), \quad \Delta_1, \Delta_2 \in \mathfrak{B}(R^1_+)$$

and spectral measure $G(\cdot)$ is defined in (1.2.15).

Remark 1.2.7. For $n = 3$, $h(m, 3) = 2m + 1$ and for each $m = 0, 1, 2, \ldots$ there exist $2m + 1$ real spherical harmonics

$$\cos \nu\phi\; P^\nu_m(\cos \phi_1)\; k(m, \nu), \quad \sin \nu\phi\; P^\nu_m(\cos \phi_1)\; k(m, \nu), \quad \nu = 0, 1, \ldots, m,$$

where

$$k(m, \nu) = \sqrt{\frac{(2m+1)\; (m-\nu)!}{2\pi(m+\nu)!}},$$

and $P_m^\nu(\cos\phi_1)$ is associated Legendre functions defined below by formula (1.2.28). The spectral representation (1.2.21) for $n = 3$ can be written in therms of complex spherical harmonics

$$Y_m^l(\phi,\phi_1) = e^{il\phi} P_m^{|l|}(\cos\phi_1)\,\frac{1}{2}\sqrt{\frac{(2m+1)(m-|l|)!}{\pi(m+|l|)!}}\,,$$

$l = 0, \pm 1, \ldots, \pm m, \quad m = 0, 1, 2, \ldots$. The complex spheric harmonics can be defined as follows

$$Y_m^l(\phi,\phi_1) = (-1)^l\sqrt{\frac{(2m+1)(m-l)!}{4\pi(m+l)!}}\,e^{il\phi}\sin^m\phi_1\left(\frac{d}{d\cos\phi_1}\right)^{m+l}\frac{(\cos^2\phi_1-1)^m}{2^m m!}$$

for $m \geq 0, \quad Y_m^{-l} = (-1)^l Y_m^l$. In particular,

$$Y_m^0 = P_m(\cos\phi_1)\sqrt{\frac{(2m+1)}{4\pi}}.$$

The functions $Y_m^l(\phi,\phi_1)$ satisfy the orthogonal property:

$$\int_{s_2(1)} Y_m^l(\phi,\phi_1)\,\overline{Y_{m'}^{l'}(\phi,\phi_1)}\,d\sigma(\phi,\phi_1) = \delta_l^{l'}\,\delta_m^{m'},$$

where $d\sigma(\phi,\phi_1) = \sin\phi_1\,d\phi_1\,d\phi, \quad 0 \leq \phi_1 < \pi, \quad 0 \leq \phi < 2\pi$. The field $\xi(x), x = (r,\phi,\phi_1)$ may represented by the series

$$\xi(r,\phi,\phi_1) = \pi\sqrt{2}\sum_{m=0}^{\infty}\sum_{l=-m}^{m} Y_m^l(\phi,\phi_1)\frac{J_{(m+1/2)}(\mu r)}{(\mu r)^{1/2}} Z_m^l(d\mu),$$

where $Z_m^l(\cdot)$ are complex random measure with control measure $G(\cdot)$ given (1.2.16).

The homogeneous field $\xi(x), x \in R^n$ has an absolutely continuous spectrum, that is, the measure F in (1.2.4) is absolutely continuous with respect to the Lebesgue measure if and only if

$$\xi(x) = \int_{R^n} b(x-y)\,W(dy), \tag{1.2.22}$$

where $b(\cdot) \in \mathbf{L}_2(R^n), \quad W(\cdot)$ is complex white noise random measure on $(R^n, \mathfrak{B}(R^n))$, that is

$$E\,|W(\Delta)|^2 = |\Delta|\,, \quad \Delta \in \mathfrak{B}(R^n) \tag{1.2.23}$$

In representation (1.2.22)

$$\gamma(\lambda) = (2\pi)^{-\frac{n}{2}} \int_{R^n} e^{-i<\lambda,y>} b(y)\, dy$$

and $f(\lambda) = |\gamma(\lambda)|^2$, $\lambda \in R^n$, is the spectral density of the homogeneous field given in (1.2.5).

If the function $b(y)$ in (1.2.22) is radial, $b(x) = b(\|x\|)$, then the field $\xi(x), x \in R^n$ is homogeneous isotropic field possessing an absolutely continuous spectrum, that is there exists a function $g(\mu)$ such that

$$G'(\mu) = |s(1)|\ \mu^{n-1} g(\mu), \qquad \mu^{n-1} g(\mu) \in \mathbf{L}_1(R^1_+), \tag{1.2.24}$$

where $G(\cdot)$ is defined in (1.2.14a), than the function $g(\mu)$ is called the isotropic spectral density function of the field $\xi(x), x \in R^n$.

In this case the representation (1.2.14) may be written as

$$B(\rho) = |s(1)| \int_0^\infty Y_n(\mu\rho)\ \rho^{n-1} g(\mu)\ d\mu. \tag{1.2.25}$$

If $B(r), r \geq 0$ falls off rapidly enough at infinity (in particular, if

$$\int_0^\infty z^{n-1}\, |B(z)|\, dz < \infty,)$$

then the function $g(\mu)$ is given by the formal expression

$$g(\mu) = (2\pi)^{-\frac{n}{2}} \int_0^\infty J_{(n-2)/2}(\mu\tau)\ (\mu\tau)^{(2-n)/2} \tau^{n-1} B(\tau)\ d\tau.$$

Accordingly, the spectral representation (1.2.21) becomes

$$\xi(r, u) = (2\pi)^{n/2} \sum_{m=0}^{\infty} \sum_{l=1}^{h(m,n)} S^l_m(u) \int_0^\infty \frac{\sqrt{\mu r}}{r^{(n-1)/2}} J_{m+\frac{n-2}{2}}(\mu r) \sqrt{g(\mu)}\ W^l_m(d\mu), \tag{1.2.26}$$

where $W^l_m(\cdot)$, $l = 1, \ldots,\ h(m,n)$, $m = 0, 1, 2, \ldots$, is a family of jointly uncorrected real white noise random measures on $(R^1_+, \mathfrak{B}(R^1_+))$, that is,

$$E\, W^l_m(\Delta_1)\, W^q_p(\Delta_2) = \delta^p_m\ \delta^q_l\, |\Delta_1 \cap \Delta_2|$$

Remark 1.2.8. If the field $\xi(x)$, $x \in R^n$, is Gaussian then all random measures in the spectral representations (1.2.3), (1.2.6), (1.2.21), (1.2.22) and(1.2.26) are Gaussian. For example, the Gaussian white noise complex random measure satisfy (1.2.23), and if $\Delta \cap (-\Delta) = \emptyset$, then r.v.s $W(\Delta)$ and $W(-\Delta)$ are independent and have the normal distribution $N(0, |\Delta|/2)$.

We provide examples of covariance functions for homogeneous isotropic random fields together with spectral representation (1.2.14) and (1.2.25). In all the examples the constant $a > 0$.

Example 1.2.3. Let the covariance function appearing in representation (1.2.14) be of the form $B(\rho) = \exp\{-a\rho\}$. Then the isotropic spectral density $g(\mu)$ (see,(1.2.24)) is of the form

$$g(\mu) = a\,\Gamma\left(\frac{n+1}{2}\right)\pi^{-\frac{n+1}{2}}(\mu^2 + a^2)^{-\frac{n+1}{2}}.$$

Example 1.2.4. Let the covariance function be of the form $B(\rho) = e^{-a\rho^2}$. Then relations (1.2.14), (1.2.24), (1.2.25) hold, with

$$g(\mu) = \frac{e^{-\mu^2/4a}}{2^n(\pi a)^{n/2}}\,.$$

Example 1.2.5. Let the covariance function $B(\tau) = (a\tau)^\nu K_\nu(a\tau)$, $\nu \geq 0$, where $K_\nu(z)$ is Bessel function of the third kind of order ν or Basset's or McDonald's function (a Bessel function of the second kind with an imaginary argument) defined by the relation

$$K_\nu(z) = \int_0^\infty e^{-z\,\mathrm{ch}\,t}\,\mathrm{ch}\,\nu t\,dt, \qquad \mathrm{ch}\,t = \frac{e^t + e^{-t}}{2}, \tag{1.2.27}$$

In this case relations (1.2.24),(1.2.25) hold with isotropic spectral density.

$$g(\mu) = 2^{\nu-1}\Gamma\left(\nu + \frac{n}{2}\right)a^{2\nu}\pi^{-n/2}(\mu^2 + a^2)^{-(\nu+n/2)}$$

Example 1.2.6. Let $B(\rho) = Y_n(a\rho)$, then the spectral function G appearing in representation (1.2.14) has the unique point of discontinuity

$$G(u) = \begin{cases} 0, & u \leq a, \\ 1, & u > a. \end{cases}$$

Example 1.2.7. Let $B(\rho) = Y_{n+2}(a\rho)$, then the spectral function G (1.2.14) is of the form

$$G(u) = \begin{cases} (u/a)^n, & u \leq a, \\ 1, & u > a. \end{cases}$$

Example 1.2.8. For $n \geq 3$ let $B(\rho) = a^{n-1}(\rho^2 + a^2)^{(2-n)/2}$. Then the representation (1.2.14) holds with

$$G'(\mu) = \frac{a^{n-1}\mu^{n-2}e^{-a\mu}}{(n-2)!}.$$

Example 1.2.9 . Let $B(\rho) = a^{n+1}(\rho^2 + a^2)^{-\frac{n+1}{2}}$, then the representation (1.2.14) holds with

$$G'(\mu) = \frac{a^{n-1}\mu^{n-1}e^{-a\mu}}{(n-1)!}.$$

Example 1.2.10. For $n = 1$ let $B(\rho) = (a\rho)^\nu K_\nu(a\rho)$, where $K_\nu(z)$ is McDonald's function defined by (1.2.27). Then the representation (1.2.5) holds with spectral density

$$f(\mu) = 2^{\nu-1}\Gamma(\nu + \frac{1}{2})\, a^{2\nu}\pi^{-1/2}(\mu^2 + a^2)^{-(\nu+1/2)}$$

Example 1.2.11. For $n = 1$ let $B(\rho) = J_\nu(a\rho)\ (a\rho)^{-\nu}$, $\nu > -\frac{1}{2}$, where $J_\nu(z)$ is a Bessel function defined by (1.2.12). Then the representation (1.2.5) holds with spectral density

$$f(\mu) = \begin{cases} \frac{(a^2-\mu^2)^{\nu-1/2}}{2^\nu\pi^{1/2}\Gamma(\nu+\frac{1}{2})a^{2\nu}}, & |\mu| < a, \\ 0, & |\mu| > a. \end{cases}$$

Note that for $\nu - \frac{1}{2} < 0$ the spectral density hasthe property $f(a) = \infty$. For instance, if

$$B(\rho) = J_0(a\rho) = \frac{1}{\pi}\int_{-1}^{1} e^{itu}(1-u^2)^{-\frac{1}{2}}du,$$

then

$$f(\mu) = \begin{cases} \pi^{-1}(a^2 - \mu^2)^{-\frac{1}{2}}, & |\lambda| < a, \\ 0, & |\lambda| > a. \end{cases}$$

and $f(a) = \infty$.

5. Vector-valued homogeneous and isotropic random fields

We emphasize the following points relating to the definition of a vector -valued random field $\xi : \Omega \times R^n \to R^m$.

Homogeneous random fields $\xi_i : \Omega \times R^n \to R^1, i = 1, 2,$ (in wide sense) with zero mean are said to be jointly homogeneous if their cross-covariance function

$$B_{12}(x, y) = E\, \xi_1(x)\, \xi_2(y) = B_{12}(x - y)$$

depends only on the difference $x - y$. Jointly homogeneous random fields are jointly isotropic if their cross-covariance function $B_{12}(x - y)$ depends only on the distance $\rho_{xy} = \|x - y\|$. A collection of m homogeneous (isotropic) fields $\xi(x) = [\xi_1(x), \ldots, \xi_m(x)]'$ which are jointly homogeneous (isotropic) is said to be a vector homogeneous (isotropic) random fields $\xi(x) \in R^m$, $x \in R^n$.

Let $B_{rs}(x) = E\, \xi_r(x+y)\, \xi_s(y)$. Then the matrix $B(x) = (B_{rs}(x))_{r,s=1,\ldots,m}$ is the covariance functional the homogeneous (isotropic) vector fields $\xi(x) \in R^m$, $x \in R^n$.

Let $B(x)$ be continuous at the point $x = 0$. Then for all r, s in $\{1, \ldots, m\}$ the functions admit the representation (1.2.4) in the case of a homogeneous field and the representation (1.2.14) in the case of homogeneous isotropic field. We denote by $B_{rs}(\cdot)$, $G_{rs}(\cdot)$ the spectral measure involved in these decompositions and refer to the matrix measures $F = (F_{rs}(\cdot))_{r,s=1,\ldots,m}$ and $G = (G_{rs}(\cdot))_{r,s=1,\ldots,m}$ respectively as the spectral measure of a vector homogeneous and homogeneous isotropic field $\xi : \Omega \times R^n \to R^m$. Then the correlation function $B(x)$ of a vector homogeneous (isotropic) random field admits a decomposition (1.2.4) (respectively (1.2.14)), where $F(\cdot)$ and $G(\cdot)$ are the matrix measure defined above. If all the elements $F_{rs}(\cdot)$ of the matrix$F$ are absolutely continuous with respect to the Lebesgue measure, that is, there exists

$$f_{rs}(\lambda) = \frac{\partial^n}{\partial \lambda_1 \ldots \partial \lambda_n} F_{rs}(\Pi(-\infty, \lambda)), \quad r, s = 1, \ldots, m,$$

then the multi-dimensional field $\xi : \Omega \times R^n \to R^m$ is said to possess the spectral density function $f(\lambda) = (f_{rs}(\lambda))_{r,s=1,\ldots,m}$. In this case the correlation function admits a decomposition (1.2.5) where $f(\lambda)$ is to be viewed as the matrix-valued function defined above. In the same manner vector analogues may be introduced for the other concepts defined above for sailer field.

Alternative generalizations to the multidimensional case occurring in turbulence theory are proposed in the works of Yaglom (1952, 1957), Monin and

Yaglom (1975), Yaglom (1987), Rosenblatt (1985), Ogura (1990).

6. Isotropic random fields on sphere

A real random field $\xi(u), u \in s(1) = s_{n-1}(1) \subset R^n$ with $E\xi^2(u) < \infty$ and $E\xi(u) = 0$ is called isotropic on a sphere if $E\xi(u)\xi(v) = B(\cos\theta)$, $u, v \in s(1)$, depends only on the geodesic (angular) distance $\cos\theta$ between u and v.

We will assume that $\xi(u), u \in s(1)$ is mean square continuous. Then $B(\cos\theta)$ is a continuous function on $[0,\pi)$.

We will need some formulas and facts from the theory of spherical function (Bateman and Erdelyi (1953), Muller(1966)).

Let $\mathbf{L}_2([-1,1],(1-u^2)^{\nu-1/2}du)$, $\nu > -\frac{1}{2}$ be a Hilbert space of a real function $f(u), u \in [-1,1]$ such that

$$\int_{-1}^{1} f^2(u)\,(1-u^2)^{\nu-1/2}du < \infty.$$

Let $P_m(u) = (d^m/du^m)(u^2-1)^m/[m!2^m]$, $u \in [-1,1], m = 0,1,\ldots,$ be a Rodrigues formula for the Legendre polynomials. Note that

$$P_0(u) = 1, \quad P_1(u) = u, \quad P_2(u) = \frac{3}{2}u^2 - \frac{1}{2}, \quad P_3(u) = \frac{5}{2}u^3 - \frac{3}{2}u, \ldots.$$

It is well known that Legendre polynomials form a complete orthogonal system in a Hilbert space $\mathbf{L}_2([-1,1],du)$ that is

$$\int_{-1}^{1} P_m(u)\,P_l(u)\,du = \delta_m^l \frac{2}{2m+1}.$$

Let $C_m^\nu(u), \nu > -\frac{1}{2}$, $\nu \neq 0$, $m = 0,1,2,\ldots,$ $u \in [-1,1]$, be a Gegenbauer (ultra-spherical) polynomial defined by its generating function

$$(1-ut+t^2)^{-\nu} = \sum_0^\infty C_m^\nu(u)\,t^m, \quad \nu > -\frac{1}{2}, \quad \nu \neq 0, \quad t \in R^1.$$

For $\nu = 0$ the Gegenbauer polynomials can be defined as Chebyshev polynomials of the first kind $C_m^0(u) = T_m(u) = \cos(\arccos u)$, $u \in [-1,1]$. The recurrence equation $T_{m+1}(u) = 2uT_m(u) - T_{m-1}(u)$ holds. Note that $T_0(u) = 1$, $T_1(u) = u$, $T_2(u) = 2u^2 - 1, \ldots.$ It is known that Chebyshev polynomials form a complete orthogonal system in a Hilbert space $\mathbf{L}_2([-1,1],(1-u^2)^{-1/2}du)$, that is

$$\int_{-1}^{1} T_m(u)\,T_k(u)\,(1-u^2)^{-\frac{1}{2}}du = J_{mk},$$

where

$$J_{mk} = 0, \quad m \neq k; \quad J_{mk} = \frac{\pi}{2}; \quad m = k > 0; \quad J_{mk} = \pi, \qquad m = k = 0.$$

Note that

$$C_m^\nu(1) = \frac{\Gamma(2m+\nu)}{m!\,\Gamma(2\nu)}, \qquad \nu \neq 0, \qquad C_m^0(1) = \frac{2}{m}$$

and

$$\begin{aligned} C_0^\nu(u) &= 1, \\ C_1^\nu(u) &= 2\nu\ u, \\ C_2^\nu(u) &= 2\nu\ (\nu+1)\left[u^2 - \frac{1}{2\nu+2}\right] \\ C_3^\nu(u) &= \frac{4}{3}\nu\ (\nu+1)(\nu+2)(u^3 - \frac{3}{2\nu+4}u), \ldots, \\ C_m^\nu(-u) &= (-1)^m C_m^\nu(u), \quad C_{2m+1}^\nu(0) = 0. \end{aligned}$$

The Gegenbauer polynomials satisfies the recurrence equations:

$$\begin{aligned} (m+1)C_{m+1}^\nu(u) &= (m+\nu)\ u\ C_m^\nu(u) - (m+2\nu-1)\ C_{m-1}^\nu(u); \\ (2\nu+m)C_m^\nu(u) &= 2\nu\ [C_m^{\nu+1}(u) - u\ C_{m-1}^{\nu+1}(u)]. \end{aligned}$$

It is known that Gegenbauer polynomials form a complete orthogonal system in a Hilbert space $\mathbf{L}_2([-1,1],(1-u^2)^{\nu-1/2}du)$, that is

$$\begin{aligned} &\int_{-1}^{1} (1-u^2)^{\nu-\frac{1}{2}} C_m^\nu(u)\ C_l^\nu(u)du \\ = &\int_0^\pi C_m^\nu(\cos\theta)\ C_l^\nu(\cos\theta)\ \sin^{2\nu}\theta d\theta \\ = &\ \delta_m^l \frac{\pi\ \Gamma(2\nu+m)}{2^{2\nu-1}(\nu+m)\ m!\ \Gamma^2(\nu)}; \quad \nu > -\frac{1}{2}, \quad \nu \neq 0. \end{aligned}$$

It is obvious that for $\nu = \frac{1}{2}$,

$$C_m^{1/2}(u) = P_m(u),$$

that is, the Gegenbauer polynomials coincide with the Legender polynomials.

Define the associated Legendre functions (of the first kind) of degree j and order m as

$$P_m^l(u) = (1-u^2)^{l/2}\frac{d^l}{du^l}P_m(u), \quad P_m^0(u) = P_m(u), \tag{1.2.28}$$

where $l = 0, 1, \ldots, m,$ $m = 0, 1, 2, \ldots$. In particular, if $u = \cos\phi,$ $0 \leq \phi \leq \pi$ then

$$\begin{aligned}
P_1^1(u) &= \sqrt{1-u^2} = \sin\phi, \\
P_2^1(u) &= \frac{3}{2}u\sqrt{1-u^2} = \frac{3}{2}\sin 2\phi, \\
P_2^2(u) &= 3(1-u^2) = \frac{3}{2}(1-\cos 2\phi), \\
P_3^1(u) &= \frac{3}{2}(5u^2-1)\sqrt{1-u^2} = \frac{3}{8}(\sin\phi + 5\sin 3\phi), \\
P_3^2(u) &= 15u(1-u^2) = \frac{15}{4}\cos\phi(\cos\phi - \cos 3\phi), \\
P_3^3(u) &= 15(1-u^2)\sqrt{1-u^2} = \frac{15}{4}(3\sin\phi - \sin 3\phi), \\
P_j^j(u) &= 1\cdot 3\cdot 5\cdots(2j-1)(1-u^2)^{j/2} = 1\cdot 3\cdot 5\cdots(2j-1)\sin^j\phi, j = 0, 1, ..
\end{aligned}$$

and as $m \to \infty$

$$P_m^j(\cos\phi) = (-1)^j\sqrt{\frac{2}{\pi m \sin\phi}}\cos\left[(m+\frac{1}{2})\phi - \frac{\pi}{4} + \frac{j\pi}{2}\right] + O(m^{-3/2}).$$

Lemma 1.2.1. (Additional theorem for spherical harmonics).
For any vectors $u, v \in s(1)$ the function

$$f(u,v) = \sum_{l=1}^{h(m,n)} S_m^l(u)\; S_m^l(v) = \frac{m+\nu}{\nu\,|s(1)|}C_m^{\nu}(\cos\theta), \quad \nu = \frac{n-2}{2} \tag{1.2.29}$$

depends only on the scalar product $< u, v >= \cos\theta$ (the angle between the unit vectors u and v), where $C_m^{\nu}, m = 0, 1, \ldots$ are the Gegenbauer polynomials.

Lemma 1.2.2. (Funk–Hecke formula).

Suppose that $f(t), t \in [-1,1]$ is continuous function. Then for any $u \in s(1)$ and integers m and l

$$\int_{s(1)} f(< u,v >)\ S^l_m(v)\ \sigma(dv) = b_m\ S^l_m(u), \tag{1.2.30}$$

with

$$b_m = \frac{|s(1)|}{C^\nu_m(1)} \int_{-1}^{1} f(t)\ C^\nu_m(t)\ (1-t^2)^{\nu-1/2} dt, \quad \nu = \frac{n-2}{2} > -\frac{1}{2} \tag{1.2.31}$$

Let $B(\cos\theta)$ be a covariance function of the isotropic random field on the sphere. Then by the Funk–Hecke formula (see lemma 1.2.2.) the coefficients b_m defined by (1.2.31) are eigenvalues and the functions S^l_m are eigenfunctions of the integral equation (1.2.30) with $f(t) = f(< u,v >) = B(\cos)$. By Merser's theorem 1.1.4. we obtain

$$B(\cos\theta) = \sum_{m=0}^{\infty} \sum_{l=1}^{h(m,n)} b_m\ S^l_m(u)\ S^l_m(v), \tag{1.2.32}$$

where $\cos\theta =< u,v >$ is the angle between the unit vectors u and v, and b_m are defined by (1.2.31) with $f(t) = f(\cos\theta) = B(\cos\theta)$.

In view of additional theorem for spherical harmonics (see lemma 1.2.1.) we arrive at the following spectral decomposition for the mean square continuous isotropic random field on the sphere

$$B(\cos\theta) = \frac{1}{|s(1)|} \sum_{m=0}^{\infty} b_m\ h(m,n) \frac{C^\nu_m(\cos\theta)}{C^\nu_m(1)}\ , \tag{1.2.33}$$

where for $\nu > -\frac{1}{2}, \quad \nu \neq 0, \quad b_m \geq 0,$

$$\sum_{m=0}^{\infty} b_m\ h(m,n)\ <\ \infty, \quad \nu = \frac{n-2}{2},$$

$$b_m\ =\ \frac{|s(1)|}{|C^\nu_m(1)|} \int_{-1}^{1} B(t)\ C^\nu_m(t)\ (1-t^2)^{\nu-1/2} dt, \tag{1.2.34}$$

Note that, $b_m \geq 0$ as a eigenvalues of non-negative and continuous function $B(\cos\theta)$.

By Karhunen's Theorem 1.1.3. the representation (1.2.33) implies the spectral decomposition of the isotropic field on the sphere $\xi(u), u \in s(1)$, itself, that is, there exists a real-valued sequence of r.v.s η_m^l such that

$$\xi(u) = \sum_{m=0}^{\infty} \sum_{l=1}^{h(m,n)} S_m^l(u)\ \eta_m^l, \tag{1.2.35}$$

where S_m^l are spherical harmonics, and

$$E\eta_m^l = 0, \qquad E\ \eta_m^l\ \eta_p^q = b_m\ \delta_m^p\ \delta_l^q. \tag{1.2.36}$$

For $n = 2$ the covariance function $B(\cos\theta)$ of isotropic random field $\xi(\phi)$, $0 \leq \phi \leq 2\pi$, on the unit circle can be expressed by (1.2.33) as follows:

$$E\xi(\phi_1)\ \xi(\phi_2) = \frac{1}{2\pi} \sum_{m=0}^{\infty} \varepsilon_m\ b_m\ \cos(\phi_1 - \phi_2), \quad \sum_{m=0}^{\infty} b_m < \infty, \tag{1.2.37}$$

where $\cos\theta$ is a geodesic (circle) distance between ϕ_1 and ϕ_2,

$$\varepsilon_m = \begin{cases} 1, & m = 0, \\ 2, & m \neq 0, \end{cases}$$

and

$$b_m = m\pi \int_0^{\pi} B(\cos\theta)\ \cos m\theta\ d\theta.$$

The isotropic field $\xi(\phi)$ admits a spectral representation (1.2.35):

$$\xi(\theta) = \sum_{m=0}^{\infty} \eta_m^{'}\ \cos m\theta + \sum_{m=1}^{\infty} \eta_m^{''}\ \sin m\theta,$$

where sequences of r.v. $\eta_m^{'}$ and $\eta_m^{''}$ such that

$$E\ \eta_m^{'}\ \eta_l^{'} = \varepsilon_m\ b_m\ \delta_m^l, \quad E\eta_m^{''}\ \eta_l^{''} = \varepsilon_m\ b_m\ \delta_m^l, \quad E\eta_m^{'}\ \eta_l^{''} = 0.$$

For $n = 3$ the covariance function $B(\cos\theta)$ of isotropic random field $\xi(u),\ u \in s_2(1) \subset R^3$ admits a spectral representation (1.2.33):

$$B(\cos\theta) = \frac{1}{4\pi} \sum_{m=0}^{\infty} (2m+1)\ b_m\ P_m(\cos\theta),$$

where $P_m(u) = C_m^{1/2}(u)$ is a Legendre polynomial, and

$$\sum_{m=0}^{\infty}(2m+1)\, b_m \ < \ \infty,$$

$$b_m \ = \ \int_{-1}^{1} B(t)\, P_m(t)\, dt = \int_{s_2} B(\cos\theta)\, P_m(\cos\theta)\, \sigma(du),$$

where $\sigma(du)$ stands for Lebesgue measure on the sphere $s_2(1)$.

The spectral representation (1.2.35) for n=3 can be written as follows

$$\xi(\phi,\phi_1) = \sum_{m=0}^{\infty}\sum_{l=0}^{m}\{\eta_m^l \ \cos m\phi + \zeta_m^l \ \sin m\phi\}\ P_m^l(\cos\phi_1),$$

where P_m^l are associated Legendre function defined by (1.2.28) and η_m^l, ζ_m^l are family of real r.v. such that

$$\begin{aligned} E\,\eta_m^l\,\eta_p^q &= \delta_m^p\,\delta_l^q\, b_m\,\frac{2}{\varepsilon_m}\,\frac{(m-l)!}{(m+l)!}, \\ E\,\zeta_m^l\,\zeta_p^q &= \delta_m^p\,\delta_l^q\, b_m\, 2\frac{(m-l)!}{(m+l)!}, \\ E\,\zeta_m^l\,\eta_p^q &= 0, \end{aligned}$$

for any m, l, p, q, and ε_m defined by (1.2.37).

The spectral representation (1.2.35) for $n = 3$ can be written in terms of complex spherical harmonics (see remark 1.2.7) as follows:

$$\xi(\phi,\phi_1) = \sum_{m=0}^{\infty}\sum_{l=-m}^{m} \eta_m^l\ Y_m^l(\phi,\phi_1),$$

where the family of complex r.v. η_m^l is such that

$$E\,\eta_m^l\,\bar{\eta}_p^q = \delta_m^p\,\delta_l^q\, b_{|m|}.$$

We provide examples of covariance functions of isotropic fields on a sphere.
Example 1.2.12. For $n > 2$ let $b_m = |s(1)|\, C_m^{\nu}(1) q^m / h(m,n)$,where

$\nu = (n-2)/2$ and $0 < q < 1$. Then the representation (1.2.33) holds with

$$B(\cos\theta) = (1 - 2q\cos\theta + q^2)^{-\nu}.$$

Example 1.2.13. For $n \geq 2$ let $b_m = q^m, \quad 0 < q < 1$. The representation (1.2.33) holds with

$$B(\cos\theta) = \frac{(1-q^2)(1-2q\cos\theta + q^2)^{n/2}}{|s(1)|}.$$

Example 1.2.14. The restriction of a homogeneous isotropic field in R^n on $s(1)$ will be a isotropic field on $s(1)$. Therefore every function

$$B(\cos\theta) = 2^{(n-2)/2}\Gamma(n/2)\int_0^\infty J_{(n-2)/2}\left(2\mu\sin\frac{\theta}{2}\right)\left[2\mu\sin\frac{\theta}{2}\right]^{(2-n)/2} dG(\mu)$$

is a correlation function of isotropic field on $s(1)$ (see (1.2.33)) with

$$b_m = c_1^2(n)\int_0^\infty J^2_{m+(n-2)/2}(\mu)\ \mu^{2-n} dG(\mu)$$

where $c_1(n)$ is defined in (1.2.19) and spectral function G is defined in (1.2.14 a)

7. Isotropic random field on Euclidean space

A real random field $\xi(x)$, $x \in R^n$, is called isotropic on n-dimensional Euclidean space R^n if $E\xi^2(x) < \infty$ and $E\xi(x) = E\xi(gx)$, $E\xi(x)\xi(y) = E\xi(gx)\xi(gy)$ for every $x, y \in R^n, \quad g \in SO(n)$. It means that its mathematical expectation and covariance function are invariant with respect to the rotation group $SO(n)$.

There exist isotropic fields which are not homogeneous. For example, the fractional Brownian motion (see, for example, Mandelbrot (1982)), that is a Gaussian random field $\zeta(x)$, $x \in R^n$, with

$$E[\zeta(x) - \zeta(y)] = 0, \quad E(\zeta(x) - \zeta(y))^2 = \|x-y\|^{2H}, \quad 0 < H \leq 1.$$

The covariance function of this field has a form:

$$cov(\zeta(x),\ \zeta(y)) = \frac{1}{2}\left\{\|x\|^{2H} + \|y\|^{2H} - \|x-y\|^{2H}\right\}. \tag{1.2.38}$$

Note that the random field $\zeta(x)$, $x \in R^n$, is self-similar with scaling parameter $H \in (0, 1]$, that is,

$$\xi(\lambda x) \stackrel{D}{=} \lambda^{2H}\xi(x)$$

for every $\lambda > 0$. If $H = 1/2$ then the fractional Brownian motion is called the Levy multi-dimensional Brownian motion.

Assume that isotropic field $\xi(x), x \in R^n$ is mean square continuous and $E\xi(x) = 0$.

For a fixed r the random field $\xi(x) = \xi(r, u)$, $r \geq 0$, $u \in s(1)$, is isotropic on the sphere $s(1)$. In view of spectral decompositions (1.2.33) and (1.2.35) we have the following expansion:

$$\xi(x) = \sum_{m=0}^{\infty} \sum_{l=0}^{h(m,n)} S_m^l \left(\frac{x}{\|x\|} \right) \eta_m^l(\|x\|), \quad x \in R^n, \tag{1.2.39}$$

where the random processes

$$\eta_m^l(r) = \int_{s(1)} \xi(ru)\ S_m^l(u)\ d\sigma(u), \quad u = \frac{x}{\|x\|} \in s(1).$$

By Lemma 1.2.2. we obtain

$$E\ \eta_m^l(r)\ \eta_p^q(s) = \delta_m^p\ \delta_l^q\ b_m(r,s), \tag{1.2.40}$$

$$\sum_{m=0}^{\infty} h(m,n)\ b_m(r,s) < \infty,$$

and

$$\begin{aligned} b_m(r,s) &= E\ \eta_m^1(r)\ \eta_p^1(s) \\ &= \frac{|s(1)|}{C_m^\nu(1)} \int_{-1}^{1} B(r,s,t)\ C_m^\nu(t)\ (1-t^2)^{\nu-1/2} dt, \quad \nu = \frac{n-2}{2}, \end{aligned}$$

where the covariance function $B(r, s, \cos\theta) = E\xi(r,u)\ \xi(s,v)$ depends on $r, s \geq 0$ and $\cos\theta$ (the spherical distance between the vectors $u \in s(1)$ and $v \in s(1)$).

Remark 1.2.9. In view of the relation (1.2.28) and (1.2.30) we obtain that an isotropic field is homogeneous if and only if

$$b_m(r,s) = c_1^2(n) \int_0^\infty \frac{J_{m+(n-2)/2}(\lambda r)}{(\lambda r)^{(n-2)/2}} \frac{J_{m+(n-2)/2}(\lambda s)}{(\lambda s)^{(n-2)/2}}\ dG(\lambda),$$

where the spectral measure G defined in (1.2.14a).

From Lemma 1.2.1. and (1.2.39), (1.2.40) we arrive at the following spectral decomposition of covariance function of isotropic random field (1.2.39):

$$\begin{aligned} \operatorname{cov}(\xi(x),\xi(y)) &= B(r,s,cos\theta) \\ &= \sum_{m=0}^{\infty}\sum_{l=1}^{h(m,n)} S_m^l(u)S_m^l(v)b_m(r,s) \\ &= \frac{1}{|s(1)|}\sum_{m=0}^{\infty} h(m,n)b_m(r,s)\frac{C_m^\nu(cos\theta)}{C_m^\nu(1)}, \qquad (1.2.41) \\ \nu &= \frac{n-2}{2.} \end{aligned}$$

where $x=(r,u)$, $y=(s,v)$, $r\geq 0$, $s\geq 0$, $u,v\in s(1)$, and $\cos\theta$ is the spherical distance between the unite vectors u and v.

Thus, for $n=3$ formula (1.2.41) becomes

$$B(r,s,\cos\theta)=\frac{1}{4\pi}\sum_{m=0}^{\infty}(2m+1)\ P_m(\cos\theta)\ b_m(r,s)$$

where $P_m(u)$ are the Legendre polynomials

$$\sum_{m=0}^{\infty}(2m+1)\ b_m(r,r)<\infty \quad \text{and} \quad b_m(r,s)=2\pi\int_{-1}^{1}B(r,s,t)\ P_m(t)dt.$$

For $n=2$ the spectral representations (1.2.39) and (1.2.41) can be written in terms of complex spherical harmonics (see remark 1.2.6) as follows:

$$\xi(r,\varphi)=\sum_{m=-\infty}^{\infty}\eta_m(r)\ e^{im\varphi}, \qquad (1.2.42)$$

where $\eta_m(r), r\geq 0$ are the family of complex-valued random processes such that

$$E\ \eta_m(r)\ \overline{\eta_k(s)}=\delta_m^k\ b_m(r,s)$$

and $b_m(r,s)$ are the sequence of non-negative definite kernels,

$$b_{-m}(r,s)=b_m(r,s),$$

and

$$\sum_{m=-\infty}^{\infty} b_m(r,r) < \infty.$$

The covariance function of the field (1.2.42) can be expanded as

$$B(r,s,\phi_1-\phi_2) = \sum_{m=-\infty}^{\infty} b_m(r,s)\, e^{im(\phi_1-\phi_2)}.$$

We provide examples of covariance functions of isotropic fields on Euclidean space. In all the examples the function $\gamma = \gamma(r,s)$ is a non-negative definite kernel on $R^1_+ \times R^1_+$.

Example 1.2.15. Using the generating function of Gegenbauer polynomials and the relation

$$\begin{aligned}\sum_{m=0}^{\infty} h(m,n)\, C_m^{\nu}(u)\, t^m &= C_m^{\nu}(1)(1-t^2)(1-2tu+t^2)^{-\frac{n}{2}},\\ \nu &= \frac{n-2}{2}, n>2, |t|<1,\end{aligned}$$

we find that

$$\begin{aligned}B(r,s,\cos\theta) &= (1-\gamma\cos\theta+\gamma^2)^{(2-n)/2},\\ B(r,s,\cos\theta) &= \gamma(1-2a\cos\theta+a^2)^{(2-n)/2}, \quad 0<a<1,\\ B(r,s,\cos\theta) &= (1-\gamma^2)(1-2\gamma\cos\theta+\gamma^2)^{-n/2}/\,|s(1)|\,,\\ B(r,s,\cos\theta) &= \gamma(1-a^2)(1-2a\cos\theta+a^2)^{-n/2}, \qquad |a|<1,\end{aligned}$$

are the covariance functions of isotropic fields with $\gamma = \gamma(r,s), n > 2$.

Example 1.2.16. Using the relation

$$\sum_{m=-\infty}^{\infty} a^{|m|} e^{im\phi} = (1-a^2)\ /\ (1-2a\cos\phi+a^2)$$

we find that for $n = 2$

$$\begin{aligned}B(r,s,\phi_1-\phi_2) &= \frac{1-\gamma^2}{1-2\gamma\cos(\phi_1-\phi_2)+\gamma^2},\\ B(r,s,\phi_1-\phi_2) &= \frac{\gamma(1-a^2)}{1-2a\cos(\phi_1-\phi_2)+a^2}, \quad |a|<1.\end{aligned}$$

are the covariance functions of isotropic field on the plane R^2, where $\gamma = \gamma(r, s)$.

1.3 Models of Random Processes and Fields with Singular Spectrum

We present some mathematical models of random processes and fields which are characterized by certain singular properties (e.g., vanishing, or unboundedness) of their spectral densities. These random functions are called by various authors the long-memory fields, random fields with long-range dependence, fields with slowly decaying correlations or strongly dependent random fields.

1. Some historical remarks

It is well known that the assumption of weak dependence is often not valid for real data. This phenomenon has been observed empirically by many scientists long before suitable mathematical models were known. Beran's (1994) book contains a pretty complete bibliography of the subject. In particular, hydrologist Hurst (1951) studying empirically the data set of the yearly minimum water level of the Nile River for the years 622–1281 found that it exhibits long-term dependence rather that short-term dependence.

For a review of long-range dependence in geophysics and hydrology, see Lawrance and Kotegoda (1977). There are many other fields of application where this type of correlation occurs. As early as 1895 the astronomer Newcomb discussed the phenomenon long-range dependence in astronomical data (see, also, Karl Pearson (1902)). Further examples are discussed, for instance, by Student (1927) for chemical data, Jeffreys (1939) for astronomical data, Smith (1938) and Whittle (1956, 1962) for agricultural data, Cox and Townsend (1948) for textile engineering data, Granger (1966, 1980), Mandelbrot (1969, 1973), Carlin, Dempster and Jonas (1985), Carlin and Dempster (1989), Porter–Hudak (1990) and Backus and Zin (1993) for economical data, Mandelbrot and Wallis (1969) for data from biology, geophysics, meteorology and hydrology, Damerau and Mandebrot (1973) for linguistic data, Graf, Hampel and Tacier (1984) for high-quality physical measurements, Haslett and Raftery (1989) for meteorological data and Willinger, Taqqu Leland and Wilson (1994) for telecommunication data. The special issue of *The Journal of Econometrics* (vol. **73**, 1996) is completely devoted to application of long-memory processes for economical data. A special issue of *The Journal of Statistical Planning and Inference* devoted to long-range dependence edited by Anh and Heyde (1999) is in press.

The first studying of processes with singular spectrum was done by Kolmogorov (19940, 1941) in the papers devoted to so called fractional Brownian motion and turbulence theory. Lamperti (1962) points to the fact that normalized sum of random variables converge to self-similar processes (he used the term 'semi-stability').

Afterwards the phenomenon of long-range dependence was studied intensively by Mandelbrot and Van Ness (1968), Mandelbrot and Wallis (1968) and many other researchers.

We shall use the term 'stationary' in place of term 'homogeneous' in all definitions of section 1.2 in the case $n = 1$.

2. Self-similar processes

A random process $Y(t), t \geq 0$ with continuous time is called self-similar with parameter $H > 0$, if for any $a > o$ $Y(at) \overset{D}{=} a^H Y(t)$.

Remark 1.3.1. If $Y(t)$, $t \geq 0$, is a self-similar process with parameter $H > 0$, and $EY(t) = 0$, $EY^2(t) < \infty$, then $\rho(t,s) = E\,Y(t)\,Y(s) < \infty$ and $\rho(at, as) = a^{2H}\rho(t,s)$. In particular,

$$\rho(t,s) = (st)^H \rho\left(\sqrt{\frac{s}{t}}, \sqrt{\frac{t}{s}}\right) = (st)^H \tilde{\rho}\left(\frac{s}{t}\right).$$

Let $w(t)$, $t \geq 0$, be a random process with continuous sample paths and such that:

(i) $w(t)$ is Gaussian;

(ii) $w(0) = 0$ a.s.;

(iii) $w(t)$ has independent increments (r.v.s $w(t_3) - w(t_2)$ and $w(t_2) - w(t_1)$ are independent for every $0 \leq t_1 < t_2 < t_3$);

(iv) $E[w(t) - w(s)] = 0$;

(v) $var[w(t) - w(s)] = \sigma^2 |t - s|, \quad \sigma^2 > 0$.

Then $w(t)$, $t \geq 0$, is called Brownian motion or a Wiener process.

Note that $Ew(t) = 0$, $Ew(t)w(s) = \sigma^2 \min\{t, s\}$.

The Wiener process $w(t)$, $t \geq 0$ is self-similar process with parameter $H = \frac{1}{2}$.

Remark 1.3.2. If $\zeta(t)$, $t \in R^1$, is a strictly stationary process, that is, $\zeta(t) \overset{D}{=} \zeta(t+\tau)$ for every $\tau \in R^1$, then the process $Y(t) = t^H \zeta(\log t)$, $t > 0$,

$Y(0) = 0$ a.s., is a self-similar process on R^1_+ with self-similar parameter $H > 0$ since

$$Y(at) \stackrel{D}{=} (at)^H \zeta\;(\log t + \log a) \stackrel{D}{=} a^H t^H \zeta\;(\log t) = a^H Y(t).$$

Conversely, if $Y(t)$, $t \geq 0$, is a self-similar process with parameter $H > 0$, then process $\zeta(t) = \exp\{-Ht\}\; Y(\exp\{t\}$ is strictly stationary, since

$$\zeta(t+a) \stackrel{D}{=} e^{-Ht} e^{-Ha} Y(e^t e^a) \stackrel{D}{=} e^{-Ht} Y(e^t) \stackrel{D}{=} \zeta(t).$$

In particular, if $w(t)$, $t \geq 0$, is a Wiener process with $\sigma^2 = 1$, then

$$U(t) = e^{-t/2} w(e^t),\;\; t \in R^1,$$

is an Ornstein–Uhlenbeck process, that is, a Gaussian process $U(t)$, $t \in R^1$, with $EU(t) = 0$ and covariance function

$$E\;U(t)\;U(s) = \exp\{-\frac{1}{2}(t+s)\}\;\exp\{\min(t,s)\} = \exp\{-\frac{1}{2}\,|t-s|\}.$$

If for any $k \geq 1$ and any points $t_1, \ldots, t_k$, the distribution of

$$(Y(t_1+a) - Y(t_1+a-1), \ldots, \quad Y(t_k+a) - Y(t_k+a-1))$$

does not depend on $a \in R^1$, then we say that the process $Y(t)$, $t \in R^1$, has stationary increments.

Assume that $EY(t) = 0$. The covariance function $\gamma(t,s) = \mathrm{cov}(Y(t), Y(s))$ of a self-similar process $Y(t), t \geq 0$, with stationary increments is of the form

$$\gamma(t,s) = \frac{\sigma^2}{2}\left(|t|^{2H} + |s|^{2H} - |t-s|^{2H}\right), \quad \sigma^2 = E[Y(1)]^2. \tag{1.3.1}$$

The increment sequence $\xi(i) = Y(i) - Y(i-1)$, $i \in Z^1$, is stationary (or homogeneous) process with covariance function

$$\begin{aligned} B(k) &= cov(\xi(i), \xi(i+k)) \\ &= \frac{\sigma^2}{2}\left(|k+1|^{2H} - 2\,|k|^{2H} - |k-1|^{2H}\right), \qquad (1.3.2) \\ \sigma^2 &= var\xi(i), \end{aligned}$$

for $k \geq 0$ and $B(k) = B(-k)$ for $k < 0$.

Note that if covariances exist and $\lim_{k\to\infty} B(k) = 0$, then $0 < H < 1$.

To simplify notation assume $\sigma^2 = 1$.

The asymptotic behavior of $B(k)$ as $k \to \infty$ follows by Taylor expansion: first note that

$$B(k) = \frac{1}{2}k^{2H}g(k^{-1}),$$

where

$$g(u) = (1+u)^{2H} - 2 + (1-u)^{2H}.$$

If $0 < H < 0$ and $H \neq \frac{1}{2}$ then the first non-zero term in the Taylor expansion of $g(u)$, expanded at the origin, is equal to $2H(2H-1)u^2$. Therefore, as $k \to \infty$, $B(k)$ is equivalent to $H(2H-1)k^{2H-2}$, i.e., $\lim_{k\to\infty} B(k)/[H(2H-1)k^{2H-2}] = 1$. If $\frac{1}{2} < H < 1$, this means that the covariance decays to zero so slowly that $\sum_{k\in Z^1} B(k) = \infty$. For $H = \frac{1}{2}$ the r.v.s $\xi(i)$ are uncorrelated. For $0 < H < \frac{1}{2}$ the covariances are summable, but $\sum_{k\in Z^1} B(k) = 0$. The spectral density of $\xi(i)$, $i \in Z^1$ can be derived from (1.3.2) (Sinai (1976)), namely,

$$f(\lambda) = 2c_f\ (1 - cos\lambda) \sum_{j\in Z^1} |2\pi j + \lambda|^{-2H-1}, \quad \lambda \in [-\pi, \pi),$$

with $c_f = (2\pi)^{-1} sin(\pi H)\ \Gamma(2H+1)$.

Under the above conditions

$$f(\lambda) = c_f\,|\lambda|^{1-2H} + O\left(|\lambda|^{\min(3-2H,2)}\right).$$

For $H \in (0,1), H \neq \frac{1}{2}$, the spectral density $f(\lambda)$ has singular properties. For $H \in (\frac{1}{2}, 1)$ $f(\lambda)$ has a pole at zero of the form $|\lambda|^{1-2H}$ ($f(0) = \infty$). For $H \in (0, \frac{1}{2})$ $f(\lambda)$ is vanishing at zero ($f(0) = 0$).

Remark 1.3.3. For $H \in (0,1)$ the fractional Brownian motion $w_H(t)$, $t \geq 0$, of the index H is a unique centered Gaussian process with covariance kernel given by (1.3.1). The sample paths of $w_H(t)$ are almost surely Hölder continuous of any order less than H, hence continuous but nowhere differentiable. So, the process $w_H(t)$ has stationary increments and is self-similar with parameter $H \in (0,1)$. The increments are independent only when $H = \frac{1}{2}$, for $H > \frac{1}{2}$, the increments are positively correlated, and for $H < \frac{1}{2}$ they are negatively correlated. Fractional Brownian motion is semi-martingale only if $H = \frac{1}{2}$ (Rogers (1997)). An interesting extention of fractional Brownian motion called fractional Riesz-Bessel motion proposed by Anh, Angulo and Ruiz-Medina (1999). The fractional Riesz-Bessel motion has stationary increments and spectral density (in generalized sence) of the form

$$f(\lambda) = \frac{c}{|\lambda|^{2\gamma}} \frac{1}{(1+\lambda^2)^{\beta}}, c > 0, 1/2 < \gamma < 3/2, \beta > 0, \lambda \in R^1.$$

The component $\frac{1}{(1+\lambda^2)^\beta}$ indicates the intermittency, which is important feature of both turbulence and finance processes.

In another paper on fractional Brownian motion, Mandelbrot and Van Ness (1968) defined the process more constructively as the integral

$$\begin{aligned} & w_H(t) - w_H(s) \\ = \; & c_H \left[\int_s^t (t-u)^{H-1/2} dw(u) \right. \\ & \left. + \int_{-\infty}^s \left((t-u)^{H-1/2} - (s-u)^{H-1/2} \right) dw(u) \right], \end{aligned}$$

where $t > s$, and $w(t), t \in R^1$ is the standard Brownian motion (two independent Brownian motions $w(t)$, $t \geq 0$, and $w(t)$, $t \leq 0$). The normalization $EZ^2(1) = 1$ is achieved with the choice

$$c_H = \sqrt{2H \frac{\Gamma\left(\frac{3}{2} - H\right)}{\Gamma\left(H + \frac{1}{2}\right) \Gamma(2 - 2H)}}.$$

Integration with respect to the fractional Brownian motion is discussed in Lin (1995), Gripenberg and Norros (1996).

Remark 1.3.4. Let $\xi(i), i \in Z^1$, be an arbitrary stationary (or homogeneous) process with covariance function $B(k) = cov(\xi(i), \xi(i+k))$, $k \in Z^1$ and spectral density $f(\lambda)$, such that $B(k) = \int_\Pi \cos(\lambda k)\, f(\lambda)\, d\lambda$, $\Pi = [-\pi, \pi)$.

Suppose that:

$$(a) \qquad \lim_{\lambda \to 0} \frac{f(\lambda)}{c_f\, |\lambda|^{1-2H}} = 1$$

for $c_f \in (0, \infty)$, $H \in (0,1)$. Some processes which satisfy (a) also satisfy the time domain property:

$$(b) \qquad \lim_{k\to\infty} \frac{B(k)}{c_B k^{2H-2}} = 1,$$

$c_B < 0$ for $0 < H < \frac{1}{2}$ and $c_B > 0$ for $\frac{1}{2} < H < 1$. With $c_B = 2c_f\Gamma(2-2H)\cos(\pi H)$, it is known that for $\frac{1}{2} < H < 1$, (b) implies (a) (Yong (1974), p. 90, Zygmund (1977)), whereas for $\frac{1}{2} < H < 1$, (a) and (b) are equivalent if the $B(k)$ are quasi monotonically convergent to zero, that is, $B(k) \to 0$ as $k \to \infty$ and for some $K > 0$,

$$B((k+1) \le B(k)(1 + K/j)$$

for all $j > 0$ large enough (Yong (1974), p. 75). In general (a) does not implying (b). In the terminology of the Fourier series the statement '(a) implies (b)' is a theorem of Abelian type and statement '(b) implies (a)' is a theorem of Tauberian type.

3. Fractional ARIMA models (FARMA models)

Fractional ARIMA (p, d, q) models or FARMA (p, d, q) models were introduced by Granger and Joyeux (1980) and Hosking (1981).

Let us first recall the definition of ARMA (p, q) processes $\xi(t)$, $t \in Z^1$ (see Brockewell and Davis (1987), ch.3). To simplify notation we assume $m = E\xi(t) = 0$. Otherwise, $\xi(t)$ must be replaced by $\xi(t) - m$ in all formulas.

In the following, B will denote the backshift operator defined by

$$B^k\xi(t) = \xi(t-k), \quad k = 0, 1, 2, \ldots .$$

Let p and q be integers. Define the polynomials

$$a(z) = 1 - \sum_{j=1}^{p} a_j z^j, \qquad b(z) = 1 + \sum_{j=1}^{q} b_j z^j.$$

Assume that $a(z)$ and $b(z)$ have no common zeros and $a(z) \neq 0$ on the complex unit circle. Furthermore, let $\varepsilon(t)$, $t \in Z^1$ be a Gaussian white noise, that is, an i.i.d. r.v. with normal distribution $\mathcal{N}(0, \sigma^2)$. An ARMA (p, q) model is defined to be the stationary solution of difference equation

$$a(B)\xi(t) = b(B)\varepsilon(t).$$

The spectral density of a ARMA (p, q) process $\xi(t)$, $t \in Z^1$ is of the form

$$f_0(\lambda) = \frac{\sigma^2}{2\pi} \frac{\left|b(e^{i\lambda})\right|^2}{\left|a(e^{i\lambda})\right|^2},$$

$\lambda \in [-\pi, \pi)$ and covariance function

$$R(k) = cov(\xi(t), \xi(t+k))$$

such that

$$|R(k)| \leq Cr^{-k}$$

for some $C > 0$ and $r \in (0,1)$. An ARIMA (p,d,q) model is defined to the solution of difference equation

$$a(B)(1-B)^d \xi(t) = b(B)\ \varepsilon(t).$$

Note that ARIMA $(p,0,q)$ process is also ARMA (p,q) process.

If $d \geq 1$ then the original process $\xi(t)$ is not stationary. To obtain a stationary process $\xi(t)$ must be differenced d times. If $d \geq 0$ is an integer, then $(1-B)^d$ can be written as

$$(1-B)^d = \sum_{k=0}^{d} \binom{d}{k} (-1)^k B^k,$$

with the binomial coefficients

$$\begin{aligned} \binom{d}{k} &= \frac{d!}{k!\ (d-k)!} \\ &= \frac{\Gamma(d+1)}{\Gamma(k+1)\Gamma(d-k+1)} . \end{aligned}$$

Here $\Gamma(s)$ denotes the gamma function, that is

$$\Gamma(s) = \begin{cases} \int_0^\infty t^{s-1} e^{-t} dt, & s > 0; \\ \infty, & s = 0; \\ s^{-1}\Gamma(1+s), & s < 0 \end{cases}$$

Note that for negative integers the gamma function has poles such that the binomial coefficient is zero if $k > d$ and d is an integer. Formally we can define $(1-B)^d$ for any real number $d > -1$ by the binomial expansion

$$\begin{aligned} (1-B)^d &= \sum_{j=0}^{\infty} \pi_j B^j \\ &= 1 - dB + \frac{d(d-1)}{2} B^2 - \cdots, \end{aligned}$$

where

$$\pi_j = \frac{\Gamma(j-d)}{\Gamma(j+1)\Gamma(-d)} = \prod_{k=1}^{j} \frac{k-1-d}{k}, \qquad j = 1, 2, \ldots; \qquad \pi_0 = 1.$$

Let $\xi(t), t \in Z^1$ be a stationary process such that

$$a(B)\ (1-B)^d \xi(t) = b(B)\ \varepsilon(t) \tag{1.3.3}$$

for some $d \in (-\frac{1}{2}, \frac{1}{2})$. Then $\xi(t)$ is called a fractional ARIMA (p, d, q) process or FARMA (p, d, q) process. In particular, if $p = q = 0$, $d \in (-\frac{1}{2}, \frac{1}{2})$ then there is a unique stationary solution $\xi(t)$, $t \in Z^1$, of difference equation (1.3.3) given by

$$\xi(t) = \sum_{j=0}^{\infty} \psi_j \varepsilon(t-j),$$

where

$$\psi_j = \frac{\Gamma(j+d)}{\Gamma(d)\Gamma(j+1)} = \prod_{k=1}^{j} \frac{k-1+d}{k}, \quad j = 0, 1, 2, \ldots.$$

From Stirling's formula

$$\Gamma(u) \sim \sqrt{2\pi} u^{-u+1} (u-1)^{u-1/2}, \quad u \to \infty,$$

it follows that

$$\psi_j \sim \frac{j^{d-1}}{\Gamma(d)}, \qquad j \to \infty,$$

and

$$\sum_{j=0}^{\infty} \psi_j^2 < \infty, \qquad \sum_{j=0}^{\infty} \psi_j = \infty.$$

Denote by $f_s(\lambda)$ and $R(k)$ the spectral density and covariance function respectively of ARMA$(0, d, 0)$ process, we have

$$f_s(\lambda) = \frac{\sigma^2/2\pi}{|1 - e^{-i\lambda}|^{2d}}, \quad \lambda \in [-\pi, \pi),$$

$$R(k) = \sigma^2 (-1)^k \frac{\Gamma(1-2d)}{\Gamma(k+1-d)\ \Gamma(1-k-d)}, \quad k = 0, \pm 1, \ldots$$

The spectral density $f(\lambda)$ of ARIMA (p,d,q) process is of the form

$$\begin{aligned} f(\lambda) &= \frac{\sigma^2}{2\pi} \frac{\left|b(e^{i\lambda})\right|^2}{\left|a(e^{i\lambda})\right|^2} \left|1-e^{-i\lambda}\right|^{-2d} \\ &= f_0(\lambda)\left(2\sin\frac{\lambda}{2}\right)^{-2d} \sim \frac{\sigma^2}{2\pi}\,[b(1)\ /\ a(1)]^2\lambda^{-2d}, \text{ as } \lambda \to 0; \end{aligned}$$

and covariance function $R(k)$ such that $R(k) \leq Ck^{2d-1}$ as $k \to \infty$ for some $C > 0$. We have

$$\sum_{k\in Z^1} R(k) = \infty \qquad \text{for } \delta \in (0, \frac{1}{2})$$

and

$$\sum_{k\in Z^1} R(k) < \infty \qquad \text{for } \delta \in (-\frac{1}{2}, 0).$$

For $d \in (-\frac{1}{2}, \frac{1}{2})$ the spectral density $f(\lambda)$ of ARIMA (p,d,q) process has singular properties. For $d \in (0, \frac{1}{2})$ $f(\lambda)$ has a pole at zero the form $|\lambda|^{-2d}$ $(f(0) = \infty)$, and for $d \in (-\frac{1}{2}, 0)$ $f(\lambda)$ is variation at zero $(f(0) = 0)$.

Comparing this with our notation in the previous sections we see that $d = H - \frac{1}{2} \in (-\frac{1}{2}, \frac{1}{2}), H \in (0,1)$. The above mentioned properties of FARMA process can be found in Brockwell and Davis (1987), ch.12.

4. Gegenbauer processes (GARMA processes)

A generalization of FARMA models was recently proposed by Hosking (1981) and later studied by Gray, Zhang and Woodward (1989) and Chung (1996 a, b).

Equation (1.3.3) is generalized in the following way:

$$a(B)\ (1 - 2\eta B + B^2)^d \xi(t) = b(B)\ \varepsilon(t), \tag{1.3.4}$$

where $\varepsilon(t), t \in Z^1$ is a independent zero r.v. with variance σ^2, the polynomials $a(z)$ and $b(z)$ have root outside the unit circle and $|\eta| \leq 1$, $|d| \leq \frac{1}{2}$, $d \neq 0$. Note that for $\eta = 1$ we obtain the original equation (1.3.3).

Gray, Zhang and Woodward (1989) call the process $\xi(t)$ defined in (1.3.4) the Gegenbauer process or GARMA (p, d, η, q) process.

The Gegenbauer polynomials (see also section 1.2.) $C_j^d(2\eta)$ are defined to be the coefficients in the power series expansion of the following function

$$(1 - 2\eta z + z^2)^{-d} = \sum_{j=0}^{\infty} C_j^d(2\eta)\ z^j, \quad |z| \leq 1, \quad |\eta| \leq 1, d \neq 0,$$

where $C_j^d(2\eta)$ can be computed in several ways.

One expression is

$$C_j^d(2\eta) = \sum_{k=0}^{[j/2]} (-1)^k \Gamma(d+j-k) \frac{(2\eta)^{j-2k}}{\Gamma(d)\,\Gamma(k+1)\,\Gamma(j-2k+1)}.$$

The easiest way to compute $C_j^d(2\eta)$ is based on the recursion formula

$$C_j^d(2\eta) = (2\eta)\left(\frac{d-1}{j}+1\right) C_{j-1}^d(2\eta) - \left(2\frac{d-1}{j}+1\right) C_{j-2}^d(2\eta)$$

with

$$C_0^d(2\eta) = 1, \quad C_1^d(2\eta) = 2d\eta, \quad C_2^d(2\eta) = 2d(d+1)\eta^2 - d, \ldots.$$

Note that

$$C_j^d(2\eta) \sim \frac{\cos[(j+d)\nu - (d\pi/2)]}{\Gamma(d)\sin^d(\nu)} \left(\frac{2}{j}\right)^{1-d} \quad \text{as} \quad j \to \infty,$$

where $\nu = \arccos \eta$. So when $d < 1$, $C_j^d(2\eta)$ also decreases at a hyperbolic rate as $j \to \infty$.

The stationary GARMA $(0, d, \eta, 0)$ process $\xi(t)$ has the moving average representation in terms of Gegenbauer polynomials

$$\begin{aligned} \xi(t) &= (1 - 2\eta\, B + B^2)^{-d} \varepsilon(t) \\ &= \sum_{j=0}^{\infty} C_j^d(2\eta)\, \varepsilon(t-j) \end{aligned}$$

for some $|\eta| \leq 1, d \neq 0$.

The Gegnbauer process is stationary if $|\eta| < 1, d < \frac{1}{2}$, when $|\eta| = 1$, then the Gegenbauer process is stationary if $d < \frac{1}{4}$. It is also straightforward to show that the spectral density of a Gegenbauer process is

$$\begin{aligned} f(\lambda) &= \frac{\sigma^2}{2\pi}\{2\,|\cos\lambda - \cos\nu|\}^{-2d} \\ &= (\sigma^2/2\pi)\left|4\sin\frac{\lambda+\nu}{2}\sin\frac{\lambda-\nu}{2}\right|^{-2d}, \end{aligned}$$

for $0 \le \lambda \le \pi$, $\cos\theta = \eta$. Moreover, we have

$$f(\lambda) \sim \left|\lambda^2 - \nu^2\right|^{-2d} \qquad \text{as} \quad \lambda \to \nu.$$

That is, the spectral density has singular properties. When $0 < d < \frac{1}{2}$ the spectral density is unbounded as $\lambda \to \nu$.

Given the spectral density we can compute the covariance function for the case $|\eta| < 1$:

$$\begin{aligned} R(j) &= \text{cov}\,(\xi(t), \xi(t+j)) \\ &= \frac{\sigma^2}{2\sqrt{\pi}}\,\Gamma(1-2d) \end{aligned}$$

$$\times [2\sin\nu]^{1/2-2d}\left[P_{j-1/2}^{2d-1/2}(\eta) + (-1)^j\; P_{j-1/2}^{2d-1/2}(-\eta)\right],$$

where $P_a^b(u)$ are associated Legendre functions. The associated Legendre functions can be calculated using the recursion

$$P_a^b(u) = \frac{2a-1}{a-b}\,uP_{a-1}^b(u) - \frac{a+b-1}{a-b}\,P_{a-2}^b(u).$$

This recursion requires the following initial terms:

$$P_{-1/2}^{2d-1/2}(\eta) = \left(\frac{1+\eta}{1-\eta}\right)^{d-1/4} \frac{1}{\Gamma(3/2-2d)}\,F\left(\frac{1}{2}, \frac{1}{2}, \frac{3}{2}, -2d, \frac{1-\eta}{2}\right),$$

$$P_{1/2}^{2d-1/2}(\eta) = \left(\frac{1+\eta}{1-\eta}\right)^{d-1/4} \frac{1}{\Gamma(3/2-2d)}\,F\left(-\frac{1}{2}, \frac{3}{2}, \frac{3}{2}, -2d, \frac{1-\eta}{2}\right),$$

where

$$F(a,b,c,z) = \sum_{n=1}^{\infty} \frac{\Gamma(c)\,\Gamma(a+n)\,\Gamma(b+n)}{\Gamma(a)\,\Gamma(b)\,\Gamma(c+n)\,\Gamma(n+1)} z^n$$

are hypergeometric functions. Finally, we have the following approximation to the covariances:

$$R(j) = \frac{2^{1-2d}\sigma^2}{\pi} \sin^{-2d}(\nu)\sin(d\pi)\Gamma(1-2d)\cos(j\nu)\frac{\Gamma(j+2d)}{\Gamma(j+1)}\left(1 + O\left(\frac{1}{j}\right)\right).$$

Then, when $|\eta| < 1$, $0 < d < \frac{1}{2}$, the correlations of the process $\xi(t)$ defined in (1.3.4) with $p = q = 0$ can be approximated by

$$\rho(j) = \frac{R(j)}{R(0)} \sim K\cos(j\nu)\; j^{2d-1} \qquad \text{as } j \to \infty,$$

where $\nu = \arccos\eta$ and K is a constant that may depend on d and ν but not on j.

The GARMA (p, d, η, q) process $\xi(t)$ is stationary if $|\eta| < 1$, $d < \frac{1}{2}$. When $|\eta| = 1$, then $\xi(t)$ is stationary $d < \frac{1}{4}$.

Given the Gegenbauer frequency $\nu = \arccos\eta$, the spectral density of the process $\xi(t)$ is

$$\begin{aligned} f(\lambda) &= \frac{\sigma^2}{2\pi}\left|\frac{b\left(e^{-i\lambda}\right)}{a\left(e^{-i\lambda}\right)}\right|^2 \left|1 - 2\eta\, e^{-i\lambda} + e^{-2i\lambda}\right|^{-2d} \\ &= \frac{\sigma^2}{2\pi}\left|\frac{b\left(e^{-i\lambda}\right)}{a\left(e^{-i\lambda}\right)}\right|^2 \{2\left|\cos(\lambda) - \cos(\nu)\right|\}^{-2d}, \ 0 \le \lambda \le \pi, \end{aligned}$$

and its covariance function $R(j)$ can be approximated by

$$R(j) \sim K\ \cos(j\nu)\ j^{2d-1} \text{ as } j \to \infty,$$

if $d \neq 0$, where K is a nonzero constant independent on j.

The spectral density has singular properties: $f(\nu) = \infty$, $0 < d < \frac{1}{2}$; and $f(\nu) = 0$ for $-\frac{1}{2} < d < 0$.

Remark 1.3.5. Viano, Deniau and Oppenheim (1995) considered the discrete time linear processes, whose spectral densities have the form

$$f(\lambda) = \left|\prod_{j=1}^{J}\{1 - \alpha_j\, e^{i\lambda j}\}^{d_j}\right|^2, \quad \lambda \in [-\pi, \pi], \tag{1.3.5}$$

where the α_j are complex numbers possibly with $|\alpha_j| = 1$, and where the d_j are necessarily real. The paper is denoted to the so called autoregressive fractional integrated seasonal moving average (ARFISMA) processes, having densities of the form (1.3.5) where the coefficients $\alpha_j = \exp\{2\pi j/s\}$ with j lying in $\{0, \ldots, s-1\}$. These densities are the product of several GARMA densities.

In particular Hassler (1994) considered the spectral densities of the form

$$f(\lambda) = \left|1 - e^{i\lambda}\right|^{2d_1} \left|1 - e^{i\lambda}\right|^{2d_2} \left|1 - e^{2i\lambda}\right|^{2d_3}, \quad \lambda \in [-\pi, \pi].$$

5. Fractional stochastic differential equations

Continuous-time fractional ARIMA models and fractional stochastic differential equations were introduced by Comte (1996), Comte and Renault (1996 a, 1996 b), Chambers (1996), Iglói and Terdik (1997).

We denote by $w_\alpha(t), t \geq 0$ a normalized fractional Brownian motion with self-similar parameter $H = \alpha + \frac{1}{2} \in (0, 1)$, where $\alpha(-\frac{1}{2}, \frac{1}{2})$, characterized by the following properties:

(i) $w_\alpha(t)$ has stationary increments;

(ii) $w_\alpha(0) = 0$, and $Ew_\alpha(t) = 0$ for all t;

(iii) $Ew_\alpha^2(t) = |t|^{2\alpha+1} = |t|^{2H}$ for all $t \geq 0$;

(iv) $w_\alpha(t)$ is Gaussian;

(v) $w_\alpha(t)$ has continuous sample paths.

Then the covariance function of a process $w_\alpha(t)$ is of the form

$$\begin{aligned}\gamma(t, s) &= \operatorname{cov}(w_\alpha(t), w_\alpha(s)) \\ &= \frac{1}{2}\{|t|^{2\alpha+1} + |s|^{2\alpha+1} - |t - s|^{2\alpha+1}\}, \quad |\alpha| < \frac{1}{2}.\end{aligned}$$

When $H \neq \frac{1}{2}$ (or $\alpha \neq 0$), the tractional Brownian motion is neither Markov process not the semimartingale (see e.g., Liptser and Shiryaev (1986), section 4.9.13). The formula for the conditional expectation $E[w_\alpha(T) \mid w_\alpha(s), \ 0 \leq s \leq t], \ 0 < t < T$, was found by Molchan (1969), Gripenberg and Norros (1996).

Since $w_\alpha(t)$ is not semimartingale, we refer to the integration theory of Gaussian processes (see e.g., Huang and Cambanis (1978)) instead of the more usual martingale approach. In fact, we need to consider deterministic integrands only, and the technical framework can be described briefly (see, for details, Lin(1995), Gripenberg and Norros (1996), Comte (1996),Comte and Renault (1996a, 1996b), Chambers (1996), Norros,Valkeila and Virtamo (1997)).

Both $\mathbf{L}_2$ and pathwise approaches are possible. For $H > \frac{1}{2}$ let A denote the integral operator

$$Af(t) = H(2H-1)\int_0^\infty f(s)\,|s-t|^{2H-2}\,ds,$$

and denote the inner product

$$< f, g >_A = < f, Ag > = H\,(2H-1)\int_0^\infty \int_0^\infty f(s)\; f(g)\; |s-t|^{2H-2}\,ds\,dt$$

where $< \cdot, \cdot >_A$ denotes the usual inner product of $\mathbf{L}_2(R^1_+)$. Denote by $\mathbf{L}_2\{A\}$ the space of equivalence classes of measurable functions f such that $< f, f >_A < \infty$. It is easy to check that the association $w_\alpha \rightarrow \chi([0, t))$

can be extended to an isometry between the Gaussian space generated by the random variables $w_\alpha(t)$, $t \geq 0$, as the smallest closed linear subspace of $\mathbf{L}_2(\Omega, \mathcal{F}, P)$ containing them, and the function space $\mathbf{L}_2\{A\}$, where $\chi(B)$ is the indicator of a set B. For $f \in \mathbf{L}_2\{A\}$, the integral $\int_0^\infty f(t)dw_\alpha(t)$ can now be defined as the image of f under this isometry. Then for $f, g \in \mathbf{L}_2\{A\}$

$$E \int_0^\infty f(t)dw_\alpha(t) \int_0^\infty f(s)dw_\alpha(s) = H(2H-1) \int_0^\infty \int_0^\infty f(s)f(g)\,|s-t|^{2H-2}\,ds\,dt.$$

For $H < \frac{1}{2}$, the integral in the above definition of A diverges, and we have to define the operator in another way. In this case an appropriate definition of A is

$$Af(t) = H \int_0^\infty |s-t|^{2H-1}\,\mathrm{sgn}(t-s)\;df(s).$$

We interpret $f(0-) = 0$ so that, for example, the indicator function $\chi_{[0,t)}$ is identified with the signed measure $\delta_0 - \delta_t$. It is easy to check that $w_\alpha(t) \to \chi([0,t))$ defines an isometry.

If f has bounded variation, the integral $\int_0^T f(t)\,dw_\alpha(t)$ can be defined ω by ω as a limit of Riemann sums, and it is easy to see the integral obtained coincides with the $\mathbf{L}_2$ integral almost surely.

Comte (1996) and Comte and Renault (1996a, 1996b) proved that singular properties of a spectrum could be found in a very large class of continuous time processes.

Consider the processes

$$\xi(t) = \int_0^t a(t-s)dw(s), \tag{1.3.6}$$

whose $w(t) = w_0(t)$ is a standard Brownian motion. They are asymptotically equivalent to the stationary process

$$\eta(t) = \int_{-\infty}^t a(t-s)dw(s) \tag{1.3.7}$$

if $\int_0^\infty a^2(u)du < \infty$. Such processes are called fractional processes if

$$a(u) = \frac{u^\alpha\,\tilde{a}(u)}{\Gamma(1+\alpha)}, \quad |\alpha| < \frac{1}{2}, \tag{1.3.8}$$

where $\tilde{a}(u)$ is continuously differentiable function on $[0, T]$ (or R^1_+).

Those processes can be also written as

$$\xi(t) = \int_0^t c(t-s)dw_\alpha(s), \qquad w_\alpha(t) = \int_0^t \frac{(t-s)^\alpha dw(s)}{\Gamma(1+\alpha)}.$$

The $a(s)$ and $c(t)$ functions are related by

$$c(u) = \frac{d}{du}\left\{\int_0^u \frac{(u-s)^{-\alpha} s^\alpha \tilde{a}(s)\, ds}{\Gamma(1-\alpha)\,\Gamma(1+\alpha)}\right\}$$

and

$$u^\alpha \tilde{a}(u) = \frac{d}{du}\left\{\int_0^u c(s)\,(u-s)^\alpha ds\right\}.$$

Those processes are known to have long memory properties if

$$\lim_{u\to\infty} u\, \tilde{a}(u) = a_\infty, \quad 0 < \alpha < \frac{1}{2}. \tag{1.3.9}$$

Those long memory properties are mainly, for $\xi(t)$ given by (1.3.6.) and satisfying (1.3.8.) and (1.3.9), as follows:

(i) $\xi(t)$ is such that

$$\gamma(h) \sim K h^{2\alpha-1}$$

for $h \to +\infty$, where

$$K = \frac{\Gamma(1-2\alpha)\,\Gamma(\alpha)\, a_\infty^2}{\Gamma(1-\alpha)\,\Gamma(1+\alpha)}$$

and $\gamma(h)$ is the covariance function of the stationary processes $\eta(t)$ associated to $\xi(t)$ (see, equation (1.3.7));

(ii) the continuous spectral density $f(\lambda)$ of $\xi(t)$ is such that

$$f(\lambda) \sim \lambda^{-2\alpha} m a_\infty^2 \quad \text{for} \quad \lambda \to 0,$$

where m is a positive scalar.

Finally, we can define a fractional derivation and integration of order α in Weyl's sense that yields a usual (asymptotically) stationary short memory process

$$\begin{aligned}\xi^{(\alpha)}(t) &= \frac{d}{dt}\int_0^t \frac{(t-s)^{-\alpha}}{\Gamma(1-\alpha)}\, \xi(t)\, dt \\ &= \int_0^t \frac{(t-s)^{-\alpha}}{\Gamma(1-\alpha)} d\xi(t).\end{aligned}$$

The process $\xi^{(\alpha)}(t)$ is well defined (at least $\mathbf{L}_2$) and if $\tilde{a}(0) \neq 0$ and $\tilde{a}$ is twice continuously differentiable, then

$$\xi^{(\alpha)}(t) = \int_0^t c(t-s)dw(s),$$

where $c(u)$ is the same as previously.

Let $\xi(t)$ be a long memory process (i.e., satisfying (1.3.8.) and (1.3.9.)). Then

a) the covariance function $\gamma(h)$ associated to $\xi(t)$ satisfies, for $h \to 0$ and k constant,

$$\gamma(h) = \gamma(0) + \frac{1}{2}k \, |h|^{2\alpha+1} + o\left(|h|^{2\alpha+1}\right);$$

b) there is a process $\zeta(t)$ equivalent to $\xi(t)$ and such that sample function of $\zeta(t)$ satisfies a Lipschitz condition of order β for every $\beta \in (0, \alpha + \frac{1}{2})$;

c) the process $\xi(t)$ is ergodic in $\mathbf{L}_2$ sense, i.e.

$$T^{-1} \int_0^T \xi(s)ds \overset{m.s.}{\to} 0, \quad T \to \infty.$$

Consider here the first-order fractional stochastic differentia equation

$$d\xi(t) = a_0 \, \xi(t) \, dt + \sigma \, dw_\alpha(t), \quad \xi(0) = 0, \quad a_0 < 0. \tag{1.3.10}$$

The solution of (1.3.10) is

$$\xi(t) = \int_0^t \exp\{a_0(t-s)\}dw_\alpha(s).$$

It is a partial case of (1.3.6) with

$$\begin{aligned} a(u) &= \frac{\sigma}{\Gamma(1+\alpha)} \frac{d}{du} \int_0^u e^{a_0 s}(u-s)^\alpha ds \\ &= \frac{\sigma}{\Gamma(1+\alpha)} \left(u^\alpha + a \, e^{a_0 u} \int_0^u e^{-a_0 s} s^\alpha ds \right). \end{aligned}$$

Therefore, the solution $\xi(t)$ of equation (1.3.10) is given by

$$\xi(t) = \int_0^t \frac{(t-s)^\alpha dw_\alpha(s)}{\Gamma(1+\alpha)}$$

and its derivative of order α is the solution of the associated standard stochastic differential equation:

$$\xi^{(\alpha)}(t) = \int_0^t \exp\{a_0(t-s)\}\sigma dw(s).$$

We can also give the (continuous time) spectral density of $\xi(t)$:

$$f_c(\lambda) = \frac{\sigma^2}{\Gamma(1+\alpha)\lambda^{2\alpha}(\lambda^2 + a_0^2)}, \quad \lambda \in R^1 \tag{1.3.11}$$

The spectral density (1.3.11) has singular properties for $|\alpha| < \frac{1}{2}$, that is $f(0) = \infty$ for $0 < \alpha < \frac{1}{2}$, and $f(0) = 0$ for $-\frac{1}{2} < \alpha < 0$. It can be checked that for $0 < \alpha < \frac{1}{2}$ the process $\xi(t)$ satisfies (1.3.8) with $a_\infty = (\alpha\sigma)/|a_0|$.

The stationary process with spectral density (1.3.11) should be called the fractional Ornstein–Uhlenbeck process.

For any step $h > 0$, $\xi_\alpha(k) = \xi(kh)$, $k \in Z^1$, is (asymptotically) a stationary Gaussian process with discrete time and spectral density (see, (1.2.8)) of the from

$$f_d(\lambda) = h^{-1} \sum_{k \in Z^1} f_c\left(\frac{\lambda + 2k\pi}{h}\right), \quad \lambda \in [-\pi, \pi),$$

which also has singular properties at zero.

Chambers (1996) present some heuristic approaches for fractional noise calculus. Consider a real-valued process $\eta(t)$, $t \in R^1$ whose derivative of order κ (where κ is not restricted to be an integer) in the noise (generalized) process $u(t)$, $t \in R^1$ with zero mean and spectral density $f_u(\lambda) = \sigma^2/2\pi$, $\lambda \in R^1$. More formally $D^\kappa\eta(t) = u(t)$, where D^κ is the Weyl fractional derivative of order κ defined by

$$D^\kappa\eta(t) = \frac{1}{\Gamma(1-\kappa)} \frac{d}{dt} \int_{-\infty}^t (t-s)^{-\kappa}\eta(s)\, ds. \tag{1.3.12}$$

The parameter κ will be referred to as long-memory parameter.

The generalized process $u(t)$ is a pure noise process in continuous time. This process has infinite variance and is not physically realizable (such assumptions have been used extensively in the study of continuous time process, see e.g., Priestley (1981)). An alternative would be to consider a random measure $\zeta(dt)$ defined over all subsets of the real line $-\infty < t < \infty$ such that

$$E\zeta(dt) = 0, \quad E\zeta(dt)^2 = \sigma^2 dt, \text{ and } \quad E\zeta(\Delta_1)\zeta(\Delta_2) = 0$$

for disjoint sets Δ_1 and Δ_2. Equation (1.3.12) could then be defined as $d[D^{\kappa-1}\eta(t)] = \zeta(dt)$. Then process corresponding to (1.3.12) could be written

$$\begin{aligned} \eta(t) &= D^{-\kappa}u(t) \\ &= \frac{1}{\Gamma(\kappa)} \int_{-\infty}^{t} (t-s)^{\kappa-1} u(s)\, ds \\ &= \frac{1}{\Gamma(\kappa)} \int_{-\infty}^{t} (t-s)^{\kappa-1} \zeta(ds). \end{aligned} \tag{1.3.13}$$

The stationary process $\xi(t) = \eta(t) - \eta(t-1)$ has the following representation

$$\xi(t) = \frac{1}{\Gamma(\kappa)} \left\{ \int_{-\infty}^{t} (t-s)^{\kappa-1}\zeta(ds) - \int_{-\infty}^{t-1} (t-1-s)^{\kappa-1}\zeta(ds) \right\}, \quad t \in R^1$$

This definition corresponds closed with the earlier definitions of fractional Brownian motion (see, Mandelbrot and van Ness (1968)), although at this stage is not being assumed that $u(t)$ is a Gaussian process. Furthermore, (1.3.13) provides a natural motivation for the fractional ARIMA model in continuous time.

The process $\xi(t)$, $t \in R^1$, is a real-valued zero-mean stationary process with spectral density

$$f_c(\lambda) = \frac{\sigma^2}{\pi}(1 - \cos\lambda)\ |\lambda|^{-2\kappa}, \quad \lambda \in R^1 \tag{1.3.14}$$

and covariance function, defined for $|\tau| \geq 0$, given by

$$\begin{aligned} \gamma(\tau) &= E\ \xi(t)\ \xi(t-\tau) \\ &= \sigma^2 V_\kappa \left[|\tau+1|^{2\kappa-1} - 2\,|\tau|^{2\kappa-1} + |\tau-1|^{2\kappa-1} \right], \end{aligned}$$

where

$$V_\kappa = -\left[2\Gamma(2\kappa)\cos(\kappa\pi)\right]^{-1} > 0, \quad \kappa \in \left(\frac{1}{2}, \frac{3}{2}\right).$$

Then for correlation function

$$\rho(\tau) = \gamma(\tau)\ /\ \gamma(0)$$

we have

$$\lim_{\tau\to\infty} \tau^{3-2\kappa}\rho(\tau) = (\kappa-1)(2\kappa-1).$$

Note the correlation for $\tau \geq 1$ are negative for $\kappa \in (\frac{1}{2}, 1)$ and positive for $\kappa \in (1, \frac{3}{2})$.

The spectral density (1.3.14) has singular properties at zero as well as spectral densities a $f_{d_1}(\lambda)$ and $f_{d_2}(\lambda)$ of a discrete-time process

$$\xi_{d_1}(t) = \xi(t), \quad t \in Z^1,$$

and

$$\xi_{d_2}(t) = \int_{t-1}^{t} \xi(s)\, ds, \quad t \in Z^1,$$

respectively. Using (1.2.8) and (1.2.11) we have

$$f_{d_1}(\lambda) = \frac{\sigma^2}{\pi}(1-\cos\lambda) \sum_{j=-\infty}^{\infty} |\lambda + 2\pi j|^{-2\kappa}, \quad \lambda \in [-\pi, \pi), \tag{1.3.15}$$

$$f_{d_2}(\lambda) = \frac{\sigma^2}{\pi}(1-\cos\lambda)^2 \sum_{j=-\infty}^{\infty} |\lambda + 2\pi j|^{-2\kappa-2}, \quad \lambda \in [-\pi, \pi). \tag{1.3.16}$$

If follows from (1.3.15) and (1.3.16) that, as $\lambda \to 0$,

$$f_{d_1}(\lambda) \sim \frac{\sigma^2}{\pi} |\lambda|^{2-2\kappa} \quad \text{and} \quad f_{d_2}(\lambda) \sim \frac{\sigma^2}{\pi} |\lambda|^{2-2\kappa}.$$

Thus, for $\kappa \in (1, \frac{3}{2})$ the spectral densities $f_{d_1}(\lambda)$ and $f_{d_2}(\lambda)$ have a pole at the origin, where as for $\kappa \in (\frac{1}{2}, 1)$ they have a zero at the zero frequency.

Note that (1.3.15) and (1.3.16) can be expressed in term of Riemann's generalized zeta function

$$\zeta(s,a) = \sum_{j=0}^{\infty}(j+a)^{-s} \tag{1.3.17}$$

as follows:

$$\begin{aligned} f_{d_1}(\lambda) &= \frac{\sigma^2}{\pi(2\pi)^{2\kappa}} (1-\cos\lambda)\, [\zeta(2\kappa, \tilde{\lambda}) + \zeta(2\kappa, -\tilde{\lambda}) - \tilde{\lambda}^{-2\kappa}], \\ f_{d_2}(\lambda) &= \frac{2\sigma^2}{\pi(2\pi)^{2+2\kappa}} (1-\cos\lambda)\, [\zeta(2+2\kappa, \tilde{\lambda}) + \zeta(2+2\kappa, -\tilde{\lambda}) - \tilde{\lambda}^{-2-2\kappa}], \end{aligned}$$

where $-\frac{1}{2} < \tilde{\lambda} = \frac{\lambda}{2\pi} \leq \frac{1}{2}$.

Consider a continuous parameter fractional ARIMA model of the form

$$a(D)D^{\kappa}\eta(t) = b(D)u(t), \ t \in R^1,$$

where $D = \frac{d}{dt}$ and $a(z) = 1 + \sum_{j=1}^{p} a_j z^j$ and $b(z) = 1 + \sum_{j=1}^{q} b_j z^j$ are polynomials such that all the root of $a(z)$ lie strictly in the left half of the complex plane. Then the stationary process $\xi(t) = \eta(t) - \eta(t-1)$ satisfies the differential equation $a(D)\xi(t) = b(D)v(t), \ t \in R^1$, where

$$v(t) = \frac{1}{\Gamma(\kappa)} \left[\int_{-\infty}^{t} (t-s)^{\kappa-1} u(s) \, ds - \int_{-\infty}^{t-1} (t-1-s)^{\kappa-1} u(s) \, ds \right].$$

The spectral density of a process $\xi(t), \ t \in R^1$, is given by

$$f_c(\lambda) = \frac{\sigma^2}{\pi} |\lambda|^{-2\kappa} (1 - \cos\lambda) \frac{|b(i\lambda)|^2}{|a(i\lambda)|^2}, \quad \lambda \in R^1. \tag{1.3.18}$$

Thus, for $\kappa \in (1, \frac{3}{2})$ the spectral density $f_c(\lambda)$ have a pole at the origin, whereas for $\kappa \in (\frac{1}{2}, 1)$ the spectral density have a zero at the zero frequency.

The spectral density of a discrete-time process $\xi_{d_1}(t), \ t \in Z^1$, is of the form

$$f_{d_1}(\lambda) = \frac{\sigma^2}{\pi} (1 - \cos\lambda) \sum_{j=-\infty}^{\infty} \frac{|b(i(\lambda + 2\pi j))|^2}{|\lambda + 2\pi j|^{2\kappa} |a(i(\lambda + 2\pi j))|^2}, \tag{1.3.19}$$

where $\lambda \in [-\pi, \pi)$.

The spectral density of a discrete-time process $\xi_{d_2}(t) = \int_t^{t-1} \xi(s)ds$, $t \in Z^1$, is of the form

$$f_{d_2}(\lambda) = \frac{2\sigma^2}{\pi} (1 - \cos\lambda)^2 \sum_{j=-\infty}^{\infty} \frac{|b(i(\lambda + 2\pi j))|^2}{|\lambda + 2\pi j|^{2+2\kappa} |a(i(\lambda + 2\pi j))|^2}, \tag{1.3.20}$$

where $\lambda \in [-\pi, \pi)$.

It is easy to see that

$$f_{di}(\lambda) \sim \frac{\sigma^2}{\pi} |\lambda|^{2-2\kappa} \quad \text{as } \lambda \to 0.$$

Both spectral densities has singularities at zero for r $\kappa \in (\frac{1}{2}, \frac{3}{2})$.

Similar model of continuous-time ARIMA-process considered by Heyde and Gay (1992).

Heyde and Gay considered equation $a(D)(D+\gamma)^{\nu}\eta(t) = b(D)u(t)$, where $u(t)$, $t \in R^1$, is white noise, $D = \frac{d}{dt}$, $a(z)$ and $a(z)$ are polynomials, and

$$(D+\gamma)^{\nu} = \sum_{j=0}^{\infty} \frac{\nu(\nu-1)\cdots(\nu-j+1)}{j!}\gamma^{\nu-j}D^j.$$

This equation has a stationary solution with spectral density

$$f_{\gamma}(\lambda) = \frac{\frac{\sigma^2}{2\pi}\left|b(i(\lambda)\right|^2}{\left|a(i(\lambda)\right|^2}(\gamma^2+\lambda^2)^{\nu}, \quad \lambda \in R^1,$$

which exhibits shot-range dependence if $\gamma > 0$. However, if $\gamma = 0$ we have long-range dependence, but when $0 < \nu < \frac{1}{2}$, $f_{\gamma}(\lambda)$ is not the spectral density of a stationary process, but of self-similar stationary increment process of order $\nu + \frac{1}{2}$. It is, however, very closely approximated by a continuous stationary process with spectral density

$$f_z(\lambda) = \frac{4\sigma^2}{2\pi}(\sin^2\lambda/2)\,\frac{\lambda^{-2-2\nu}\left|b(i\lambda)\right|^2}{\left|a(i\lambda)\right|^2}, \quad \lambda \in R^1,$$

derived from

$$z(t) = \int_{t-1/2}^{t+1/2} \xi(s)ds,$$

where the process $\xi(t), t \in R^1$, has stationary increments and spectral density $f_{\gamma}(\lambda)$.

Another approach was proposed by Viano, Deniau and Oppenheim (1994).

Remark 1.3.6. Inoue (1993) investigated a class of Langevin equations with delay. The unique of this equation is long-memory stationary Gaussian process. Let $B(t)$ be the covariance function of the solution, then

$$\int_0^{\infty} B(t)dt = \infty.$$

6. Fractional Laplace equation

Gay and Heyde (1990, 1992) proposed some model of continuous parameter random field with singular spectrum via the stochastic Laplace equation.

Consider the stochastic operational Laplace equation

$$(\nabla - \beta^2)^\nu \eta(x) = u(x), \quad \nu > 0, \quad x \in R^n \tag{1.3.21}$$

where $u(x)$ is a white noise, β is real, ∇ is Laplacian (see, section 1.1.), and the operator $(\nabla - \beta^2)^\nu$ is defined by

$$(\nabla - \beta^2)^\nu = (-1)^\nu \sum_{j=0}^{\infty} \binom{\nu}{j} (-\nabla)^j \, \beta^{2(\nu-j)}.$$

Then a homogeneous isotropic solution to equation (1.3.21) has spectral density

$$f_\beta(\lambda) = \frac{\sigma^2}{(2\pi)^n} \left(\|\lambda\|^2 + \beta^2\right)^{-2\nu}, \quad \lambda \in R^n.$$

For $\beta^2 > 0$ the field $\eta(x)$ in (1.3.21) exhibits short-range dependence. However, for

$$(n-1)/4 < \nu < n/4, \qquad , f_0(\lambda) \sim \sigma^2 (2\pi)^{-n} \|\lambda\|^{-4\nu}$$

is the n-dimensional spectral density of an isotropic self-similar random field $\xi(x)$, $x \in R^n$, say, with homogeneous increments. Also, the field

$$Z(x) = \int_{A_x} \xi(y) \, dy,$$

$$A_x = \{x_i - 1/2 \leq y_i \leq x_i + 1/2, i = 1, \dots, n\}$$

which has spectral density

$$f_Z(\lambda) = \sigma^2 \pi^{-n} \left\{ \sum_{i=1}^{n} \lambda_i^{-2} \sin^2 \frac{\lambda_i}{2} \right\} (\lambda_1^2 + \ldots + \lambda_n^2)^{-2\nu}, \quad \lambda \in R^n \tag{1.3.22}$$

is homogeneous with asymptotic covariance $B(\tau) \sim \text{const} \|\tau\|^{4\nu - n}$, $\tau \in R^n$, as $\|\tau\| \to \infty$, (see, Sinai (1976)).

The spectral density (1.3.22) has singular properties at zero.

Remark 1.3.6. The spectral densities of the form

$$f(\lambda_1, \dots, \lambda_n) = \frac{\sigma^2 \prod_{j=1}^{p} \left| \|\lambda\|^2 + \theta_j \right|^{2k_j}}{(2\pi)^n (\|\lambda\|^2 + \beta^2)^{2\nu} \prod_{j=1}^{q} \left| \|\lambda\|^2 + \phi_j \right|^{2m_j}}, \quad \lambda \in R^n,$$

where k_j and m_j are positive integers satisfying $\sum m_i \geq \sum k_i + 1$, while θ_j, $1 \leq j \leq p$, are real and ϕ_i, $1 \leq i \leq q$, are positive, covers a comprehensive range of short- and long-range behavior.

7. Long memory shot noise

Giraitis, Molchanov and Surgailis (1993) introduced long memory shot noise processes and fields.

Consider the shot noise random process $\xi(t), t \in R^1$, of the form

$$\begin{aligned} \xi(t) &= \sum_j \eta_j \, g\left(\frac{t-\tau_j}{\zeta_j}\right) \\ &\stackrel{D}{=} \int_X f_z(s,y,z)\, q(ds,dy,dz), \end{aligned}$$

$g(t)$, $t \in R^1$, is a non-random function such that $\int_{R^1} g^2(x)\,dx < \infty$; τ_j, $j \in Z^1$, is a stationary Poisson point process with constant rate $\lambda > 0$, and (η_j, ζ_j), $j \in Z^1$, is an i.i.d. sequence of $R^1 \times R^1_+$-valued random vectors, independent of τ_j, $j \in Z^1$, and $q(ds,dy,dz)$ is the centered Poisson random measure on $X = R^1 \times R^1 \times R^1_+$ with variance

$$\mu(ds,dy,dz) = \lambda ds P(\eta \in dy,\ \zeta \in dz),$$

and

$$f_t(s,y,z) = yg\left(\frac{t-s}{z}\right).$$

The characteristic function of $\xi(t)$ is equal to

$$E \exp\left\{ i \sum_{k=1}^{r} a_k \, \xi(t_k) \right\} = \exp\left\{ \lambda \int_{R^1} \langle e^{i\phi(s)} - 1 - i\phi(s) \rangle \, ds \right\},$$

where

$$\phi(s) = \eta \sum_{k=1}^{r} a_k \, g\left(\frac{t_k - s}{\zeta}\right), \quad a_1, \ldots, a_r \in R^1, \quad t_1, \ldots, t_r \in R^1_+.$$

By $\langle \cdot \rangle$ we denote the expectation corresponding to a generic pair (η, ζ) of the sequence $(\eta_j, \zeta_j), j \in Z^1$. Suppose that $\langle \eta g(t/\zeta) \rangle = 0$, $t \in R^1$, $\langle \eta^2 \zeta \rangle < \infty$.

Then random process $\xi(t)$, $t \in R^1$, is well defined under the above conditions. More precisely, the series converge a.s. and defines a strictly stationary process $\xi(t)$ ($\xi(t) \stackrel{D}{=} \xi(t+\tau)$ for all $t, \tau \in R^1$) with $E\xi(t) = 0$ and covariance function

$$B(t) = E\xi(0)\xi(t) = \left\langle \eta^2 \zeta B_0\left(\frac{t}{\zeta}\right)\right\rangle,$$

where

$$B_0(t) = \lambda \int\limits_{R^1} g(s)g(t+s)ds.$$

If $B_0(t) \sim t^{-\alpha}$, $0 < \alpha < 1$, as $t \to \infty$ and $\langle \eta^2 \zeta^{1+\alpha}\rangle < \infty$, then $B(t) \sim t^{-\alpha}$, as $t \to \infty$ (see, Giraitis, Molchanov and Surgailis (1993)). These results can be extended on random field.

Remark 1.3.7. Barndorff-Nielsen (1998) introduced Ornstein-Uhlenbeck type processes with long-range dependence and marginal normal inverse Gaussian (or inverse Gaussian) distributions. These processes are interesting both for finance and turbulence theory.

1.4 Tauberian and Abelian Theorems for Correlation Function of Homogeneous Isotropi Random Fields

We present theorems of Tauberian and Abelian types for nonintegrated covariance functions of homogeneous isotropic random fields.

1. Tauberian and Abelian Theorems

Consider the spectral decomposition of covariance function of mean-square continuous homogeneous and isotropic random field $\xi(x)$, $x \in R^n$, with mean zero and unite variance (see, (1.2.14))

$$B_n(r) = \int\limits_0^\infty Y_n(\lambda r)\, dG(\lambda), \qquad \int\limits_0^\infty dG(\lambda) = B_n(0) = 1, \tag{1.4.1}$$

where G is spectral function of the field defined by (1.4.14 a), that is bounded nondecreasing function, and Y_n is a spherical Bessel function defined by (1.2.13)

Remark 1.4.1. We use the spectral function G in the spectral decomposition (1.4.1) instead of spectral measure in the formula (1.2.14). We use the same letter G for both spectral function and spectral measure. Hope that there is no danger of misunderstanding.

Remark 1.4.2. In this section we write $B_n(r)$ in (1.4.1) specifying the dimension $n \geq 1$ of the field.

Let $v(r) = \{x \in R^n : \|x\| < r\}$ and $s(r) = \{x \in R^n : \|x\| = r\}, n \geq 2$ be a ball and sphere of radius $r > 0$, and

$$\eta(r) = \int_{v(r)} \xi(x)\, dx, \qquad \zeta(r) = \int_{s(r)} \xi(x)\, d\sigma(x) \tag{1.4.2}$$

where $d\sigma(x)$ is an element of Lebesgue measure on the sphere s(r) (see, (1.1.4)) we shall use the following know formulas (see, Gikhman and Skorokhod (1974))

$$\int\limits_{v(r)} e^{i<x,y>} dx = \left(\frac{2\pi r}{\|y\|}\right)^{n/2} J_{n/2}(r\,\|y\|),$$

$$\int\limits_{s(r)} e^{i<x,y>} d\sigma(x) = \left(\frac{2\pi r}{\|y\|}\right)^{n/2} \|y\|\;\; J_{(n-2)/2}(r\,\|y\|), \tag{1.4.3}$$

where J_ν is the Bessel function of the first kind of order ν defined (1.2.12).

Denote the incomplete beta function by

$$I_\mu(p,q) = \frac{\Gamma(p+q)}{\Gamma(p)\,\Gamma(q)} \int\limits_0^\mu t^{p-1}(1-t)^{q-1}dt, \quad \mu \in [0,1], \quad p > 0,\ q > 0 \tag{1.4.4}$$

The following two assertion holds.

Lemma 1.4.1. Let β and γ be two independent and uniformly distributed inside the ball $v(r)$ random vectors. Then the density function $\psi_r(u)$, $u \geq 0$ of the distance $\rho_{\beta\gamma} = \|\beta - \gamma\|$ between β and γ is

$$\psi_r(u) = n\, u^{n-1} r^{-n} I_{1-(u/2r)^2}\left(\frac{n+1}{2}, \frac{1}{2}\right), \quad 0 \leq u \leq 2r \tag{1.4.5}$$

Remark 1.4.3. For $n = 1,2,3$ relation (1.4.5) yields respectively the expressions:

$$\frac{1}{r}\left(1 - \frac{u}{2r}\right); \quad 8\frac{u}{2r}\left[\arccos\frac{u}{2r} - \frac{u}{2r}\sqrt{1-\left(\frac{u}{2r}\right)^2}\,\right]\frac{1}{\pi r},$$

$$6\left(\frac{u}{2r}\right)^2\left(1-\frac{u}{2r}\right)^2\left(2+\frac{u}{2r}\right)\frac{1}{r}.$$

Lemma 1.4.2. Les $s(r)$ be a sphere in R^n, $n \geq 2$. If β and γ are independent random vectors selected on $s(r)$ according to the uniform law, that is,

$$\begin{aligned} P\{\beta \in \Delta\} &= P\{\gamma \in \Delta\} \\ &= \int\limits_{\Delta \cap s(r)} \frac{d\sigma(x)}{r^{n-1}\,|s(1)|}, \quad \Delta \in s(r), \end{aligned}$$

then the density function of the distance $\rho_{\beta\gamma}$ between β and γ is

$$q_r(u) = \frac{\Gamma(n/2)}{\sqrt{\pi}\,\Gamma((n-1)/2)} r^{1-n} u^{n-2}\left[1-\left(\frac{u}{2r}\right)^2\right]^{(n-3)/2}, \quad 0 \leq u \leq 2r. \tag{1.4.6}$$

The proof of lemmas 1.4.1. and 1.4.2. utilize geometrical probabilities (see Kendall and Moran (1963) or Santalo(1976)).

By randomization of the covariance function we obtain from (1.4.1), (1.4.2), (1.4.3) and (1.4.5):

$$\begin{aligned} b_n(r) &= \operatorname{var}\eta(r) \\ &= \int\limits_{v(r)}\int\limits_{v(r)} B_n(\|x-y\|)dx\,dy \\ &= (2\pi)^n r^{2n} \int\limits_0^\infty J^2_{n/2}(\lambda r)\,(\lambda r)^{-n} dG(\lambda) \\ &= |v(1)|^2\, r^{2n} E\, B_n(\|\beta-\gamma\|) \\ &= \frac{4\pi^n}{n\,\Gamma^2(n/2)} r^n \int\limits_0^{2r} z^{n-1} B_n(z)\, I_{1-(\frac{z}{2r})^2}\left(\frac{n+1}{2},\frac{1}{2}\right) dz, \quad (1.4.7) \end{aligned}$$

where $\|\beta-\gamma\|$ is the distance between independent and uniformly distributed inside the ball $v(r)$ random vectors β and γ, and $I_\mu(p,q)$ is incomplete beta-function defined in (1.4.4).

Similarly for $n \geq 2$ we obtain from (1.4.1), (1.4.2), (1.4.3) and (1.4.5):

$$
\begin{aligned}
l_n(r) &= \operatorname{var}\zeta(r) \\
&= \int\limits_{s(r)}\int\limits_{s(r)} B_n(\|x-y\|)d\sigma(x)d\sigma(y) \\
&= (2\pi)^n r^{2(n-1)} \int\limits_0^\infty J^2_{(n-2)/2}(\lambda r)\,(\lambda r)^{2-n} dG(\lambda) \\
&= r^{2n-2}\,|s(1)|^2\,EB_n\left(\left\|\tilde{\beta}-\tilde{\gamma}\right\|\right) \\
&= \frac{2^n\pi^{n-1}}{(n-2)!} r^{n-1} \int\limits_0^{2r} z^{n-2}\left[1-(\frac{z}{2r})^2\right]^{(n-3)/2} B_n(z)dz, \quad (1.4.8)
\end{aligned}
$$

where $\left\|\tilde{\beta}-\tilde{\gamma}\right\|$ is the distance between independent and uniformly distributed random vectors $\tilde{\beta}$ and $\tilde{\gamma}$ on $s(r)$ (see lemma 1.4.2.).

Recall that the function $L:(0,\infty)\to(0,\infty)$ is called slowly varying at infinity if for all $\lambda>0$,

$$\lim_{t\to\infty}\frac{L(\lambda t)}{L(t)}=1.$$

For example, the functions

$$L(t)\equiv 1, \quad L(t)=[\log(1+t)]^\alpha, \quad L(t)=\log(t+e),$$

$$L(t)=\log\log(t+e^e), \quad L(t)=k_1(1+k_2t^{-\beta})+o(t^{-\beta}), \quad t\to\infty,$$

$\beta>0$, $k_1>0$, $k_2>0$ vary slowly at infinity.

Let $\mathcal{L}$ be a class of functions that are slowly varying at infinity and bounded on each finite interval.

In this section we are interested in the connection between the behavior of the function $G(\lambda)$ as $\lambda\to 0+$ and the behavior of the function $B_n(r)$, $b_n(r)$ and $l_n(r)$ as $r\to\infty$.

We state the main results of this section

Theorem 1.4.1. Let $0<\alpha<n-1$, $n\geq 2$, $L\in\mathcal{L}$. The following two statements are equivalent:

a) $G(\lambda)/\lambda^\alpha \sim L(\frac{1}{\lambda})$, $\lambda\to 0+$;

b) $l_n(r)/r^{2n-\alpha-2} \sim L(r)c_2(n,\alpha), \quad r \to \infty,$
where and the functions G and l_n are defined in (1.4. 1) and (1.4.8) and

$$c_2(n,\alpha) = \frac{\alpha\pi^n 2^{\alpha+1}\Gamma(n-\alpha-1)\,\Gamma\left(\frac{\alpha}{2}\right)}{\Gamma^2\left(\frac{n-\alpha}{2}\right)\Gamma\left(n-1-\frac{\alpha}{2}\right)}. \tag{1.4.9}$$

Theorem 1.4.2. Let $\quad 0 < \alpha < n, \quad L \in \mathcal{L}$. The following two statement are equivalent:
a) $G(\lambda)\ /\ \lambda^\alpha \sim L(\frac{1}{\lambda}), \quad \lambda \to 0+;$
b) $b_n(r)\ /\ r^{2n-\alpha} \sim L(r)\ c_3(n,\alpha), \quad r \to \infty$, where

$$c_3(n,\alpha) = \frac{\alpha\ \pi^n 2^\alpha\Gamma(n-\alpha-1)\,\Gamma\left(\frac{\alpha}{2}\right)}{\Gamma^2\left(\frac{n-\alpha+2}{2}\right)\Gamma\left(\frac{2n-\alpha+2}{2}\right)} \tag{1.4.10}$$

and the functions G and b_n are defined in (1.4. 1) and (1.4.7) respectively.
Theorem 1.4.3. Let $\quad L \in \mathcal{L}$.
a) $G(\lambda)\ /\ \lambda^\alpha \sim L(\frac{1}{\lambda}), \quad \lambda \to 0+;$
b) $B_n(r)\ r^\alpha \sim c_4(n,\alpha)\ L(r), \quad r \to \infty, \quad$ where

$$c_4(n,\alpha) = \frac{2^\alpha\Gamma\left(\frac{\alpha+2}{2}\right)\Gamma\left(\frac{n}{2}\right)}{\Gamma\left(\frac{n-\alpha}{2}\right)} \tag{1.4.11}$$

Then for $0 < \alpha < (n-3)/2, \quad n \geq 4$, the statement a) implies the statement b); and for $0 < \alpha < n$ the statement b) implies the statement a).
The proof of the theorems 1.4.1.–1.4.3. are present below.
Remark 1.4.4. For $n = 1$ the constant (1.4.11) can be simplified:

$$c_4(n,\alpha) = \cos\frac{\alpha\pi}{2}\Gamma(1+\alpha).$$

Remark 1.4.5. In the terminology of the integral transforms a statement 'a) implies b)' is a theorem of Abelian type and statement 'b) implies a)' is a theorem of Tauberian type. We preserve this historical notation.
Remark 1.4.6. Bingham (1972) presents an alternative variant of Tauberian and Abelian Theorems for Hunkel type transform (1.4.1) (see, also, remark 1.2.3.): let $0 < \gamma < 2$, $L \in \mathcal{L}$, then the following two statements are equivalents:
a) $1 - B_n(r) \sim r^\gamma L(\frac{1}{r}), \quad r \to 0+;$
b) $1 - G(\lambda) \sim \frac{L(\lambda)}{\lambda^\gamma}\frac{2^\gamma\Gamma(\frac{n+\gamma}{2})}{\Gamma(\frac{n}{2})\Gamma(1-\frac{\gamma}{2})}, \quad \lambda \to \infty.$

If $\gamma = 2$, the statement a) is equivalent to

$$\int_0^\lambda \mu[1 - G(\mu)]\, d\mu \sim n\, L(\lambda), \quad \lambda \to \infty,$$

or

$$\int_0^\lambda \mu^2 dG(\mu) \sim 2n\, L(\lambda), \quad \lambda \to +\infty$$

If $\gamma = 0$, the statement b) implies a) (with $\gamma = 0$) if $1 - G(\lambda)$ is convex for λ sufficiently large, but not in general. Using the inversion formula (1.2.14 b) and Bingham's result, we obtain the following result:

$$G(\lambda)/\lambda^\alpha \sim \text{const}\, L(\lambda), \quad \lambda \to 0+,$$

if and only if

$$B_n(r) r^\alpha \sim \text{const}\, L(r), \quad r \to \infty,$$

where $\alpha > n$. However, the method of Bingham (1972) cannot be used to derive similar results in the case $\alpha \in (0, n)$ and $n \geq 2$. To study the case $\alpha \in (0, n)$ we use the ideas described by Laue (1987) (where she assumes that $n = 1$) and look for the relation between the behavior of $G(\lambda)/\lambda^\alpha L(1/\lambda)$ as $\lambda \to 0+$and $b_n(r)/r^{2n-\alpha} L(r)$ as $r \to \infty$, where $\alpha \in (0, n)$. The case an $\alpha = n$ is more delicate and open for investigation. The Tauberian results can be find in Bingham at all (1989) for $n = 1$, $\alpha \in (0, 1)$, in Pitman (1968) for $n = 1$, $\alpha > 1$, and Miroshin (1981) $n \geq 1, \alpha > n$.

We state the following assumption.

A. Let $\xi(x), x \in R^n$ be a real mean-square continuous homogeneous isotropic random field with $E\xi(x) = 0$, $E^2(x) = 1$ and with the covariance function

$$B(\|x\|) = \frac{L(\|x\|)}{\|x\|^\alpha}, \quad 0 < \alpha < n, \quad \text{as} \quad \|x\| \to \infty,$$

where $L \in \mathcal{L}$. Suppose that there exist a spectral density $f(\lambda) = f(\|\lambda\|)$, $\lambda \in R^n$ (see (1.2.5) or (1.2.25)) and $f(\|\lambda\|)$ is decreasing for $\|\lambda\| \geq k > 0$ and some constant $k > 0$.

Then from Theorem 1.4.3. and (1.2.25) we obtain

Theorem 1.4.4. Under the assumption A the following asymptotic behavior of the spectral density $f(\|\lambda\|)$ as $\|\lambda\| \to 0$ holds:

$$f(\|\lambda\|) \sim L\left(\frac{1}{\|\lambda\|}\right) \|\lambda\|^{\alpha - n} c_5(n, \alpha), \quad 0 < \alpha < n, \tag{1.4.12}$$

where

$$\begin{aligned} c_5(n,\alpha) &= \frac{\alpha}{|s(1)|\, c_4(n,\alpha)} \\ &= \frac{\alpha\Gamma\left(\frac{n-\alpha}{2}\right)}{2^{\alpha+1}\pi^{n/2}\Gamma\left(1+\frac{\alpha}{2}\right)} \\ &= \frac{\Gamma\left(\frac{n-\alpha}{2}\right)}{2^{\alpha}\pi^{n/2}\Gamma\left(\frac{\alpha}{2}\right)}, \end{aligned} \tag{1.4.13}$$

where the constant $c_4(n,\alpha)$ is given by (1.4.11). For $n = 1$ constant

$$c_5(1,\alpha) = 2\Gamma(\alpha)\ \cos\frac{\alpha\pi}{2}.$$

Example 1.4.1. Let $B(\|x\|) = (1 + \|x\|^2)^{-\alpha/2}$, $0 < \alpha < n$. Then there exists a spectral density (see, Donoghue (1969), p.293) of the form

$$f_\alpha(\lambda) = \pi^n 2^{n+(\alpha-2)/2}\Gamma(\alpha/2)K_{(n-\alpha)/2}(\|\lambda\|)\,\|\lambda\|^{(\alpha-n)/2},$$

where $K_\nu(z)$ is the modified Bessel function of the third kind of order ν (see, (1.2.27)). Using the asimptotic relation:

$$K_\nu(z) \sim \Gamma(\nu)2^{\nu-1}z^{-\nu}, z \to 0,$$

we obtain that

$$f(\|\lambda\|) = c_5(n,\alpha)\,\|\lambda\|^{\alpha-n}\,(1-\theta(\|\lambda\|),$$

where $\theta(\|\lambda\|) \to 0$ as $\|\lambda\| \to 0$. The radial function $f_\alpha(\lambda) = g(\rho), \rho = \|\lambda\|$, is called Bessel potential of order α (see, , Donoghue (1969), p.294). Note that

$$K_\nu(z) = (\pi/2)^{1/2}z^{-1/2}e^{-z}(1 - \frac{4\nu^2-1}{8z} + \cdots)$$

as $z \to \infty$. In this example the slowly varying function $L = 1$.

2. Auxiliary results

Before proving Theorems 1.4.1.-1.4.3., we mention some well-known results.

Remark 1.4.7. We shall show that functions $G(\lambda)/\lambda^\alpha$ and $b_n(r)/r^{2n-\alpha}L(r)$ (or $l_n(r)/r^{2n-\alpha-2}L(r)$) have limits at zero and infinity, respectively. The proof will show that each of these limits depends only on the other. Therefore, these limits and the relation between them can be taken from the special case cited in Yadrenko (1983) and Ivanov and Leonenko (1989).

Lemma 1.4.3. Consider the integral

$$S(r) = \int_0^\infty T(\lambda r)\lambda^\delta g(\lambda)\, d\lambda,$$

and suppose that:

a) the function $g(\lambda)$ is continuous in a neighborhood of zero, $g(0) \neq 0$ and $g(\lambda)$ is bounded on $[o, \infty)$;

b) $\int_0^\infty T(z) z^\delta dz = \gamma(\delta)$, $\int_0^\infty |T(z)|\, z^\delta dz = \tilde{\gamma}(\delta) < \infty$.

Then

$$S(r) = r^{-\delta-1} g(0)\, \gamma(\delta)\, (1 + o(1)) \quad \text{as} \quad r \to \infty.$$

Proof

Note that

$$\int_0^\infty T(\lambda r)\, \lambda^\delta d\lambda = r^{-\delta-1}\gamma(\delta).$$

Choose $\beta(r) \to 0$ such that $r\beta(r) \to \infty$ as $r \to \infty$.

Note the identity

$$S(r) = r^{-\delta-1} g(0)\, \gamma(\delta) + f(r), \qquad f(r) = \int_0^\infty T(\lambda r)\, \lambda^\delta [g(\lambda) - g(0)]\, d\lambda.$$

We shall prove that under the conditions a) and b) $f(r) = o(r^{-\delta-1})$ as $r \to \infty$. Indeed

$$|f(r)| \leq \int_0^{\beta(r)} |T(\lambda r)|\, |g(\lambda) - g(0)|\, \lambda^\delta d\lambda + \int_{\beta(r)}^\infty |T(\lambda r)|\, |g(\lambda) - g(0)|\, \lambda^\delta d\lambda \leq$$

$$\max_{0 \leq \lambda \leq \beta(r)} |g(\lambda) - g(0)|\, \tilde{\gamma}(\delta)\, r^{-\delta-1} + 2 \sup_{0 \leq \lambda \leq \infty} |g(\lambda)| \cdot r^{-\delta-1} \int_{r\beta(r)}^\infty |T(z)|\, z^\delta dz.$$

Hence,

$$f(r) = o(r^{-\delta-1}) \quad \text{as } r \to \infty.$$

Lemma 1.4.4. Let (1.4.1) hold, and

$$G'(\lambda) = |s(1)|\, \lambda^{\delta}\, g(\lambda), \quad \delta \in (-1, n). \tag{1.4.14}$$

Then, as $r \to \infty$

$$b_n(r) = (2\pi)^n\, |s(1)|\, c_6(n, \delta)\; g(0)\; r^{2n-\delta-1}(1 + o(1)), \tag{1.4.15}$$

where

$$c_6(n, \delta) = \frac{\Gamma(n-\delta)\, \Gamma\left(\frac{1+\delta}{2}\right)}{2^{n-\delta}\Gamma^2\left(\frac{n-\delta+1}{2}\right)\, \Gamma\left(\frac{2n-\delta+1}{2}\right)}, \tag{1.4.16}$$

and $b_n(r)$ is defined in (1.4.7).

Proof. We substitute (1.4.14) into (1.4.7) and in Lemma 1.4.3. set

$$T(z) = J^2_{n/2}(z) z^{-n}.$$

Utilizing relation 6.574. 2 from Gradstein and Ryzhik (1962):

$$\int_0^{\infty} J_v(\rho t) J_\mu(\rho t) t^{-\lambda} dt = \frac{\rho^{\lambda-1}\Gamma(\lambda)\, \Gamma\left(\frac{\nu+\mu-\lambda+1}{2}\right)}{2^{\lambda}\Gamma\left(\frac{\mu-\nu+\lambda+1}{2}\right)\Gamma\left(\frac{\nu+\mu+\lambda+1}{2}\right)\Gamma\left(\frac{\nu-\mu+\lambda+1}{2}\right)}$$

$\rho > 0$, $\mathrm{Re}(\nu + \mu + 1) > \mathrm{Re}\lambda > 0$.

For $-1 < \delta < n$ we set

$$\int_0^{\infty} J^2_{n/2}(z)\; z^{\delta-n} dz = c_6(n, \delta).$$

Then (1.4.15) follows from Lemma 1.4.3.

Lemma 1.4.5. Let (1.4.1) and (1.4.14) hold with $\delta \in (-1, n-2)$, $n \geq 2$. Then as $r \to \infty$

$$l_n(r) = (2\pi)^n g(0)\;\; |s(1)|\;\; c_6(n-2, \delta)\; r^{2n-\delta-3}(1 + o(1)), \tag{1.4.17}$$

where $c_6(n, \delta)$ is defined in (1.4.16) and $l_n(r)$ is defined in (1.4.8).

Proof

We substitute (1.4.14) into (1.4.8) and in Lemma 1.4.3. set

$$T(z) = J^2_{(n-2)/2}(z)z^{2-n} \qquad \text{for} \quad n \geq 2.$$

Then we obtain (1.4.17). For such a choice of $T(z)$, to ensure convergence of the integral $\int_0^\infty T(z)z^\delta dz$ it is necessary to assume that $\alpha \in (-1, n-2)$.

Lemma 1.4.6. Let (1.2.25) hold with

$$g(\lambda) = \frac{h(\lambda)}{\lambda^{n-\alpha}}, \quad \alpha \in (0, \frac{1}{2}(n-1)), \quad n \geq 2, \tag{1.4.18}$$

where $h(\lambda), \lambda \geq 0$, is a function continuous in a neighborhood of zero, $h(0) \neq 0$ and $h(\lambda)$ is bounded on $[0, \infty)$.

Then as $r \to \infty$

$$B_n(r) = r^{-\alpha} h(0) \frac{1}{c_5(n, \alpha)} (1 + o(1)),$$

where $c_5(n, \alpha)$ is defined in (1.4.13).

Proof

We substitute (1.4.18) into (1.2.25). Utilizing the relation (see Bateman and Erdelyi (1958))

$$\int_0^\infty J_\mu(\rho t)\, t^{\gamma-1} dt = \frac{2^{\gamma-1}\Gamma\left(\frac{\mu+\gamma}{2}\right)}{\rho^\gamma \Gamma\left(1 + \frac{\mu-\gamma}{2}\right)},$$

$-\mathrm{Re}\mu < \mathrm{Re}\gamma < \frac{1}{2}$, $\rho > 0$, we obtain from Lemma 1.4.4.

$$\begin{aligned} B_n(r) &= |s(1)| \int_0^\infty Y_n(\rho r)\, h(\rho)\, \rho^{\alpha-1} d\rho \\ &= \frac{h(0)}{r^\alpha c_5(n, \alpha)} (1 + o(1)), \end{aligned}$$

as $r \to \infty$.

We formulate a theorem of Abelian and Tauberian type for Laplace–Stiltjes transforms.

Lemma 1.4.7. (Feller (1971)) Let $U(t), t \geq 0$, be bounded nondecreasing function such that its Laplace–Stiltjes transform

$$w(\lambda) = \int_0^\infty e^{-\lambda t} dU(t)$$

exists for $\lambda > 0$. Then for every $0 \le \rho < \infty$, $L \in \mathcal{L}$:

a) $w(\lambda) \sim \lambda^{-\rho} L(\frac{1}{\lambda})$, $\lambda \to 0+$, if and only if

$$U(t) \sim \frac{t^{\rho} L(t)}{\Gamma(1+\rho)}, t \to \infty;$$

b) $w(\lambda) \sim \lambda^{-\rho} L(\lambda)$, $\lambda \to \infty+$, if and only if

$$U(t) \sim \frac{t^{\rho} L(\frac{1}{t})}{\Gamma(1+\rho)}, t \to 0+.$$

Lemma 1.4.8. (Widder (1969)) Let

$$w(\lambda) = \int_0^\infty e^{-\lambda t} F(t)\, dt, \quad \lambda > 0$$

be a Laplace transform of a function $F(t)$, then $w'(\lambda)$ is a Laplace transform of a function $-tF(t)$.

Lemma 1.4.9. (Bateman and Erdelyi (1953), relation (8.2.5)) Let

$$H_\nu(f, \lambda) = \int_0^\infty f(u)\ J_\nu(u\lambda)\ (u\lambda)^{1/2} du, \quad \lambda > 0$$

be a Hankel transform of a function $f(u)$. Then for $f(u) = u^{1/2}(a^2 + y^2)^{-3/2}$, $\mathrm{Re} a > 0$ we obtain $H_\nu(f, \lambda) = \lambda^{-1/2} e^{-a\lambda}$.

Let

$$g(s) = \int_0^\infty f(u)\ u^{s-1} du, \quad s \in \mathbf{C}, \tag{1.4.19}$$

be a Mellin transform of a function $f(u)$.

Lemma 1.4.10. (Bateman and Erdelyi (1969)) Let $g(s)$ and $f(u)$ are defined in (1.4.19). Then the following table is true

N	$f(u)$	$g(s)$
(1)	$\begin{cases}(1-u^h)^{\nu-1}, & 0<u<1\\ 0, & u\geq 1,\end{cases}$ $h>0, \operatorname{Re}\nu>0$	$h^{-1}B(\nu, \frac{s}{h})$, $\operatorname{Re}s>0$
(2)	$u^\alpha f(u)$	$g(s+\alpha)$
(3)	$f(u^h), \quad h>0$	$h^{-1}g(\frac{s}{h})$
(4)	$e^{-\alpha u}, \quad \operatorname{Re}\alpha>0$	$\alpha^{-s}\Gamma(s)$
(5)	$u^\alpha \int\limits_0^\infty t^\beta f_1(ut)\, f_2(t)dt$	$g_1(s+\alpha)\, g_2(1-s-\alpha+\beta)$
(6)	$J_\nu(au), \quad a>0$	$2^{s-1}\Gamma(s/2+\nu/2)/a^s\Gamma(\nu/2-s/2+1)$ $\operatorname{Re}\nu<\operatorname{Re}s<3/2$
(7)	$J_\mu(au)\, J_\nu(au),\ a>0$	$\frac{2^{s-1}a^{-s}B(1-s,\mu/2+\nu/2+s/2)}{\Gamma(\nu/2-\mu/2-s/2+1)\Gamma(\mu/2-\nu/2-s/2+1)}$

Lemma 1.4.11. Let

$$f(t) = \int\limits_0^\infty e^{i\lambda t} dG(\lambda), \quad t \in R^1$$

be a characteristic function of distribution function G, such that $G(0+) = 0$. Then for $y > 0$

$$f(iy) = \frac{2y}{\pi} \int\limits_0^\infty \frac{\operatorname{Re}\ f(u)}{u^2+y^2} du, \quad y > 0 \tag{1.4.21}$$

and

$$f(iy) = A(n)\, y \int\limits_0^\infty \frac{u^{n-1}}{(u^2+y^2)^{(n+1)/2}} B_n(u) du, \quad y > 0 \tag{1.4.22}$$

where $A(n)$is a constant and B_n is defined in (1.4.1).

Proof

The formula (1.4.21) is known (see Laue (1975), Rossberg, Jesiak and Siegel (1985)). An integration by parts produces

$$\begin{aligned} f(iy) &= \frac{2y}{\pi}\left[\frac{1}{u^2+y^2}\int\limits_0^u \operatorname{Re} f(v)dv\ \Big|_0^\infty + \int\limits_0^\infty \frac{2u}{(u^2+y^2)^2}\int\limits_0^u \operatorname{Re} f(v)dvdu\right] \\ &= \frac{2y}{\pi}\int\limits_0^\infty \frac{2u}{(u^2+y^2)^2}\int\limits_0^u \operatorname{Re}\ f(v)dvdu. \end{aligned}$$

Since f is the characteristic function of distribution function G, we have

$$\begin{aligned}\int_0^u \mathrm{Re} f(v)dv &= \int_0^u\int_0^\infty \cos(\lambda v)dG(\lambda)dv \\ &= \int_0^\infty\int_0^u \cos(\lambda v)dvdG(\lambda) \\ &= \int_0^\infty \frac{\sin(\lambda u)}{\lambda}dG(\lambda).\end{aligned}$$

The order of integration can be changed, since

$$\sup_{v\in[0,u]}\int_A^\infty cos(\lambda v)dG(\lambda)\to 0,\quad A\to\infty.$$

Therefore,

$$f(iy)=\frac{2y}{\pi}\int_0^\infty \frac{2u^2}{(u^2+y^2)^2}\int_0^\infty J_{1/2}(\lambda u)(\lambda u)^{-1/2}dG(\lambda)du.$$

It is easy to see that if $n=3$ then

$$f(iy)=L(3)\ y\int_0^\infty \frac{u^2B_3(u)}{(u^2+y^2)^2}du.$$

We use formulas for differentiating Bessel function derived in Betamen and Erdelyi (1953) to obtain

$$\begin{aligned}\left[J_{(n-2)/2}(\lambda u)\ (\lambda u)^{(2-n)/2}\right]' &= J_{(n-4)/2}(\lambda u)\ \left[u\ (\lambda u)^{(n-4)/2}\right]^{-1} \\ &= -(n-2)\ \frac{J_{(n-2)/2}(\lambda u)}{u\ (\lambda u)^{(n-2)/2}}.\end{aligned}$$

Therefore,

$$\int_0^u J_{(n-4)/2}(\lambda v)\ (\lambda v)^{(4-n)/2}dv = \int_0^u (n-2)\ J_{(n-2)/2}(\lambda v)\ (\lambda v)^{(2-n)/2}dv$$

$$+\int_0^u v\ (J_{(n-2)/2}(\lambda v)\ (\lambda v)^{(2-n)/2})' dv$$

$$= (n-3)\int_0^u J_{(n-2)/2}(\lambda v)\ (\lambda v)^{(2-n)/2} dv$$
$$+u\ J_{(n-2)/2}(\lambda u)\ (\lambda u)^{(2-n)/2}. \quad (1.4.23)$$

Let $n = 2k+1$. We use induction to show that

$$f(iy) = L(n)\ y\int_0^\infty u^{2k} B_n(u)(u^2+y^2)^{-(k+1)} du. \quad (1.4.24)$$

The above equation has been proved before in the case $n = 3$. Using (1.4.23) we obtain

$$\int_0^\infty u^{2k} B_{2k+1}(u)(u^2+y^2)^{-(k+1)} du = -\int_0^\infty (u^{2k}(u^2+y^2)^{-(k+1)})' \times \int_0^u B_{2k+1}(v) dv du$$

$$= -\int_0^\infty (u^{2k}(u^2+y^2)^{-(k+1)})' \times [u B_{2(k+1)+1}(u) + 2k\int_0^u B_{2(k+1)+1}(v) dv] du$$

$$= -\int_0^\infty [(u^{2k}(u^2+y^2)^{-(k+1)})' u - 2k u^{2k}(u^2+y^2)^{-(k+1)}] \times B_{2(k+1)+1}(u) du$$

$$= 2(k+1)\int_0^\infty (u^{2k+1}(u^2+y^2)^{-(k+2)}) \times B_{2k+3}(u) du.$$

Therefore, formula (1.4.12) holds for odd n. We now show that (1.4.22) holds for even n.

In the case $n = 2$ we have

$$\int_0^\infty u(u^2+y^2)^{-3/2} B_2(u)du = \int_0^\infty u(u^2+y^2)^{-3/2} \int_0^\infty J_0(\lambda u)dG(\lambda)du \quad (1.4.25)$$

Changing the order of integration is formula (1.4.25), we obtain

$$\begin{aligned}\int_0^\infty\int_0^\infty u(u^2+y^2)^{-2} J_{1/2}(\lambda u)(\lambda u)^{-1/2}dudG(\lambda) &= \int_0^\infty \frac{dG(\lambda)}{\lambda}\int_0^\infty u(u^2+y^2)^{-2} \\ &\quad \times J_{1/2}(\lambda u)(\lambda u)^{1/2}du \\ &= \sqrt{\frac{\pi}{8}}\int_0^\infty y^{-1}e^{-y\lambda}dG(\lambda),\end{aligned}$$

$$\begin{aligned}\int_0^\infty\int_0^\infty u(u^2+y^2)^{-3/2} J_0(\lambda u)dudG(\lambda) &= \int_0^\infty \frac{dG(\lambda)}{\lambda^{1/2}}\int_0^\infty u^{1/2}(u^2+y^2)^{-3/2} \\ &\quad \times J_0(\lambda u)(\lambda u)^{1/2}du \\ &= \int_0^\infty y^{-1}e^{-y\lambda}dG(\lambda).\end{aligned}$$

Here we have used Lemma 1.4.9. Thus formula (1.4.22) holds in the case $n = 2$. Carrying out the calculations performed in the case of odd n, we obtain the formula (1.4.22) for all $n \geq 1$.

Lemma 1.4.12. (Bingham, Goldie and Teugels (1989)) Let $L \in \mathcal{L}$ be a function which is slowly varying at infinity. Then:

1) relation $L(t\lambda)/L(t) \to 1$ holds uniformly for $\lambda \in [a, b]$, $0 < a < b < \infty$ as $t \to \infty$;

2) for any $\delta > 0$, $t^\delta L(T) \to \infty$,

$$t^{-\delta}L(t) \to 0, \qquad \text{as } t \to \infty;$$

3) for any $0 < a < b < \infty$,

$$\lim_{t\to\infty}\int_a^b \frac{L(ts)}{L(t)}ds = \int_a^b \lim_{t\to\infty}\frac{L(ts)}{L(t)}ds = b - a;$$

4) for any $0 < a < b < \infty$,

$$\lim_{t\to\infty} \sup_{at<s<bt} \left| \frac{L(s)}{L(t)} - 1 \right| = 0;$$

5) if L is a function integrable in the Riemann or Lebesgue sense (in the latter case the measurability of the function is assumed in its definition) over any finite interval $[a, b]$, then L admits the canonical representation:

$$L(t) = c(t) \exp\{\int_a^t \varepsilon(s) s^{-1} ds\},$$

where $c(t), \varepsilon(t)$ are functions such that $c(t) \to c > 0$, $\varepsilon(t) \to 0$ as $t \to \infty$;

6) if $f(t) \sim L(t)$ as $t \to \infty$ where $L \in \mathcal{L}$, then there exists $\tilde{L} \in \mathcal{L}$ such that $\tilde{L}(t) \sim L(t)$ as $t \to \infty$ and $f(t) = \tilde{L}(t)$;

7) let $L(t) \in \mathcal{L}$, $t \in [A, \infty)$, $A > 0$, be a function such that Lebesgue integral $\int_\alpha^\infty f(t) t^\eta dt$, $\alpha \in [0, \infty)$, is defined for a fixed $\eta \geq 0$, then integral $\int_\alpha^\infty f(t) L(st) dt$ is defined a) for $\eta \geq 0$; b) for nondecreasing at infinity functions $L(s)$, if $\eta = 0$; in both cases

$$\int_\alpha^\infty f(t) L(st) dt \sim L(s) \int_\alpha^\infty f(t) dt, \quad s \to \infty;$$

8) let $L(t) \in \mathcal{L}$ and $\int_0^\beta t^{-\eta} f(t) dt < \infty$ for some $0 < \beta < \infty$ and $\eta \geq 0$, then

$$\int_0^\beta f(t) L(st) dt \sim L(s) \int_\alpha^\beta f(t) dt, \quad s \to \infty \qquad \text{for} \quad \eta > 0$$

(for $\eta = 0$ the L should be nondecreasing on $[0, \infty)$).

Lemma 1.4.13. Let condition b) of Theorem 1.4.1. hold.

Then $\int_0^\infty \lambda^{\delta-\alpha} dG(\lambda)$ exists for $0 < \delta < \alpha$.

Proof

Consider the integral $\int_0^R (l_n(r)/r^{2n-1-\alpha+\delta}) dr$. From condition b) we obtain

$$f(r) = l_n(r)/r^{2n-1-\alpha+\delta} \sim L(r)/r^{1+\delta} \quad \text{as} \quad r \to \infty.$$

From Lemma 1.4.12. (2)) we obtain $L(r)/r^\gamma \to 0$ as $r \to \infty$ for any $\gamma > 0$. Thus the function $f(r)$ is integrable on $[A, \infty)$, $A > 0$. Consider the behavior of a function $f(r)$ at zero. By definition

$$f(r) \leq \text{const} \frac{r^{2n-2}}{r^{2n-1-\alpha+\delta}} = \frac{\text{const}}{r^{1-\alpha+\delta}}.$$

Therefore for $\delta \in (0, \alpha)$ the integral $\int_0^\infty f(r)dr$ is finite, and using (1.4.8) we obtain

$$\int_0^R f(r)\,dr = (2\pi)^n \int_0^R r^{\alpha-1-\delta} \int_0^\infty J^2_{(n-2)/2}(r\lambda)\,(r\lambda)^{2-n} dG(\lambda)$$
$$= \int_0^\infty (2\pi)^n \lambda^{2-n} \int_0^R r^{1+\alpha-n-\delta} J^2_{(n-2)/2}(r\lambda)\,dr\,dG(\lambda).$$

From Lemma 1.4.10. (formula 7)) we obtain

$$\lim_{R\to\infty} \int_0^R r^{1+\alpha-n-\delta} J^2_{(n-2)/2}(r\lambda)\,dr\,dG(\lambda) = \text{const}\ \lambda^{n-2+\delta-\alpha}.$$

Therefore

$$\int_0^\infty f(r)\,dr = \text{const} \int_0^\infty \lambda^{\delta-\alpha} dG(\lambda).$$

Lemma 1.4.14. Let condition b) of Theorem 1.4.2. hold. Then $\int_0^\infty \lambda^{\delta-\alpha} dG(\lambda)$ exists for $0 < \delta < \alpha$.

Proof

Consider the integral

$$U_R = \int_0^R f(r)\,dr,$$

where

$$f(r) = b_n(r)/r^{2n-\alpha+1+\delta}.$$

From condition b) we obtain

$$f(r) \sim L(r)/r^{1+\delta} \quad \text{as} \quad r \to \infty.$$

From Lemma 1.4.12. (statement 2)) we obtain

$$L(r) r^{-\gamma} \to 0, \quad r \to \infty$$

for any $\gamma > 0$. Consider the behavior of a function $f(z)$ at zero. We obtain by definition

$$\begin{aligned} \frac{b_n(r)}{r^{2n-\alpha+1+\delta}} &\leq \text{const}\, \frac{r^{2n}}{r^{2n-\alpha+1+\delta}} \\ &= \text{const}\ r^{1-\alpha+\delta}. \end{aligned}$$

Thus the integral U_R is convergence for $\alpha > \delta > 0$. Using (1.4.7) we obtain

$$\begin{aligned} U_R &= (2\pi)^2 \int_0^R \frac{1}{r^{1-\alpha+\delta}} \int_0^\infty J_{n/2}^2(r\lambda)\ (r\lambda)^{-n} dG(\lambda) dr \\ &= (2\pi)^2 \int_0^\infty \lambda^{-n} \int_0^R r^{\alpha-n-1-\delta} J_{n/2}^2(r\lambda)\ dr dG(\lambda). \end{aligned}$$

From Lemma 1.4.10. (formula 7)) we obtain

$$\lim_{R\to\infty} \int_0^R r^{\alpha-n-1-\delta} J_{n/2}^2(r\lambda)\ dr = \text{const}\ r^{n+\delta-\alpha}, \quad \alpha > \delta,$$

and

$$\int_0^\infty f(r)\ dr = \text{const} \int_0^\infty \lambda^{\delta-\alpha} dG(\lambda).$$

3. Proofs

Proof of Theorem 1.4.1.

Let condition a) hold. Form Lemma 1.4.12. (statement 6)) we obtain that there exists a function $\tilde{L} \in \mathcal{L}$ such that

$$G(\lambda)/\lambda^\alpha = \tilde{L}(\frac{1}{\lambda}) \qquad \text{as} \quad \lambda \to 0+.$$

Then from (1.4.8) we have

$$\frac{l_n(r)}{r^{2n-\alpha-2}} = (2\pi)^n r^\alpha \int_0^\infty J_{(n-2)/2}(\lambda r)(\lambda r)^{2-n} dG(\lambda)$$

$$
\begin{aligned}
&= (2\pi)^n \int_0^\infty \tilde{L}(\frac{r}{u}) \left[\frac{(n-2) J^2_{(n-2)/2}(u)}{u^{n-1-\alpha}} \right. \\
&\quad \left. - \frac{2J_{(n-2)/2}(u)}{u^{n-2-\alpha}} \left(J_{(n-4)/2}(u) + J_{n/2}(u) \right) \right] du \\
&= (2\pi)^n \left[\int_{1/A}^{\infty} + \int_0^{1/A} \right] \tilde{L}(zr) \\
&\quad \times \left[\frac{(n-2) J^2_{n/2}(1/z)}{z^{\alpha+1-n}} - \frac{2J_{(n-2)/2}(1/z)}{z^{\alpha+2-n}} \right. \\
&\quad \left. \times \left(J_{(n-4)/2}(1/z) + J_{n/2}(1/z) \right) \right] \frac{dz}{z^2} \qquad (1.4.26)
\end{aligned}
$$

It is well-known that

$$
J_\nu(z) = \left(\frac{2}{\pi z}) \right)^{1/2} \cos(z - \frac{\nu\pi}{2} - \frac{\pi}{4}) + O(z^{-3/2}), \quad z \to \infty; \qquad (1.4.27)
$$

$$
J_\nu(z) = \frac{z^\nu}{2^\nu \Gamma(\nu+1)} + O(z^{\nu+1}), \quad z \to 0.
$$

From (1.4.27) we obtain that expression in the square brackets of (1.4.26) has an order $z^{1-\alpha}$ at infinity and an order $z^{n-\alpha}$ at zero. Then there exists $\eta > 0$ such that next integrals of Lebesgue are finite:

$$
\int_{1/A}^{\infty} z^{\eta-2} U_n(z)\, dz, \qquad \int_0^{1/A} z^{-\eta-2} U_n(z)\, dz,
$$

where

$$
\begin{aligned}
U_n(z) &= (n-2)\ J^2_{n/2}(1/z) z^{n-\alpha-1} - 2J_{(n-2)/2}(1/z) \\
&\quad \times \left(J_{(n-4)/2}(1/z) + J_{n/2}(1/z) \right)\ z^{n-\alpha-2}.
\end{aligned}
$$

Using the statements 6)-8) from Lemma 1.4.12. we obtain from (1.4.26):

$$
\frac{l_n(r)}{r^{2n-\alpha-2}} \sim \tilde{L}(r) \int_0^\infty \left[\frac{(n-2) J^2_{n/2}(u)}{u^{n-1-\alpha}} - \frac{2J_{(n-2)/2}(u)}{u^{n-2-\alpha}} \left(J_{(n-4)/2}(u) + J_{n/2}(u) \right) \right] du
$$

$$\sim L(r)\,\text{const}, \quad r \to \infty$$

Thus Abelian part of Theorem 1.4.1. is proved (see Remark 1.4.7.).

Let us prove the Tauberian part of Theorem 1.4.1, that is, b) implies a). Suppose that $n = 2k + 1$ is odd. Then we shall show that there exists a function $g_n(u, y)$ such that

$$\int_0^\infty u^{n-1}(u^2+y^2)^{-(n+1)/2} B_n(u)du = \int_0^\infty g_n(u,y) \times \int_0^\infty z^{n-2}\left(1-(z/u)^2\right)^{(n-3)/2} B_n(z)dzdu. \tag{1.4.28}$$

Note that for odd n the expression $(1-(z/u)^2)^{(n-3)/2}$ is polynomial of degree $n-3$ from z/u. An integration by parts of (1.4.28) produces

$$u^{n-1}(u^2+y^2)^{-(n+1)/2} = \int_0^\infty v^{n-2} g_n(v,y)\ \left(1-(u/v)^2\right)^{(n-3)/2} dv.$$

Continuing the differentiation, we obtain

$$g_n(u,v) = A_1\ u^{2(k-1)} \underbrace{\left(\left(\cdots\left(\left(\frac{u}{(u^2+y^2)^{-(n+1)/2}}\right)'\frac{1}{u}\right)'\cdots\right)'\frac{1}{u}\right)'}_{k},$$

where A_1 is some constant.

We easily see that at infinity $g_n(u, y)$ has order u^{-n-1}, and at zero it has order constant. Therefore there exists $\eta > 0$ such that $t^\eta t^{n-\alpha-1} g_n(t, 1)$ is integrable on $[A, \infty)$, where $A > 0$, and $t^{-\eta} t^{n-\alpha-1} g_n(t, 1)$ is integrable on $[0, A]$. From condition b) and lemma 1.4.12. (statement 6)) we obtain that there exists a function $\tilde{L}(t) \in \mathcal{L}$ such that

$$\frac{l_n(y)}{y^{2n-\alpha-2}} = \tilde{L}(y).$$

From (1.4.22) we obtain

$$y^\alpha f(iy) = A_2 \int_0^\infty t^{n-1-\alpha} g_n(t,1)\ l_n(ty)\ (ty)^{2+\alpha-2n} dt, \quad y > 0 \tag{1.4.28a}$$

Using the statement 7) and 8) of lemma 1.4.12. we obtain

$$y^{\alpha} f(iy) \sim L(y)\,\text{const}\,.$$

Then our statement for odd u follows from Lemma 1.4.7. Suppose that n is even. We define the following functions:

$$\begin{aligned}\varphi_1(u) &= \int_0^u v^{n-2} B_n(v)\,\left(1-(v/u)^{(n-3)/2}\right)\,dv \\ &= u^{n-1} \int_0^1 t^{n-1} B_n(tu)\,(1-t^2)^{(n-3)/2} dt,\end{aligned}$$

and

$$\varphi_2(u) = u^{n-1} \int_0^1 t^{n-2} B_{n+1}(tu)\,(1-t^2)^{(n-3)/2} dt,$$

where B_{n+1} corresponds the same spectral function G as B_n (see (1.4.1)).

Let $g(s)$ and $\tilde{g}(s)$ be Mellin transforms of a functions $\varphi_1(u)$ and $\varphi_2(u)$, respectively, $g_1(s)$ and $\tilde{g}_1(s)$ be a Mellin transforms of a functions B_n and B_{n+1} and $g_2(s)$ be a Mellin transform of a function

$$\begin{cases}(1-u^2)^{(n-3)/2}, & |u| \le 1, \\ 0, & |u.| > 1.\end{cases}$$

For a definition Mellin transform, see (1.4.19). From Lemma 1.4.10. (formula (5)) we obtain

$$g(s) = g_1(s+n-1)\, g_2(-s), \qquad \tilde{g}(s) = \tilde{g}_1(s+n-1)\, g_2(-s). \tag{1.4.29}$$

Let us investigate for what s all these transforms are defined correctly.

From Lemma 1.4.10. (formula (1)) we obtain that $g_2(s)$ is defined correctly for Re$s < 0$.

Note that

$$g_1(s) = \int_0^{\infty} u^{s-1} B_n(u) du$$

$$= \text{const} \int_0^\infty u^{s-1-(n-2)/2} du$$
$$\times \int_0^\infty J_{(n-2)/2}(\lambda u)\ \lambda^{(2-n)/2} dG(\lambda)$$

From lemma 1.4.10. (formula (5)) we obtain

$$g_1(s) = \text{const}\, g_0(s - \frac{1}{2}(n-2))\ \tilde{g}_0(1 - s + \frac{1}{2}(n-2)), \qquad (1.4.30)$$

where $g_0(s)$ is Mellin transform of a function $J_{(n-2)/2}(u)$ and $\tilde{g}_0(s)$ is Mellin transform of a function $\lambda^{(2-n)/2} dG(\lambda)$. From Lemma 1.4.10. (formula (6)) we have that Mellin transform $g_0(s - \frac{1}{2}(n-2))$ is defined correctly for

$$s : \frac{2-n}{2} < \text{Re}(s - \frac{1}{2}(n-2)) < 3/2 \quad \text{or} \quad 0 < \text{Re}s < \frac{1}{2}(n+1).$$

From Lemma 1.4.13. we obtain that Mellin transform

$$\tilde{g}_0\left(1 + \frac{n-2}{2} - s\right) = \int_0^\infty \lambda^{-s} dG(\lambda)$$

is defined correctly for $0 < \text{Re}s < \alpha$. Therefore the function $g_1(s)$ is defined for $s : 0 < \text{Re}s < \min((n+1)/2, \alpha)$ and $g_1(s+n-1)$ for $s : -(n-1) < \text{Re}s < -(n-1) + min((n+1)/2, \alpha)$. The function $g(s)$is defined for the same s. Similarly, the Mellin transform $\tilde{g}(s)$ is defined correctly for $s : 1-n < \text{Re}s < 1 - n + \min((n+1)/2, \alpha)$.

From (1.4.30) we have

$$g(s) = g_2(-s) g_0(s + n/2) \tilde{g}_0(1 - s - n/2).$$

Similarly

$$\tilde{g}(s) = g_2(-s) \breve{g}_0(s + n/2 - 1/2) \tilde{g}_0(1 - s - n/2),$$

where $g_0(s)$ is Mellin transform of a function $J_{(n-2)/2}(u)$. From Lemma 1.4.10. (formula (6)) we obtain

$$g_0(s + n/2) = \frac{\Gamma\left(\frac{s+n-1}{2}\right) 2^{s+n/2-1}}{\Gamma(1 - s/2)},$$

$$\breve{g}_0\left(s + \frac{n-1}{2}\right) = \frac{\Gamma\left(\frac{s+n-1}{2}\right) 2^{s+n/2-3/2}}{\Gamma(1-s/2)}. \tag{1.4.31}$$

Using formula 6) from Lemma 1.4.10. we choose functions f and $\tilde{f}$ with Mellin transforms g_3 and $\tilde{g}_3$ such that $g(s)g_3(1-s+\beta_1)$ and $\tilde{g}(s)\tilde{g}_3(1-s+\beta_2)$ are equal up to constant. Let $f(u) = \tilde{f}(u) = \exp(-u^2)$, $\beta_1 = 0$, $\beta_2 = 1$. From Lemma 1.4.10. (statements 3) and 4)) we obtain

$$g_3(1-s) = \frac{1}{2}\Gamma\left(\frac{1-s}{2}\right), \qquad \tilde{g}_3(2-s) = \frac{1}{2}\Gamma\left(1 - \frac{s}{2}\right). \tag{1.4.32}$$

From (1.4.31) and (1.4.32) we have

$$g(s)\; g_3(1-s+\beta_1) = \tilde{g}(s)\; \tilde{g}_3(1-s+\beta_2)$$

Therefore the Mellin transforms of the functions

$$\int_0^\infty e^{-u^2}\varphi_1(tu)\; du, \qquad \int_0^\infty u\; e^{-u^2}\varphi_2(tu)\; du \tag{1.4.33}$$

are equal up to constant. These transforms are defined correctly for

$$1 - n < \mathrm{Re} s < 1 - n + \min((n+1)/2, \alpha).$$

Since the functions (1.4.33) are equal up to constant.

Note that

$$\int_0^\infty e^{-u^2}\varphi_1(tu)\; du = \frac{1}{2t}\int_0^\infty e^{-z/t^2} z^{-1/2}\varphi_1(\sqrt{z})\; dz \tag{1.4.34}$$

The last integral is Laplace transform of a function $\varphi_1(\sqrt{z})/\sqrt{z}$. We study the asymptotic behavior of $\varphi_1(z)$. From the definition of $l_n(r)$ we easily obtain

$$l_n(r/n)/(r/n)^{n-1} = A_3\varphi_1(r), \quad A_3 > 0.$$

Therefore

$$\varphi_1(\sqrt{z})/\sqrt{z} \sim A_3 z^{(n-2-\alpha)/2} L(\sqrt{z}), \quad z \to \infty.$$

By Lemma 1.4.7.

$$\int_0^\infty e^{-u^2}\varphi_1(tu)\,du \sim A_4 t^{n-\alpha-1}L(t), \quad A_4>0, \quad t\to\infty. \tag{1.4.35}$$

Now we consider the second function from (1.4.33):

$$\int_0^\infty u\,e^{-u^2}\varphi_2(tu)\,du = \frac{1}{2t^2}\int_0^\infty e^{-z/t^2}\varphi_2(\sqrt{z})\,dz. \tag{1.4.36}$$

From (1.4.35) and (1.4.36) we obtain

$$\int_0^\infty e^{-zp}\varphi_2(\sqrt{z})\,dz \sim A_5 p^{-(n-\alpha+1)/2}L\left(\frac{1}{\sqrt{p}}\right), \quad A_5>0, \quad p\to 0+.$$

The function B_{n+1} is a covariance function of homogeneous isotropic random field on R^n. Let $\tilde{l}_n(r)$ is defined from (1.4.8) with B_{n+1} instead of is B_n. Similar to φ_1 we obtain

$$\tilde{l}_n(r/n)/(r/n)^{n-1} = A_3\varphi_2(r).$$

Therefore

$$F_1(z) = \int_0^z \varphi_2(\sqrt{u})du$$

is a measure.

From Lemma 1.4.7. we obtain

$$\int_0^z \varphi_2(\sqrt{u})\,du \sim A_6 z^{(n+1-\alpha)/2}L(\sqrt{z}), \quad A_6>0, \quad z\to\infty \tag{1.4.37}$$

Now we consider the functions φ_2 and φ_3, where

$$\varphi_3(u) = u^{n-1}\int_0^1 t^{n-1}(1-t^2)^{(n-2)/2}\,B_{n+1}(tu)\,du.$$

From Lemma 1.4.10. (formula (5)) we obtain that Mellin transform of a function φ_3 is equal to $\tilde{g}_1(s+n-1)\tilde{g}_2(1-s)$, where g_2 is a Mellin transform of a function

$$f(t) = \begin{cases} (1-t^2)^{(n-2)/2}, & |t| < 1, \\ 0, & |t| \geq 1. \end{cases}$$

From Lemma 1.4.10. (formula (1)) we obtain

$$\tilde{g}_2(1-s) = \frac{1}{2}\,B\left(\frac{n}{2}, \frac{1-s}{s}\right), \qquad g_2(-s) = \frac{1}{2}\,B\left(\frac{n-1}{2}, \frac{-s}{2}\right).$$

As before for the functions φ_1 and φ_2, we take the functions

$$\begin{aligned} f_1(u) &= e^{-u^2}, \quad \beta_1 = 0; \\ \tilde{f}_1(u) &= \frac{n-1}{2}e^{-u^2} + u^2 e^{-u^2}; \quad \beta_2 = -1. \end{aligned}$$

From Lemma 1.4.10. (statements 1), 2) and 3))we obtain

$$g_4(1-s) = \frac{1}{2}\Gamma\left(\frac{1-s}{2}\right), \qquad \tilde{g}_4(-s) = \frac{1}{2}\left(\frac{n-1}{2} - \frac{s}{2}\right)\Gamma\left(-\frac{s}{2}\right),$$

where g_4 and $\tilde{g}_4$ are Mellin transforms of the functions f_4 and $\tilde{f}_4$ respectively. The functions $\tilde{g}(s)g_4(1-s)$ and $\tilde{g}_1(s)\tilde{g}_2(1-s)\tilde{g}_4(-s)$ are equal up to constant. As before we obtain that there exists s such that all these functions are defined correctly. Using inversion formula for Mellin transform we obtain that the functions

$$\int_0^\infty e^{-u^2}\varphi_2(tu)du, \qquad \int_0^\infty \frac{e^{-u^2}}{u}\left(\frac{n-1}{2} + u^2\right)\varphi_3(tu)du \tag{1.4.38}$$

are equal up to the constant.

Consider the fist function from (1.4.38):

$$\begin{aligned} \int_0^\infty e^{-u^2}\varphi_2(tu)\,du &= \frac{1}{2t}\int_0^\infty e^{-z/t^2}\frac{\varphi_2(\sqrt{z})}{\sqrt{z}}dz \\ &= \frac{\sqrt{p}}{2}\int_0^\infty e^{-zp}\frac{\varphi_2(\sqrt{z})}{\sqrt{z}}dz \\ &= \frac{\sqrt{p}}{2}g(p). \end{aligned}$$

From Lemma 1.4.8. we obtain

$$-g'(p) = \int_0^\infty e^{-zp}\sqrt{z}\varphi_2(\sqrt{z})\, dz.$$

Consider the measures

$$F_1(z) \quad \text{and} \quad F_2(z) = \int_0^z \sqrt{u}\varphi_2(\sqrt{u})\, du.$$

Note that F_2 is not necessarily finite measure. We obtain:

$$F_1'(z)\sqrt{z} = F_2'(z),$$

so therefore

$$\begin{aligned} F_2(z) &= \int_0^z F_2'(u)\, du \\ &= \int_0^z F_1'(u)\, \sqrt{u}\, du. \end{aligned}$$

Integrating by parts, we obtain

$$\begin{aligned} F_2(z) &= \sqrt{z}\, F_1(z) - \frac{1}{2}\int_0^z \frac{F_1(u)}{\sqrt{u}}\, du \\ &= \sqrt{z}\, F_1(z) - \frac{1}{2}\int_0^z \frac{F_1(zu)}{\sqrt{zu}}\, z\, du. \end{aligned}$$

From (1.4.37) and Lemma 1.4.12. (statement 6)) we have

$$F_1(z) = A_6(z^{(n-\alpha+1)/2}\, \tilde{L}(\sqrt{z}), \quad z \to \infty.$$

The integral $\int_0^1 u^{(n-\alpha)/2-\eta} du$ is exists for $\eta < \frac{1}{2}(n-\alpha)+1$. From Lemma 1.4.12. (statement 8)) we obtain

$$\begin{aligned} z^{(\alpha-n-1)/2}\int_0^1 \frac{F_1(zu)}{\sqrt{u}}\, du &\sim A_6\, \tilde{L}(\sqrt{z})\int_0^1 u^{(n-\alpha)/2} du \\ &\sim \frac{2A_6}{n-\alpha+2}\, L(\sqrt{z}), \quad z \to \infty, \end{aligned}$$

and

$$F_2(z) \sim \left(1 - \frac{1}{n-\alpha+2}\right) A_6\, z^{(n-\alpha+z)/2} L(\sqrt{z}), \quad z \to \infty.$$

From Lemma 1.4.7.

$$\int_0^\infty e^{-zp}\sqrt{z}\, \varphi_2(\sqrt{z})\, dz \sim A_7\, p^{-(n-\alpha+2)/2} L\left(\frac{1}{\sqrt{p}}\right), \quad p \to +0.$$

Therefore,

$$g'(p) \sim A_7\, p^{-(n-\alpha+2)/2}\, L\left(\frac{1}{\sqrt{p}}\right), \quad p \to +0$$

We easily obtain

$$\begin{aligned} g(p) &= -\int_p^\infty g'(t)\, dt \\ &= -\int_0^1 g'\left(\frac{1}{zm}\right) \frac{dz}{mdz^2} \\ &= -m^{(n-\alpha)/2} \int_0^1 \frac{g'\left(\frac{1}{zm}\right)\, z^{(n-\alpha+2)/2}}{\left(\frac{1}{zm}\right)^{-(n-\alpha+2)/2} z^2} dz, \quad p = \frac{1}{m} \end{aligned}$$

and

$$g(p) = m^{(n-\alpha)/2} \int_0^1 \tilde{L}(\sqrt{zm})\, z^{(n-\alpha)/2-1} dz.$$

From Lemma 1.4.12. (statement 7) we obtain for $\eta < (n-\alpha)/2$:

$$\begin{aligned} g(p) &\sim A_8\, p^{-(n-\alpha)/2} \tilde{L}\left(\frac{1}{\sqrt{p}}\right) \\ &\sim A_8\, p^{-(n-\alpha)/2} L\left(\frac{1}{\sqrt{p}}\right), \quad p \to +0. \end{aligned}$$

Therefore

$$\int_0^\infty e^{-x^2} \varphi_2(tx)\, dx \sim \frac{1}{2} A_8\, t^{n-\alpha+1} L(t), \quad t \to +\infty$$

and

$$\int_0^\infty \frac{1}{x} e^{-x^2} (\frac{n-1}{2} + x^2)\ \varphi_3(tx)\ dx \sim A_9\ t^{n-\alpha-1} L(t), \quad t \to +\infty.$$

For the second function from (1.4.38):

$$\begin{aligned}\int_0^\infty \frac{1}{x} e^{-x^2} \left(\frac{n-1}{2} + x^2\right)\ \varphi_3(tx)\ dx &= \frac{1}{2}\int_0^\infty e^{-u} \left(\frac{n-1}{2} + u\right) \frac{\varphi_3(t\sqrt{u})}{u}\ du \\ &= \frac{1}{2}\int_0^\infty e^{-z/t^2} \left(\frac{n-1}{2} + \frac{z}{t^2}\right) \frac{\varphi_3(\sqrt{z})}{z}\ dz.\end{aligned}$$

Consider

$$\psi(p) = \int_0^\infty e^{-zp} \frac{\varphi_3(\sqrt{z})}{z}\ dz, \quad p = \frac{1}{t^2}.$$

From Lemma 1.4.8. we obtain that the second function in (1.4.38) is equal to

$$\frac{n-1}{4}\psi(p) - \frac{p}{2}\psi'(p) = \psi_1(p).$$

A solution of this equation is a function

$$\psi(p) = p^{(n-1)/2} \int_p^\infty \frac{2\psi_1(t)}{t^{(n-1)/2+1}} dt.$$

We note that

$$\begin{aligned}\psi(p) &= p^{(n-1)/2} \int_0^{1/p} 2\psi_1\left(\frac{1}{z}\right)\ z^{(n-1)/2-1} dz \\ &= m^{1-(n-1)/2} \int_0^1 2\psi_1\left(\frac{1}{zm}\right)\ (zm)^{(n-1)/2-1} dz\end{aligned}$$

and

$$\psi_1(p) = A_{10}\ p^{-(n-\alpha-1)/2} \tilde{L}\left(\frac{1}{\sqrt{p}}\right).$$

From Lemma 1.4.12. (statement 8)) $(n - \frac{\alpha}{2} + 2 > -1)$ we obtain

$$\begin{aligned}\psi(p) &= m^{(n-\alpha-1)/2} \int_0^1 2A_{10}\psi_1\left(\frac{1}{zm}\right)\left(\frac{1}{zm}\right)^{(n-\alpha-1)/2-1} z^{n-\alpha/2-2}dz \\ &\sim A_{11}\, m^{(n-\alpha-1)/2}\tilde{L}(\sqrt{m}) \\ &\sim A_{11}\, p^{-(n-\alpha-1)/2}\tilde{L}\left(\frac{1}{\sqrt{p}}\right), \quad p \to +0.\end{aligned}$$

From Lemma 1.4.7.

$$\psi(p) \sim A_{11}\, p^{-(n-\alpha-1)/2} L\left(\frac{1}{\sqrt{p}}\right), \quad p \to +0 \tag{1.4.39}$$

By definition of φ_3 we obtain

$$\frac{\varphi_3(x)}{x} = A_{12}\frac{l_{n+1}\left(\frac{x}{2}\right)}{\left(\frac{x}{2}\right)^{n+2}}, \tag{1.4.40}$$

where l_{n+1} corresponds to covariance function corresponds to $B_{n+1.}$ Since $\varphi_3 \geq 0$, $\varphi_3(x)/x \to 0$, $x \to 0+$, we obtain from Lemma 1.4.7. and (1.4.39):

$$\int_0^z \frac{\varphi_3(\sqrt{r})}{r}dr \sim A_{13}\, z^{(n-\alpha-1)/2} L(\sqrt{z}), \quad z \to +\infty.$$

Using the identity:

$$\int_0^z \frac{\varphi_3(\sqrt{r})}{r}\, dr = 2\int_0^{\sqrt{z}} \frac{\varphi_3(r)}{r}\, dr,$$

we obtain from (1.4.40):

$$\int_0^z \frac{l_{n+1}(r)}{r^{n+2}}dr \sim A_{14}\, z^{n-\alpha-1} L(z), \quad z \to +\infty. \tag{1.4.41}$$

From (1.4.29) we obtain $(n+1$ is odd):

$$y^\alpha f(iy) = A_2 \int_0^\infty g_{n+1}(t,1)\, t^{n-\alpha}\frac{l_{n+1}(ty)}{(ty)^{2n-\alpha}}\, dt.$$

Integrating by parts we obtain

$$y^\alpha f(iy) = A_2\, t^2 \int_0^t \frac{l_{n+1}(xy)}{(xy)^{n+2}}\, dx\; g_{n+1}(t,1)\; y^{\alpha+2-n}\, |_0^\infty$$

$$-A_2 \int_0^\infty y^{\alpha+2-n}\, \left(t^2 g_{n+1}(t,1)\right)' \int_0^t \frac{l_{n+1}(xy)}{(xy)^{n+2}}\, dx\, dt.$$

By definition of and $g_{n+1}(t,1)$ we have $g_{n+1}(t,1) \sim t^{-n-2}$ as $t \to \infty$ and $g_{n+1}(t,1) \sim$ const as $t \to 0$. Therefore from (1.4.41) we obtain

$$g_{n+1}(t,1) \cdot t^2 \int_0^t \frac{l_{n+1}(xy)}{(xy)^{n+2}}\, dx \to 0 \text{ as } \quad t \to \infty.$$

Thus,

$$y^\alpha f(iy) = -A_2 y^{\alpha+2-n} \int_0^\infty \left(t^2 g_{n+1}(t,1)\right)' \int_0^t \frac{l_{n+1}(xy)}{(xy)^{n+2}} dx dt$$

$$= -A_2\; y^{\alpha+2-n} \int_0^\infty \left(2\, t\; g_{n+1}(t,1) + t^2 g'_{n+1}(t,1)\right)$$

$$\times \frac{1}{y} \int_0^t \frac{l_{n+1}(x)}{x^{n+2}}\, dx dt.$$

From definition $g_{n+1}(t,1)$ we obtain that $t g_{n+1}(t,1)$ has order t^{-n-1} as $t \to \infty$, the function $t g_{n+1}(t,1)$ has order t as $t \to 0$, and the function $t^2 g_{n+1}(t,1)$ has order t^{-n-1} as $t \to \infty$, and $t^2 g'_{n+1}(t,1) \to 0$ as $t \to 0$. The function $\int_0^t l_{n+1}(x) x^{-n-2} dx$ is bounded on every finite interval from $(0,\infty)$. From Lemma 1.4.12. (statement 6),7),8)) we obtain

$$y^\alpha f(iy) \sim A_{16}\; L(y), \quad y \to \infty.$$

Then from Lemma 1.4.7. we obtain the statement a) of Theorem 1.4.1.

Theorem 1.4.1. is proved.

Proof of Theorem 1.4.2.

The proof of Theorem 1.4.2. is analogous to that of Theorem 1.4.1. with used the function $b_n(r)$ (see 1.4.7)) in place of a function $l_n(r)$ (see (1.4.8)). So, we shall omit some details.

Let condition a) of Theorem 1.4.2. holds. From Lemma 1.4.12. (statement 6)) we obtain that there exists a function $\tilde{L} \in \mathcal{L}$ such that

$$G(\lambda)/\lambda^\alpha = \tilde{L}(\frac{1}{\lambda}) \quad \text{as} \quad \lambda \to 0+.$$

From (1.4.7) we obtain

$$\begin{aligned}
\frac{b_n(r)}{r^{2n-\alpha}} &= (2\pi)^n r^\alpha \int_0^\infty J^2_{n/2}(\lambda r)(\lambda r)^{-n} dG(\lambda) \\
&= (2\pi)^n r^\alpha \int_0^\infty J^2_{n/2}(u) u^{-n} dG\left(\frac{u}{r}\right) \\
&= (2\pi)^n r^\alpha J^2_{n/2}(u)\, u^{-n} G\left(\frac{u}{r}\right)\Big|_0^\infty \\
&\quad -(2\pi)^n r^\alpha \int_0^\infty \left[2J_{n/2}(u)\left(J_{n/2-1}(u) + J_{n/2+1}(u)\right) u^n \right. \\
&\quad \left. - n\, u^{n-1} J^2_{n/2}(u)\right] u^{-2n} G\left(\frac{u}{r}\right) du \\
&= (2\pi)^n \int_0^\infty G\left(\frac{u}{r}\right)\left(\frac{u}{r}\right)^{-\alpha} \left[\frac{nJ^2_{n/2}(u)}{u^{n+1-\alpha}}\right. \\
&\quad \left. - \frac{2J_{n/2}(u)}{u^{n-\alpha}}\left(J_{n/2-1}(u) + J_{n/2+1}(u)\right)\right] du,
\end{aligned}$$

and

$$\begin{aligned}
\frac{b_n(r)}{r^{2n-\alpha}} &= (2\pi)^n \int_0^\infty \tilde{L}\left(\frac{r}{u}\right) \left[\frac{nJ^2_{n/2}(u)}{u^{n+1-\alpha}}\right. \\
&\quad \left. - \frac{2J_{n/2}(u)}{u^{n-\alpha}}\left(J_{n/2-1}(u) + J_{n/2+1}(u)\right)\right] du \\
&= (2\pi)^n \int_{1/A}^\infty \tilde{L}(zr) \left[nJ_{n/2}\left(\frac{1}{z}\right) z^{n+1-\alpha}\right.
\end{aligned}$$

$$- 2J_{n/2}\left(\frac{1}{z}\right) z^{n-\alpha} \left(J_{n/2-1}\left(\frac{1}{z}\right) + J_{n/2+1}\left(\frac{1}{z}\right)\right)\Bigg] z^{-2}dz$$

$$+(2\pi)^n \int_0^{1/A} \tilde{L}(zr) \Bigg[nJ^2_{n/2}\left(\frac{1}{z}\right) z^{n+1-\alpha}$$

$$- 2J_{n/2}\left(\frac{1}{z}\right) z^{n-\alpha} \left(J_{n/2-1}\left(\frac{1}{z}\right) J_{n/2+1}\left(\frac{1}{z}\right)\right)\Bigg] z^{-2}dz.$$

From (1.4.27) we obtain that there exists $\eta > 0$, such that the integrals

$$\int_{1/A}^{\infty} z^n U_n(z)\, dz, \qquad \int_0^{1/A} z^n U_n(z)\, dz$$

exist, where

$$U_n(z) = \left[nJ^2_{n/2}\left(\frac{1}{z}\right) z^{n+1-\alpha} - 2\, J_{n/2}\left(\frac{1}{z}\right) z^{n-\alpha}\left(J_{n/2-1}\left(\frac{1}{z}\right) + J_{n/2+1}\left(\frac{1}{z}\right)\right)\right] z^{-2}.$$

From Lemma 1.4.12. (statements 6), 7), 8)) we obtain

$$\frac{b_n(r)}{r^{2n-\alpha}} \sim \tilde{L}(r) \left[n\ J^2_{n/2}(u)\ u^{\alpha-1-n} - 2J_{n/2}(u)\ u^{\alpha-n}\left(J_{n/2-1}(u) + J_{n/2+1}(u)\right)\right] du \sim$$

$$\tilde{\ }L(r)\ \text{const}, \qquad r \to \infty.$$

Thus the Abelian part is proved. Let us prove the Tauberian part of Theorem 1.4.2. Integrating by parts in (1.4.22) we obtain

$$f(iy) = A(n)\ (n+1) \int_0^{\infty} u\ \left(u^2 + y^2\right)^{-(n+1)/2-1} \int_0^{u} B_n(\nu)\ d\nu\ du.$$

Suppose that $n = 2k+1$ is odd. Then we shall show that there exists a function $g_n(u, y)$ such that

$$\int_0^{\infty} u \left(u^2 + y^2\right)^{-(n+1)/2-1} \int_0^{u} \nu^{n-1} B_n(\nu) d\nu du$$

$$= \int_0^{\infty} g_n(u, y) \int_0^{u} \nu^{n-1} B_n(\nu) I_{1-(\nu/u)^2}\left(\frac{n+1}{2}, \frac{1}{2}\right) d\nu du. \quad (1.4.42)$$

By the formula (1.4.4) we obtain for odd n:

$$I_{1-(\nu/u)^2}\left(\frac{n+1}{2},\frac{1}{2}\right) = B^{-1}\left(\frac{n+1}{2},\frac{1}{2}\right)\int_0^{1-(\nu/u)^2} t^{(n-1)/2}(1-t)^{-1/2}dt$$

$$= B^{-1}\left(\frac{n+1}{2},\frac{1}{2}\right)\int_{(\nu/u)^2}^{1} x^{-1/2}\left(1-x\right)^{(n-1)/2}dx$$

$$= B^{-1}\left(\frac{n+1}{2},\frac{1}{2}\right)\int_{\nu/u}^{1}(1-x)^{(n-1)/2}dx. \qquad (1.4.43)$$

is a polynomial of degree n from ν/u. Integrating by parts we obtain from (1.4.42) and (1.4.43) the following equation for $g_n(u,y)$:

$$\int_0^u \nu\,(\nu^2+y^2)^{-(n+1)/2-1}d\nu = \int_0^u g_n(\nu,y)\;I_{1-(\nu/u)^2}\left(\frac{n+1}{2},\frac{1}{2}\right)\,d\nu.$$

Differentiating it, we obtain:

$$\frac{u}{(u^2+y^2)^{(n+1)/2+1}} = \int_0^u \frac{g_n(\nu,y)}{\nu}\left(1-(u/\nu)^2\right)^{(n-1)/2}d\nu.$$

Continuing the differentiation, we obtain

$$g_n(u,v) = A_1\;u^{2k+1}\underbrace{\left(\left(\cdots\left(\left(\frac{u^2}{(u^2+y^2)^{(n+1)/2+1}}\right)'\frac{1}{u}\right)'\cdots\right)'\frac{1}{u}\right)'}_{k}, \qquad A_1>0.$$

We can easily see that at infinity $g_n(u,y)$ has order u^{-n-2}, and at zero it has order u. Therefore there exists $\eta>0$ such that $t^{\eta}t^{n-\alpha}g_n(t,1)$ is integrable on $[A,\infty)$ and $t^{-\eta}t^{n-\alpha}g_n(t,1)$ is integrable on $[A,\infty)$, where $A>0$. By the definitions of $b_n(r)$ and $g_n(t,1)$ we obtain from Lemma 1.4.12.:

$$y^{\alpha}f(iy) = A_2\int_0^{\infty} t^{n-\alpha}g_n(t,1)\;b_n(ty)\;(ty)^{\alpha-2n}dt$$

$$\sim \text{const}\;L(y) \quad \text{as} \quad y\to\infty.$$

Then by Lemma 1.4.7.

$$G(\lambda)/\lambda^{\alpha} \sim L(\frac{1}{\lambda}), \quad \lambda \to 0+.$$

So, our statement for odd n follows from Lemma 1.4.7.

Suppose that n is even. We define the following functions:

$$\begin{aligned}\varphi_1(u) &= \int_0^u v^n B_n(v) \left(1-\left(\frac{v}{n}\right)\right)^{(n-1)/2} dv \\ &= u^{n+1} \int_0^1 t^n B_n(tu)(1-t^2)^{(n-1)/2} dt;\end{aligned}$$

$$\varphi_2(u) = u^{n+1} \int_0^1 t^{n+1} B_n(tu)(1-t^2)^{n/2} dt, \tag{1.4.44}$$

where B_{n+1} corresponds the same spectral function G in (1.4.1) as B_n. From Lemma 1.4.10. (formula 5)) we obtain

$$g(s) = g_1(s+n+1)\, g_2(-s), \qquad \tilde{g}(s) = \tilde{g}_1(s+n+1)\, g_2(-s),$$

where $g(s)$ and $\tilde{g}(s)$ are Mellin transforms of a functions $\varphi_1(u)$ and $\varphi_2(u)$, g_1 and $\tilde{g}_1(s)$ are Mellin transforms of a functions B_n and B_{n+1} and g_2 is Mellin transform of a function

$$\begin{cases} (1-t^2)^{(n-1)/2}, & |t| \le 1 \\ 0, & |t| > 1 \end{cases}$$

From Lemma 1.4.10. (formula 5), 6)) we obtain that the function g_2 is defined correctly for Re$s < 0$ and the function

$$\begin{aligned}g_1(s) &= \int_0^\infty x^{s-1} B_n(x)\, dx \\ &= \text{const} \int_0^\infty \frac{x^{s-1}}{x^{(n-2)/2}} \int_0^\infty J_{(n-2)/2}(\lambda)\, \frac{dG(\lambda)}{\lambda^{(n-2)/2}} \\ &= \text{const}\ g_0\left(s - \frac{n-2}{2}\right)\, \tilde{g}_0\left(1 + \frac{n-2}{2} - s\right),\end{aligned}$$

where g_0 is Mellin transform of $J_{(n-2)/2}$ and $\tilde{g}_0$ is Mellin transform of a function $\lambda^{(n-2)/2}dG(\lambda)/d\lambda$.

From Lemma 1.4.10. (formula 6)) we obtain that Mellin transform $g_0(s-\frac{1}{2}(n-2))$ is defined correctly for

$$-\frac{n-2}{2} < \mathrm{Re}\left(s-\frac{n-2}{2}\right) < \frac{3}{2} \quad \text{or} \quad 0 < \mathrm{Re}\, s < \frac{n+1}{2}.$$

Note that

$$\tilde{g}_0(1+\frac{1}{2}(n-2)-s) = \int_0^\infty \lambda^{-s} dG(\lambda)$$

and from Lemma 1.4.14. we obtain that this transform is defined correctly for $0 < \mathrm{Re} s < \alpha$. So that the function $g_1(s)$ is defined for $0 < \mathrm{Re}\, s < \min\left(\frac{1}{2}(n+1), \alpha\right)$ and $g_1(s+n+1)$ for $-(n+1) < -(n+1)+\min\left(\frac{1}{2}(n+1), \alpha\right)$. Finally for $g(s)$ we obtain the same domain of definition. Similar for the function $\tilde{g}(s)$ we obtain the following domain of definition: $-(n+1) < \mathrm{Re}\, s < -(n+1) + \min\left(\frac{1}{2}(n+1), \alpha\right)$.

From Lemma 1.4.10. (formulas 5) 6)) we obtain

$$g_1(s+n+1) = \text{const } g_0\left(s+\frac{n}{2}+2\right) \tilde{g}_0\left(-\frac{1}{2}-s-\frac{n}{2}\right),$$

$$\tilde{g}_1(s+n+1) = \text{const } \bar{g}_0\left(s+\frac{n}{2}+\frac{1}{2}\right) \tilde{g}_0\left(-\frac{1}{2}-s-\frac{n}{2}\right),$$

where $g_0, \bar{g}_0$ are Mellin transform of $J_{(n-2)/2}$ and $J_{(n-1)/2}$, and

$$g_0\left(s+\frac{n}{2}+2\right) = \frac{\Gamma\left(\frac{s}{2}+\frac{n}{2}+\frac{1}{2}\right) 2^{s+\frac{n}{2}+1}}{\Gamma\left(-\frac{1+s}{2}\right)},$$

$$\bar{g}_0\left(s+\frac{n}{2}+\frac{1}{2}\right) = \frac{\Gamma\left(\frac{s}{2}+\frac{n}{2}+\frac{1}{2}\right) 2^{s+\frac{n}{2}+1}}{\Gamma\left(-\frac{s}{2}\right)}. \tag{1.4.45}$$

From Lemma 1.4.10. (formula 5)) we chose f and $\tilde{f}$ with Mellin transform g_3 and $\tilde{g}_3$ respectively such that $g(s)\ g(1-s+\beta_1)$ and $\tilde{g}(s)\ \tilde{g}_3(1-s+\beta_2)$ are equal up to constant. Let

$$f(x) = \tilde{f}(x) = \exp(-x^2), \quad \beta_1 = -2, \quad \beta_2 = -1.$$

From Lemma 1.4.7. (formulas 3) and 4)) and (1.4.45) we obtain that $g(s)\, g_3(1-s+\beta_1)$ and $\tilde{g}(s)\, \tilde{g}_3(1-s+\beta_2)$ are equal. Therefore the Mellin transforms of a functions

$$\int_0^\infty e^{-x^2} \varphi_1(tx)\, x^{-2} dx, \qquad \int_0^\infty e^{-x^2} \varphi_2(tx)\, x^{-1} dx, \tag{1.4.46}$$

are equal up to constant. For

$$-(n+1) < \operatorname{Re} s < (n+1) + \min\left(\frac{n+1}{2}, \alpha\right)$$

these transforms are defined correctly, then by the inversion formula for Mellin transforms the functions (1.4.16) are equal up to constant. Note that functions (1.4.46) are equal to:

$$\frac{t}{2} \int_0^\infty e^{-z/t^2} \frac{\varphi_1(\sqrt{z})}{z^{3/2}}\, dz, \qquad \frac{1}{2} \int_0^\infty e^{-z/t^2} \frac{\varphi_2(\sqrt{z})}{z}\, dz.$$

From (1.4.7) and (1.4.44) we obtain:

$$\frac{\varphi_1(r)}{r^2} = A_{14} \left(\frac{b_n\left(\frac{r}{2}\right)}{\left(\frac{r}{2}\right)^n} \right)'.$$

Denote

$$g(p) = \int_0^\infty e^{-zp} \frac{\varphi_1(\sqrt{z})}{z^{3/2}}\, dz.$$

From Lemma 1.4.8. we obtain

$$-g'(p) = \int_0^\infty e^{-zp} \frac{\varphi_1(\sqrt{z})}{\sqrt{z}}\, dz.$$

Consider the function:

$$\bar{K}_2(z) = \frac{1}{A_{17}} \int_0^\infty \frac{\varphi_1(\sqrt{u})}{\sqrt{u}}\, du, \quad \bar{K}_1(z) = \frac{b_n(\sqrt{z}/2)}{(\sqrt{z}/2)^n} = \int_0^\infty \frac{\varphi_1(\sqrt{t})}{A_{17} t} dt.$$

Therefore $\bar{K}_2'(z) = \sqrt{z}\bar{K}_1'(z)$. Similarly to the proof of Theorem 1.4.1. we obtain:

$$\bar{K}_2(z) = \sqrt{z}\,\bar{K}_1(z) - \frac{1}{2}\int_0^z \frac{\bar{K}_1(r)}{\sqrt{r}}\,dr. \tag{1.4.47}$$

From definition of $\bar{K}_1(z)$ and condition b) we obtain

$$\bar{K}_1(z) \sim A_{18}\, z^{(n-\alpha)/2} L(\sqrt{z}), \quad z \to \infty.$$

Therefore from Lemma 1.4.12. (statement 8)) and (1.4.47) we obtain

$$g'(p) \sim \mathrm{const}\, p^{-(n-\alpha+1)/2} L(\frac{1}{\sqrt{p}}), \quad p \to 0+.$$

Similarly to proof of Theorem 1.4.1. we find

$$g(p) \sim \mathrm{const}\, p^{-(n-\alpha+1)/2} L(\frac{1}{\sqrt{p}}), \quad p \to 0+.$$

Thus

$$\frac{t}{2}\int_0^\infty e^{-z/t^2} \frac{\varphi_1(\sqrt{z})}{z^{3/2}}\,dz = A_{18}\, t^{n-\alpha} L(t), \quad t \to \infty,$$

$$\frac{1}{2}\int_0^\infty e^{-z/t^2} \frac{\varphi_2(\sqrt{z})}{z}\,dz \sim A_{19}\, t^{n-\alpha} L(t), \quad t \to \infty, \tag{1.4.48}$$

or

$$\int_0^\infty e^{-x^2} \frac{1}{x}\varphi_2(tx)\,dx = A_{19}\, t^{n-\alpha} L(t), \quad t \to \infty. \tag{1.4.49}$$

Consider functions φ_2 and φ_3, where

$$\varphi_3(u) = u^{n+1} \int_0^1 t^{n+1} B_{n+1}(t)\,(1-t^2)^{n/2} dt.$$

From Lemma 1.4.10. (formula 5)) Mellin transforms of these functions are equal to $\tilde{g}_1(s+n+1)\, g_2(-s)$ and $\tilde{g}_1(s+n+1)\tilde{g}_2(1-s)$, where $\tilde{g}_2$ is Mellin transform of a function

$$\begin{cases} (1-t^2)^{n/2}, & |t| \le 1, \\ 0, & |t| > 1. \end{cases}$$

From Lemma 1.4.10. (formula 1))

$$\begin{aligned}
g_2(-s) &= \frac{1}{2}\,B\left(\frac{n+1}{2},-\frac{s}{2}\right)\\
&= \frac{\Gamma\left(-\frac{s}{2}\right)\Gamma\left(\frac{n+1}{2}\right)}{2\Gamma\left(\frac{n+1}{2}-\frac{s}{2}\right)},\\
\tilde{g}_2(1-s) &= \frac{1}{2}\,B\left(\frac{n+2}{2},\frac{1-s}{2}\right)\\
&= \frac{\Gamma\left(\frac{1-s}{2}\right)\Gamma\left(\frac{n+2}{2}\right)}{2\Gamma\left(\frac{n+1}{2}-\frac{s}{2}\right)\Gamma\left(\frac{n+1-s}{2}\right)}.
\end{aligned}$$

Consider the functions:

$$f_1(x)=e^{-x^2},\quad \beta_1=0;\quad \tilde{f}_1(x)=\frac{n+1}{2}e^{-x^2}+x^2e^{-x^2},\quad \beta_2=-1.$$

Arguments similar to those given above lead us to conclude that expressions

$$\int\limits_0^\infty e^{-x^2}\varphi_2(tx)\,dx,\qquad \int\limits_0^\infty e^{-x^2}\left(\frac{n+1}{2}+x^2\right)\frac{1}{x}\,\varphi_3(tx)\,dx \tag{1.4.50}$$

are equal up to constant. The first function from (1.4.50) is equal to

$$\frac{1}{2t}\int\limits_0^\infty e^{-z/t^2}\frac{\varphi_2(\sqrt{z})}{\sqrt{z}}dz.$$

By definition of $\varphi_2(\sqrt{z})$ we obtain

$$\frac{\varphi_2(\sqrt{z})}{z}=A_{17}\left(\frac{\tilde{b}_n\left(\frac{\sqrt{r}}{2}\right)}{\left(\frac{\sqrt{r}}{2}\right)^n}\right)',$$

where $\tilde{b}_n$ is given by (1.4.7) with B_{n+1} in place of B_n. Therefore

$$\begin{aligned}
\int\limits_0^z \frac{\varphi_2(\sqrt{t})}{t}\,dt &= A_{17}\,\frac{\tilde{b}_n\left(\frac{\sqrt{z}}{2}\right)}{\left(\frac{\sqrt{z}}{2}\right)^n}\\
&\geq 0.
\end{aligned}$$

As $b_n(r)/r^n \to 0$, $r \to 0$, the measure

$$F_3(z) = \int_0^z \int_0^u \frac{\varphi_2(t)}{t}\, dt\, du,$$

is defined. From ((1.4.48) and (1.4.49) we obtain

$$\begin{aligned}\int_0^\infty e^{-zp}\, dF_3(z) &= \frac{1}{p}\int_0^\infty e^{-zp}\frac{\varphi_2(\sqrt{z})}{\sqrt{z}}\, dz \\ &\sim A_{20}\, p^{-(n-\alpha+2)/2} L(\frac{1}{\sqrt{p}}), \quad p \to 0.\end{aligned}$$

From Lemma 1.4.7.

$$F_3(z) \sim A_{21}\, z^{(n-\alpha+2)/2} L(\sqrt{z}), \quad z \to \infty. \tag{1.4.51}$$

We consider the function

$$F_4(z) = \int_0^z \int_0^t \frac{\varphi_2(\sqrt{z})}{\sqrt{z}}\, dz\, dt.$$

Note that $F_3''(z)\sqrt{z} = F''(z)$. Similarly to proof of Theorem 1.4.1. we obtain

$$\begin{aligned}F_4'(z) &= \sqrt{z}\, F_3'(z) - \frac{1}{2}\int_0^z \frac{F_3'(r)}{\sqrt{r}}\, dr, \\ F_4(z) &= \sqrt{z}\, F_3(z) - \frac{1}{2}\int_0^z \frac{F_3(r)}{\sqrt{r}}\, dr - \int_0^z\int_0^t \frac{F_3'(r)}{\sqrt{r}}\, dr\, dt \\ &= \sqrt{z}\, F_3(z) - \int_0^z \frac{F_3(r)}{\sqrt{r}}\, dr + \frac{1}{4}\int_0^z\int_0^t \frac{F_3(r)}{r^{3/2}}\, dr\, dt.\end{aligned}$$

From (1.4.51) and Lemma 1.4.12. (statement 8)) we obtain

$$\frac{1}{2t}\int_0^\infty e^{-z/t^2}\frac{\varphi_2(\sqrt{z})}{\sqrt{z}}\, dz \sim A_{22}\, t^{n-\alpha} L(t), \quad t \to \infty.$$

Thus for a second function from (1.4.50):

$$\int_0^\infty e^{-x^2}\left(\frac{n+1}{2}+x^2\right)\frac{1}{x}\,\varphi_3(tx)\,dx \sim A_{23}\,t^{n-\alpha}L(t), \quad t\to\infty. \qquad (1.4.52)$$

The second function from (1.4.50) is equal to

$$\frac{1}{2t}\int_0^\infty e^{-z/t^2}\left(\frac{n+1}{2}+x^2\right)\frac{\varphi_2(\sqrt{z})}{z}\,dz.$$

Let

$$\psi(p)=\int_0^\infty e^{-zp}\frac{\varphi_2(\sqrt{z})}{z}\,dz, \quad p=\frac{1}{t^2}.$$

From Lemma 1.4.8. we obtain that second function from (1.4.50) is equal to

$$\frac{1}{2}\left(\frac{n+1}{2}\,\psi(p)-p\,\psi'(p)\right).$$

Similar to the proof of Theorem 1.4.1. we obtain

$$\psi(p)=A_{24}\,p^{-(n-\alpha)/2}L\left(\frac{1}{\sqrt{p}}\right), \quad p\to 0+. \qquad (1.4.53)$$

By definition of φ_3 we obtain

$$\frac{\varphi_3(r)}{r}=A_{17}\left(\frac{b_{n+1}\left(\frac{r}{2}\right)}{\left(\frac{r}{2}\right)^{n+1}}\right),$$

where b_{n+1} is defined by (1.4.7) with spectral function G. Therefore

$$\begin{aligned} A_{17}\int_0^t \frac{\varphi_3(\sqrt{z})}{z}dz &= \int_0^t \frac{A_{17}}{\sqrt{r}}\frac{\varphi_3(\sqrt{z})}{\sqrt{z}}dz \\ &= \int_0^t \frac{1}{\sqrt{z}}\left(\frac{b_{n+1}\left(\frac{\sqrt{z}}{2}\right)}{\left(\frac{\sqrt{z}}{2}\right)^{n+1}}\right)'dz \end{aligned}$$

$$\begin{aligned}
&= \frac{1}{\sqrt{z}}\left(\frac{b_{n+1}\left(\frac{\sqrt{z}}{2}\right)}{\left(\frac{\sqrt{z}}{2}\right)^{n+1}}\right)\Bigg|_0^t \\
&\quad +\frac{1}{2}\int_0^t \frac{1}{z^{3/2}}\left(\frac{b_{n+1}\left(\frac{\sqrt{z}}{2}\right)}{\left(\frac{\sqrt{z}}{2}\right)^{n+1}}\right) dz \\
&= \frac{1}{\sqrt{t}}\left(\frac{b_{n+1}\left(\frac{\sqrt{t}}{2}\right)}{\left(\frac{\sqrt{t}}{2}\right)^{n+1}}\right) \\
&\quad +\frac{1}{2}\int_0^t \frac{1}{z^{3/2}}\left(\frac{b_{n+1}\left(\frac{\sqrt{z}}{2}\right)}{\left(\frac{\sqrt{z}}{2}\right)^{n+1}}\right) dz \\
&\geq 0.
\end{aligned}$$

Let

$$\begin{aligned}
F_5(0,x) &= F_5(x) \\
&= \int_t^x \int_0^t \frac{\varphi_3(\sqrt{z})}{z}\, dz\, dt
\end{aligned}$$

be a measure. We obtain

$$\frac{\psi(p)}{p} = \int_0^\infty e^{-zp} \int_0^z \frac{\varphi_3(\sqrt{t})}{t}\, dt\, dz.$$

From (1.4.53) we obtain

$$\int_0^\infty e^{-zp} \int_0^z \frac{\varphi_3(\sqrt{t})}{t}\, dt\, dz \sim A_{24}\, p^{-(n-\alpha+2)/2} L\left(\frac{1}{\sqrt{p}}\right), \quad p \to 0+,$$

and by Lemma 1.4.8.

$$\int_t^x \int_0^t \frac{\varphi_3(\sqrt{z})}{z}\, dz\, dt \sim A_{25}\, x^{-(n-\alpha+2)/2} L(\sqrt{x}), \quad x \to \infty. \tag{1.4.54}$$

Consider

$$F_6(x) = \int_t^x \int_0^t \frac{\varphi_3(\sqrt{z})}{z}\, dz\, dt$$
$$= A_{17} \int_0^x \frac{b_{n+1}\left(\frac{\sqrt{t}}{2}\right)}{\left(\frac{\sqrt{t}}{2}\right)^{n+1}}\, dt.$$

Note that $F_5''(x)\sqrt{x} = F_6''(x)$, similarly to the measure F_3 and F_4 we obtain

$$F_6(x) = \sqrt{x}\, F_5(x) - \int_0^x \frac{F_5(t)}{\sqrt{t}}\, dt + \frac{1}{4}\int_t^x \int_0^t \frac{F_5(z)}{z^{3/2}}\, dz\, dt.$$

From Lemma 1.4.12. (statement 8)) and (1.4.54) we obtain

$$\frac{1}{8}\int_0^x \frac{b_{n+1}\left(\frac{\sqrt{t}}{2}\right)}{\left(\frac{\sqrt{t}}{2}\right)^{n+1}} dt = \int_0^{\sqrt{x}/2} \frac{b_{n+1}(z)}{z^{n+1}}\, z\, dz$$
$$\sim A_{25}\, x^{-(n-\alpha+3)/2} L(\sqrt{x}), \quad x \to \infty$$

or

$$\int_0^x \frac{b_{n+1}(z)}{z^n}\, z\, dz \sim A_{26}\, x^{(n-\alpha+3)} L(x), \quad x \to \infty. \tag{1.4.55}$$

We have ($n+1$ is odd)

$$y^\alpha f(iy) = A_2 \int_0^\infty g_{n+1}(t,1)\, t^{n-\alpha+1} \frac{b_{n+1}(ty)}{(ty)^{2n-\alpha+2}} dt$$
$$= \frac{A_2\, g_{n+1}(t,1)}{y^{n-\alpha+2} t} \int_0^t \frac{b_{n+1}(xy)}{(xy)^n}\, dx \Bigg|_0^\infty$$
$$-\frac{A_2}{y^{n-\alpha+2}} \int_0^\infty \left(\frac{g_{n+1}(t,1)}{t}\right)' \int_0^t \frac{b_{n+1}(xy)}{(xy)^n}\, dx\, dt.$$

The function

$$\frac{g_{n+1}(t,1)}{t}\int\limits_0^t \frac{b_{n+1}(xy)}{(xy)^n}dx$$

has order $t^{-\alpha-1}$ as $t \to \infty$ and tends to zero as $t \to 0$. Therefore

$$y^\alpha f(iy) = \frac{A_2}{y^{n-\alpha+3}}\int\limits_0^\infty \left[\frac{g_{n+1}(t,1)}{t^2} - \frac{g'_{n+1}(t,1)}{t}\right]\int\limits_0^{ty} \frac{b_{n+1}(z)}{z^n}\,dz\,dt.$$

The expression in square brackets has on order t^{-n-5} as $t \to \infty$ and order constant as $t \to 0$. From Lemma 1.4.12. (statements 6)-8)) and (1.4.55) we obtain

$$y^\alpha f(iy) \sim A_{27} L(y), \quad y \to \infty$$

and statement a) holds.

Proof of Theorem1.4.3.

Integrating by parts in (1.4.1) we obtain

$$\begin{aligned} r^\alpha B_n(r) &= r^\alpha 2^{(n-2)/2}\Gamma\left(\frac{n}{2}\right) \\ &\quad\times \left[J_{(n-2)/2}(\lambda r)(\lambda r)^{(n-2)/2}G(\lambda)\Big|_0^\infty \right. \\ &\quad \left. + r^2\int\limits_0^\infty J_{(n-2)/2}(\lambda r)\ (\lambda r)^{-n/2}\ \lambda\ G(\lambda)\ d\lambda\right] \\ &= 2^{(n-2)/2}\Gamma\left(\frac{n}{2}\right)\int\limits_0^\infty r^{\alpha+2}J_{n/2}(\lambda r)(\lambda r)^{-n/2}\lambda G(\lambda)d\lambda \\ &= 2^{(n-2)/2}\Gamma\left(\frac{n}{2}\right)\int\limits_0^\infty J_{n/2}(u)u^{-n}u^{\alpha+1}G\left(\frac{u}{r}\right)\left(\frac{u}{r}\right)^{-\alpha}du. \end{aligned}$$

From (1.4.27) we obtain that $J_{n/2}(u)\ u^{\alpha-n/2+1}$ has order $u^{\alpha+1}$ at zero and $u^{\alpha+1-(n+1)/2}$ at infinity. Thus $J_{n/2}(u)\ u^{\alpha-n/2+1}$ is integrable for $-1 < \alpha < (n-3)/2$. From Lemma 1.4.12. (statement 8)) and Lemma 1.4.4. we obtain $r^\alpha B_n(r) \sim c_4(n,\alpha)\ L(r), \quad r \to \infty$. Thus the statement a) implies b) for $-1 < \alpha < (n-3)/2$, $n \geq 4$.

Let condition b) is hold. From Lemma 1.4.11.

$$y^\alpha f(iy) = A(n) \int_0^\infty t^{n-\alpha-1}(1+t^2)^{-(n+1)/2} B_n(ty)\ (ty)^\alpha dt.$$

The function $t^{n-\alpha-1}/(1+t^2)^{(n+1)/2}$ has order $t^{n-1-\alpha}$ at zero and $t^{-2-\alpha}$ at infinity. From Lemma 1.4.12. (statement 8)) we obtain $y^\alpha f(iy) \sim L(y)$ const. The statement of Theorem is follows from Lemma 1.4.4.

Proof of Theorem 1.4.4.

Follows from a properties of slowly varying functions (see, Bingham, Goldie and Teugels (1989)).

Chapter 2

LIMIT THEOREMS FOR NON-LINEAR TRANSFORMATIONS OF RANDOM FIELDS

2.1 Some Properties of Gaussian and χ-Squared Random Fields

We introduce a class of random fields with given one-dimensional distributions and given covariance function.

Let $\mathbf{L}_2((a,b), p(u)\, du)$ be a Hilbert space of equivalence classes of Lebesgue measurable functions G: $(a,b) \to R^1$ satisfying the inequality

$$\int_a^b G^2(u)\, p(u)\, du < \infty, \quad p(u) \geq 0,$$

and let $\{e_k(u)\}_{k=0}^{\infty}$ be a complete orthonormal system of functions in $\mathbf{L}_2((a,b), p(u)\, du)$, that is,

$$\int_a^b e_k(u)\, e_m(u)\, p(u) du = \delta_k^m. \tag{2.1.1}$$

A. Let $\xi(x)$, $x \in R^n$, be a mean square continuous homogeneous isotropic random field with $E\xi(x) = k$ =const and covariance function

$$B(\|x\|) = \mathrm{cov}(\ \xi(0), \xi(x)), \quad x \in R^n.$$

Suppose that there exist densities

$$\begin{aligned} p(u) &= \frac{d}{du} P\{\xi(x) \le u\}, \quad u \in (a,b) \subseteq R^1; \\ p(u,w,\|x-y\|) &= \frac{\partial^2}{\partial u \partial w} P\{\xi(x) \le u,\ \xi(y) \le w\}, \\ (u,w) &\in S = (a,b) \times (a,b) \subseteq R^2 \end{aligned}$$

such that bilinear expansion

$$p(u,w,\|x\|) = p(u)\ p(w)\ \left[1 + \sum_{k=1}^{\infty} r_k(\|x\|)\ e_k(u)\ e_k(w)\right], \tag{2.1.2}$$

hold, where

$$|r_k(\|x\|)| \le 1, \quad \sum_{k=1}^{\infty} r_k^2(\|x\|) < \infty \quad \text{for all } \|x\| > 0,$$

$r_k(0) = 1$, and $\{e_k(u)\}$ is a complete orthogonal system in Hilbert space $\mathbf{L}_2((a,b), p(u)\ du)$. Assume also that e_k is orthogonal (with respect to $p(u)$) to every constant function and define $e_0(u) \equiv 1$ and $r_0(\|x\|) \equiv 1$. The series (2.1.2) converge in mean square (with weight $1/p(u)\ p(w)$) .

Remark 2.1.1. In the most cases $r_k(\|x\|) = B^k(\|x\|)$, $k = 1, 2, \ldots$, where $B(\|x\|)$ is covariance function of random field $\xi(x)$, $x \in R^n$ with given one-dimensional densities $p(u)$, $u \in (a,b)$.

Remark 2.1.2. If $p(u) > 0$, $u \in (a,b)$, $p(u,w,\|x\|) = p(w,u,\|x\|)$, $(u,w) \in S$, and

$$\int\int_S K^2(u,w,\|x\|) p(u) p(w) du dw < \infty,$$

for $\|x\| > 0$, where

$$K(u,w,\|x\|) = \frac{p(u,w,\|x\|)}{\sqrt{p(u)\ p(w)}}$$

is a symmetric kernel on S, then there exists a sequence of eigenvalues

$$r_k(\|x\|), \quad k = 1, 2, \ldots, \quad 1 \ge |r_1(\|x\|)| \ge \cdots \ge |r_k(\|x\|)| \ge \cdots,$$

and a sequence of eigenfunctions $\{e_k(u, \|x\|)\}_{k=1}^{\infty}$ such that expansion in eigenfunctions (2.1.2) hold with $e_k(u, \|x\|)$ in place of $e_k(u)$.

The condition A implies

$$Ee_k(\xi(x)) = 0, \qquad Ee_k(\xi(x))e_m(\xi(y)) = \delta_m^k \; r_m(\|x-y\|). \tag{2.1.3}$$

In fact

$$\begin{aligned} Ee_k(\xi(x))e_m(\xi(y)) &= \int\int_S e_k(u)e_k(w)p(u,w, \|x\|)dudw \\ &= \sum_\nu r_\nu(\|x-y\|) \int_{R^1} e_k(u)e_\nu(u)p(u)du \\ &\quad \times \int_{R^1} e_m(w)e_\nu(w)p(w)dw \\ &= \delta_m^k \; r_m(\|x-y\|). \end{aligned}$$

Example 2.1.1. Gaussian random field. Let $(a,b) = R^1$, and

$$p(u) = \phi(u) = \frac{e^{-u^2/2}}{\sqrt{2\pi}}, \qquad u \in R^1, \tag{2.1.3 a}$$

be a standard normal density function given by (1.1.5) with $a = 0$, $\sigma^2 = 1$.

It is well known that Chebyshev–Hermite polynomials with leading coefficient equal to 1 form a complete orthogonal system in the Hilbert space $\mathbf{L}_2(R^1, \phi(u)\, du)$ that is

$$\int_{-\infty}^{\infty} H_m(u)\; H_q(u)\; \phi(u)\; du = \delta_m^q \; m! \tag{2.1.4}$$

for all m, $q \geq 0$.

Note that (2.1.4) is partial case of (2.1.1) with $e_k(u) = H_k(u)/\sqrt{k!}$, $u \in R^1$.

Using the representation

$$H_m(u) = (-1)^m e^{u^2/2} \frac{d^m}{du^m} e^{-u^2/2}, \qquad m = 0, 1, 2 \ldots \tag{2.1.5}$$

one can derive expressions for the first few polynomials:

$$H_0(u) = 1, \quad H_1(u) = u, \quad H_2(u) = u^2 - 1,$$

$$H_3(u) = u^3 - 3u, \quad H_4(u) = u^4 - 6u + 3, \tag{2.1.6}$$

and so on.

Let $B(\|x - y\|)$, $(x, y) \in R^n \times R^n$, be a continuous non-negative definite kernel on $R^n \times R^n$ depends on $\|x - y\|$ only. Put

$$r_k(\|x\|) = B^k(\|x\|), \ k = 1, 2, \ldots.$$

Then by Meler's formula (see Bateman and Erdelyi (1953), ch.10) for $B(\|x\|) < 1$

$$\begin{aligned} &\phi(u)\phi(w) \left[1 + \sum_{k=1}^{\infty} B^k(\|x\|) \left(\frac{H_k(u)}{\sqrt{k!}}\right) \left(\frac{H_k(w)}{\sqrt{k!}}\right)\right] \\ = \ &\frac{1}{2\pi\sqrt{1 - B^2(\|x\|)}} \\ &\times \exp\left\{-\frac{u^2 + w^2 - 2u\ w\ B(\|x\|)}{2(1 - B^2(\|x\|))}\right\} \\ = \ &\phi(u, w, B(\|x\|)), \end{aligned} \tag{2.1.7}$$

The proof of (2.1.7) can be found in Berman (1992), p. 34.

The right hand side of (2.1.7) is defined the standard bivariate density $\phi(u, w, B(\|x\|))$, $(u, w) \in R^2$, of homogeneous isotropic Gaussian random field $\xi(x)$, $x \in R^n$, with

$$E\xi(x) = 0, \quad E\xi^2(x) = 1, \quad E\xi(0)\ \xi(x) = B(\|x\|).$$

The formula (2.1.7) is particular case of (2.1.2), and (2.1.3) becomes

$$E\ H_k(\xi(x)) = 0, \quad E\ H_k(\xi(x))\ H_m(\xi(y)) = \delta_m^k\ m!\ B^m(\|x - y\|). \tag{2.1.8}$$

Remark 2.1.3. In the natations of example 2.1.1. let us put

$$S = R^2, \quad p(u) = \phi(u), \quad e_k(u) = \frac{H_{2k}(u)}{\sqrt{(2k)!}}, \quad k = 0, 1, 2, \ldots.$$

Then the right hand side of the formula

$$\begin{aligned} \phi(u)\phi(w) \left[1 + \sum_{k=1}^{\infty} B^{2k} \frac{(\|x\|) H_{2k}(u) H_{2k}(w)}{(2k)!}\right] \quad = \quad &\frac{1}{2}\left[\varphi(u, w, B(\|x\|))\right. \\ &\left. + \varphi(u, w, -B(\|x\|))\right], \\ (u, w) \in \ &R^2, \end{aligned}$$

is defined the bivariate densities of homogeneous isotropic non-Gaussian random field $\xi(x)$, $x \in R^n$, with

$$E\xi(x) = 0, \quad E\xi^2(x) = 1, \quad E\xi(0)\,\xi(x) = B^2(\|x\|).$$

One-dimensional distributions of this field are Gaussian but two-dimensional distributions are non-Gaussian. Note that the system $\{H_{2k}(u)/\sqrt{(2k)!}\}_{k=1}^{\infty}$ is not complete in $\mathbf{L}_2(R^1, \phi(u)\,du)$) but the diagonal expansion (2.1.2) holds.

Example 2.1.2. Gamma-correlated random field. We consider the standard Gamma density

$$p(u) = p_\beta(u) = \frac{u^{\beta-1}e^{-u}}{\Gamma(\beta)}, \quad u > 0, \quad \beta > 0. \tag{2.1.9}$$

It is well known that the complete orthonormal system of functions in a Hilbert space $\mathbf{L}_2((0,\infty),\, p_\beta(u)du)$ has the form

$$\begin{aligned} e_k(u) &= e_k^{(\beta)}(u) \\ &= L_k^{(\beta-1)}(u)\left\{\frac{k!\Gamma(\beta)}{\Gamma(\beta+k)}\right\}^{1/2}, \quad k = 0, 1, 2, \ldots, \end{aligned} \tag{2.1.10}$$

where $L_k^{(\beta)}(u)$ are the generalized Laguerre polynomials of index β for $k \geq 0$ (see, Bateman and Erdelyi (1953), ch. 10). The functions (2.1.10) are orthogonal with respect to the density $p_\beta(u)$, $u > 0$, that is, formula (2.1.1) holds.

Using the representation

$$L_k^{(\beta)}(u) = (k!)^{-1}u^{-\beta}e^{u}\frac{d^k}{du^k}\{e^{-u}u^{\beta+k}\}$$

one can derive for the first few polynomials:

$$e_0^{(\beta)}(u) \equiv 1, \quad e_1^{(\beta)}(u) = (\beta - u)\,\beta^{-1/2},$$

$$e_2^{(\beta)}(u) = (u^2 - 2(\beta+1)\,u + (\beta+1)\,\beta)\,[2(\beta+1)\,\beta]^{-1/2}, \tag{2.1.11}$$

and so on.

We introduce the modified Bessel function of the first kind of order $\nu \in R^1$

$$\begin{aligned} I_\nu(z) &= \sum_{k=0}^{\infty}\left(\frac{z}{2}\right)^{2k+\nu}\frac{1}{k!\,\Gamma(k+\nu+1)} \\ &= \frac{(z/2)^\nu}{\sqrt{\pi}\,\Gamma(\nu+1/2)}\int_{-1}^{1}(1-t^2)^{\nu-1/2}e^{zt}dt, \quad z > 0 \end{aligned} \tag{2.1.12}$$

and two-dimensional density

$$p_\beta(u,w,\gamma) = \left(\frac{uw}{\gamma}\right)^{(\beta-1)/2} \exp\left\{-\frac{u+w}{1-\gamma}\right\} \times I_{\beta-1}\left(2\frac{\sqrt{uw\gamma}}{1-\gamma}\right)\frac{1}{\Gamma(\beta)\,(1-\gamma)}, \qquad (2.1.13)$$

where $(u,w)\in R_+^2$, $0\le\gamma<1$.

Note that martingal densities of a two-dimensional density (2.1.13) are defined by (2.1.9) and corresponding characteristic function of (2.1.9) is of the form (see, Wicksell (1933))

$$\varphi_\beta(t,s,\gamma) = [1-it-is-t\;s\;(1-\gamma)]^{-\beta}. \qquad (2.1.14)$$

Then by Myller–Lebedeff's formula or Hille–Hardy's formula (see, Bateman and Erdelyi (1953), ch. 10)

$$p_\beta(u,w,\gamma) = p_\beta(u)\;p_\beta(w)\;\left[1+\sum_{k=1}^{\infty}\gamma^k\;e_k^{(\beta)}(u)\;e_k^{(\beta)}(w)\right]. \qquad (2.1.15)$$

Suppose that

$$p_\beta(u,w,0) - p_\beta(u)\;p_\beta(w) = 0. \qquad (2.1.16)$$

The formula (2.1.15) is a particular case of (2.1.2).

Let $\xi(x)$, $x\in R^n$, be homogeneous isotropic random field with one dimensional densities (2.1.9) and two-dimensional densities of the form (2.1.13), and let $\gamma=\gamma(\|x-y\|)$ be a continuous non-negative definite kernel on $R^n\times R^n$. Then (2.1.3) becomes

$$Ee_k^{(\beta)}(\xi(x)) = 0, \quad E\;e_m^{(\beta)}(\xi(x))\;e_k^{(\beta)}(\xi(y)) = \delta_m^k\;\gamma^m(\|x-y\|). \qquad (2.1.17)$$

The homogeneous isotropic field $\xi(x)$, $x\in R^n$, is called Gamma correlated random field. As we shall see below this field can be given constructive for every $\alpha=p/2$, where $p\ge 1$ is integer. So the class of Gamma correlated random fields is not empty.

Note that the bivariate density $p_\beta = p_\beta(u,w,\gamma)$, $\beta>1$, given by (1.2.13), satisfies the following partial differential equation (see Berman (1984))

$$\frac{\partial p_\beta}{\partial\gamma} = \beta\frac{\partial^2 p_{\beta+1}}{\partial u\;\partial w}. \qquad (2.1.18)$$

Example 2.1.3. χ -squared random field . Consider the random field of the form

$$\xi_p(x) = \frac{1}{2}(\eta_1^2(x) + \cdots + \eta_p^2(x)), \quad x \in R^n, \quad p \geq 1, \tag{2.1.19}$$

where $\eta_1(x), \ldots, \eta_p(x)$ are independent copies of homogeneous Gaussian field $\eta(x)$, $x \in R^n$, with

$$E\eta(x) = 0, \quad E\eta^2(x) = 1, \quad \mathrm{cov}(\eta(0),\ \eta(x)) = B(x), \quad x \in R^n,$$

(see example 2.1.1.).

Let $(X_1, Y_1), \ldots, (X_p, Y_p)$ be i.r.v. with common standard normal bivariate distribution with correlation coefficient ρ. Then it can be shown that the r.v. $\frac{1}{2}(X_1^2, Y_1^2)$ has the characteristic function (2.1.14) with $\gamma = \rho^2$ and $\beta = \frac{1}{2}$. Consequently, the function (2.1.14) is the characteristic function of $\frac{1}{2}(X_1^2 + \cdots + X_p^2,\ Y_1^2 + \cdots + Y_p^2)$ with $\beta = p/2$.

In what follows, the Gamma correlated random field $\xi(x), x \in R^n$ (see example 2.1.2.) may be realized for $\beta = p/2$, $\gamma(\|x\|) = B^2(\|x\|)$, as χ - squared random field (2.1.19).

Note that

$$E\xi_p(x) = \frac{p}{2}, \quad \mathrm{var}\,\xi_p(x) = \frac{p}{2}, \quad \mathrm{cov}(\xi_p(0),\ \xi_p(x)) = \frac{p}{2}B^2(\|x\|). \tag{2.1.20}$$

The one-dimensional and two-dimensional densities of χ -squared random field $\xi_p(x)$ are given by (2.1.9) and (2.1.13) respectively with $\beta = p/2$. We obtain from (2.1.17)

$$Ee_k^{(p/2)}(\xi_p(x))\ e_m^{(p/2)}(\xi_p(y)) = \delta_m^k\ B^{2m}(\|x - y\|). \tag{2.1.21}$$

Remark 2.1.4. Many other examples of the diagonal expansion (2.1.2) can be found in a paper of Sarmanov (1961). In particular, if $\bar{P}_m(u) = P_m(u)\sqrt{(2m+1)/2}$, $u \in [-1,1]$, is an orthonormal Legendre polynomials (see section 1.2), then (2.1.2) with $r_k(\|x\|) = [r_1(\|x\|)]^k = t^k$ becomes

$$\begin{aligned}\sum_{k=0}^{\infty}(k + \frac{1}{2})\ t^k \bar{P}_k(\cos\theta)\bar{P}_k(\cos\psi) \quad &= \quad \frac{1}{2\pi}\int_0^{\pi}(1 - t^2) \\ &\times[1 - 2t\ (\cos\theta\cos\psi \\ &+ \sin\theta\sin\psi\sin w) + t^2]^{-3/2}dw, \\ &|t| < 1.\end{aligned}$$

Similar formulas exists for Gegenbauer, Chebyshev, and Jacobi orthogonal polynomials (see Sarmanov (1961)). In these formulas $r_k(\|x\|)$ is not necessary equal to $B^k(\|x\|)$.

2.2 Reduction Theorems for the Local Functionals of Random Fields with Slowly Decaying Correlations

In this section reduction conditions are provided under which the limit distributions of a local functionals of random fields coincide with corresponding distributions of the functionals of more simple structure. Reduction conditions for random processes and fields with slowly decaying correlations were first derived by Taqqu (1975, 1979) Dobrushin and Major (1979), Berman (1979, 1984), Ivanov and Leonenko (1989) and others.

We shall assume that

B. The random field $\xi(x)$, $x \in R^n$, satisfies assumptions A of section 2.1.and a non-random function $G_r(u)$, $u \in R^1$, $r > 0$, such that $G_r(u) \in \mathbf{L}_2((a,b), p(u)\ du)$.

Under assumption B, $G_r(u)$ may be expanded in the series

$$G_r(u) = \sum_{q=0}^{\infty} C_q(r)\ e_q(u), \qquad C_q(r) = \int_a^b G_r(u)\ e_q(u)\ p(u)\ du, \tag{2.2.1}$$

$q = 0, 1, 2, \ldots$, which converges in the Hilbert space $\mathbf{L}_2(R^1, p(u)\ du)$, and in view of the Parseval equality,

$$\begin{aligned}\sum_{q=0}^{\infty} C_q^2(r) &= \int_a^b G_r^2(u)\ p(u)\ du \\ &< \infty\end{aligned} \tag{2.2.2}$$

for every fixed $r > 0$.

Under the assumption B, we shall study functionals of the form

$$\eta_r = \int_{v(r)} G_r(\xi(x))\ dx - C_0(r)\ r^n\ |v(1)|, \tag{2.2.3}$$

where $v(r)$ is ball of radius $r > 0$ and $|v(1)|$ is the volume of a unit ball in R^n (see section 1.1)

In view of (2.1.3) we obtain

$$E\int_{v(r)} e_m(\xi(x))\,dx \int_{v(r)} e_q(\xi(y))\,dy = \delta_m^q\,\sigma_m^2(r), \tag{2.2.4}$$

where similarly to (1.4.7)

$$\begin{aligned}\sigma_m^2(r) &= \operatorname{var}\left[\int_{v(r)} e_m(\xi(x))\,dx\right]\\ &= \left[4\pi^n/\,n\Gamma^2(n/2)\right] r^n\\ &\quad\times\int_0^{2r} z^{n-1} r_m(z) I_{1-(z/2r)^2}\left(\frac{n+1}{2},\frac{1}{2}\right) dz.\end{aligned} \tag{2.2.5}$$

Consider under assumptions B the random processes

$$\eta_{m,r} = \operatorname{sgn}\{C_m(r)\}\int_{v(r)} e_m(\xi(x))\,dx.$$

C. There exists an integer $m \geq 1$ such that

$$\overline{\lim_{r\to\infty}}\frac{\operatorname{var}\,\eta_r}{C_m^2(r)\,\sigma_m^2(r)} \leq 1. \tag{2.2.6}$$

Lemma 2.2.1. Under the assumptions B and C the limiting distributions of the r.v.

$$X_r = \frac{\eta_r}{|C_m(r)|\,\sigma_m(r)}$$

and the r.v.

$$X_{r,r} = \frac{\eta_{r,m}}{\sigma_m(r)}$$

coincide as $r \to \infty$.

Remark 2.2.1. In the statement of a theorem of this type, the coincidence of the limiting distributions means the following: if the limiting distributions of one collection of r.v.s exists, then so does the limiting distribution of the other, and they are equal.

Proof of Lemma 2.2.1.

Set $Q_r = X_r - X_{m,r}$. In view of (2.2.4), $\operatorname{var}\eta_r \geq C_m^2(r)\,\sigma_m^2(r)$ and assumption (2.2.6) imply that $\lim_{r\to\infty}\operatorname{var} X_r = 1$, therefore $\lim\operatorname{var} Q_r = 0$.

B'. The random field $\xi(x)$, $x \in R^n$, satisfies assumptions A of section 2.1 and non-random function $G_x(u) \in \mathbf{L}_2((a,b),p(u)\,du)$ for every $x \in R^n$.

Then the function $G_x(u)$ can be expanded in the space $\mathbf{L}_2((a,b),p(u)du)$ in the series

$$G_x(u) = \sum_{q=0}^{\infty} C_q(x)\ e_q(u), \qquad C_q(x) = \int_a^b G_x(u)\ e_k(u)\ p(u)\ du. \qquad (2.2.7)$$

Using (2.1.3) we obtain for $m \geq 1$, $q \geq 1$

$$E \int_{v(r)} C_q(x)\ e_q(\xi(x))\ dx \int_{v(r)} C_m(y)\ e_m(\xi(y))\ dy = \delta_m^q\ \tau_m^2(r),$$

where

$$\tau_m^2(r) = \int_{v(r)} \int_{v(r)} C_m(x)\ C_m(y)\ r_m(\|x-y\|)\ dx\ dy.$$

Let

$$\kappa_r = \int_{v(r)} G_x(\xi(x))\ dx - \int_{v(r)} C_0(x)\ dx, \quad \kappa_{m,r} = \int_{v(r)} C_m(x)\ e_m(\xi(x))\ dx$$

C'. There exists an integer $m \geq 1$ such that

$$\overline{\lim_{r\to\infty}} \frac{\text{var}\ \kappa_r}{\tau_m^2(r)} \leq 1. \qquad (2.2.8)$$

Lemma 2.2.2. Under the assumptions B' and C' the limiting distributions of the r.v. $Y_r = \kappa_r/\tau_m(r)$ and the r.v. $Y_{r,m} = \kappa_{r,m}/\tau_m(r)$ coincide as $r \to \infty$.

The proof is analogous to the preceding one.

B". The random field $\xi(x)$, $x \in R^n$, satisfies assumptions A and non-random function $G(u)$, $u \in R^1$, is such that $G(u) \in \mathbf{L}_2((a,b),p(u)du)$.

Then the function $G(u)$ can be expanded in this series

$$G(u) = \sum_{q=0}^{\infty} C_q\ e_q(u), \quad C_q = \int_a^b G(u)\ e_k(u)\ p(u)\ du, \quad \sum_{q=0}^{\infty} C_q^2 < \infty. \quad (2.2.9)$$

Consider under assumptions B" the random processes

$$\zeta_r = \int_{v(r)} G(\xi(x))\ dx - C_0\ r^n\,|v(1)|, \qquad \zeta_{r,m} = \text{sgn}\{C_m\} \int_{v(r)} e_m(\xi(x))\ dx.$$

C".There exists $\delta \in (0,1)$ and an integer $m \geq 1$ such that $C_1 = \cdots = C_{m-1} = 0$, $C_m \neq 0$, and

$$\lim_{r\to\infty} \frac{\sigma_m^2(r)}{r^{n(1+\delta)}} = \infty, \tag{2.2.10}$$

where $\sigma_m^2(r)$ is defined in (2.2.5).

Lemma 2.2.3.Let assumptions B", C" hold.Suppose that

1) $r_k(z) \geq 0, \quad k = 1,2,\ldots;$

2) there exists a continuous function $\psi(z) \in (0,1]$ such that

$$r_{m+1}(z) \leq \psi(z)\, r_m(z)$$

and

$$\sup\{\psi(z),\ z \geq (2r)^\delta\} \to 0 \quad \text{as } r \to \infty,$$

with $m \geq 1$ given in assumption C", and $r_k(z)$ given in expansion (2.1.2).

Then the limiting distributions of the r.v.

$$Z_r = \frac{\zeta_r}{|C_m|\, \sigma_m(r)}$$

and

$$Z_{m,r} = \frac{\zeta_{m,r}}{\sigma_m(r)}$$

coincide as $r \to \infty$.

Proof

We write $\zeta_r = W_1(r) + W_2(r)$, where

$$W_1(r) = C_m \int_{v(r)} e_m(\xi(x))\, dx, \quad W_2(r) = \sum_{q=m+1}^{\infty} C_q \int_{v(r)} e_q(\xi(x))\, dx.$$

The expansion is in the sense of convergence in the Hilbert space $\mathbf{L}_2(\Omega)$. By (2.2.4),

$$\operatorname{var} \zeta_r = \operatorname{var} W_1(r) + \operatorname{var} W_2(r),$$

and

$$\operatorname{var} W_1(r) = C_m^2 \sigma_m^2(r).$$

It is suffices to show that

$$\lim_{r\to\infty} \frac{\operatorname{var}\ W_2(r)}{C_m^2\ \sigma_m^2(r)} = 0. \tag{2.2.11}$$

Utilizing (2.2.9) and conditions 1), 2) we obtain

$$\begin{aligned}
\operatorname{var} W_2(r) &= \sum_{q=m+1}^{\infty} C_q^2 \, \sigma_q^2(r) \\
&\leq \sigma_{m+1}^2(r) \sum_{q=m+1}^{\infty} C_q^2, \\
\sigma_{m+1}^2(r) &= \frac{4\pi^n}{n\Gamma^2(n/2)} \, r^n \left[\int_0^{(2r)^\delta} + \int_{(2r)^\delta}^{2r} \right] \\
&\quad \times z^{n-1} r_{m+1}(z) \, I_{1-(z/2r)^2}\left(\frac{n+1}{2}, \frac{1}{2} \right) dz
\end{aligned}$$

Utilizing the bound $r_{m+1}(z) \leq 1$ for $0 \leq z \leq (2r)^\delta$ and using condition 2) we obtain

$$\begin{aligned}
\sigma_{m+1}^2(r) &\leq k_1 \left\{ r^{n(1+\delta)} + \sup\{\psi(z), z \geq (2r)^\delta\} \right. \\
&\quad \times r^n \int_{(2r)^\delta}^{2r} z^{n-1} r_m(z) \, I_{1-(z/2r)^2}\left(\frac{n+1}{2}, \frac{1}{2} \right) dz,
\end{aligned}$$

where $k_1 > 0$.

Divide the right hand side of the above inequality by $C_m^2 \sigma_m^2(r)$ we have

$$\frac{\operatorname{var} W_2(r)}{C_m^2 \, \sigma_m^2(r)} \leq \frac{k_2 r^{n(1+\delta)}}{\sigma_m^2(r)} + k_3 T_r \sup\{\psi(z), \; z \geq (2r)^\delta\}, \tag{2.2.12}$$

where

$$\begin{aligned}
T_r &= \frac{\int_{(2r)^\delta}^{2r} z^{n-1} r_m(z) \, I_{1-(z/2r)^2}\left(\frac{n+1}{2}, \frac{1}{2}\right) \, dz}{\int_0^{2r} z^{n-1} r_m(z) \, I_{1-(z/2r)^2}\left(\frac{n+1}{2}, \frac{1}{2}\right) dz} \\
&\leq 1
\end{aligned}$$

and $k_2 > 0$, $k_3 > 0$.

In view of the assumption (2.2.10) and assumption 2) from (2.2.12) we obtain (2.2.11).

D. Let $\xi(x)$, $x \in R^n$, be a mean square continuous homogeneous isotropic Gaussian random field with

$$E\xi(x) = 0, \quad E\xi^2(x) = 1, \quad B(\|x\|) = cov(\xi(0), \; \xi(x)) \searrow 0 \quad \text{as } \|x\| \to \infty.$$

Under assumption D we derive

$$\begin{aligned}\sigma_m^2(r) &= var\left[\int_{v(r)} H_m(\xi(x))\,dx\right]\\ &= \frac{4m!\,\pi^n}{n\Gamma^2(n/2)} r^n\\ &\times \int_0^{2r} z^{n-1} B^m(z) I_{1-(z/2r)^2}\left(\frac{n+1}{2},\frac{1}{2}\right) dz, \qquad (2.2.13)\end{aligned}$$

where $H_m(u)$ is a Chebyshev–Hermite polynomial defined in (2.1.5).

E. For some $\alpha > 0$, $B(\|x\|) = \|x\|^{-\alpha} L(\|x\|)$, where $\|x\| > 0$ and $\|x\| \to \infty$ and $L \in \mathcal{L}$ (see section 1.4 for definition of a class $\mathcal{L}$ of slowly varying functions).

F. Let $G \in \mathbf{L}_2(R^1, \phi(u), du)$ be a non-random function, and let an integer $m \geq 1$ exist such that $C_1 = \ldots = C_{m-1} \neq 0$, where C_q is defined in (2.2.17). If F holds, m is called the Hermitian rank of G, denoted as $m = H\,\mathrm{rank}\,G$. For example, the function $G(u) = H_m(u)$, $m \geq 1$, is of Hermitian rank m.

Lemma 2.2.4. Let D, E hold and $r \to \infty$. Then

1) for $\alpha \in (0, n/m)$

$$\sigma_m^2(r) = c_7(n,m,\alpha) r^{2n-m\alpha} L^m(r)\ (1+o(1)), \qquad (2.2.14)$$

where $\sigma_m^2(r)$ is defined in (2.2.13) and

$$c_7(n,m,\alpha) = \frac{m!\ 2^{n-m\alpha+1}\pi^{n-1/2}\Gamma\left(\frac{n-m\alpha+1}{2}\right)}{(n-m\alpha)\ \Gamma\left(\frac{n}{2}\right)\ \Gamma\left(\frac{2n-m\alpha+2}{2}\right)}; \qquad (2.2.15)$$

2) If F holds and $\alpha \in (0, n/m)$, $m = H\,\mathrm{rank}\,G$, then

$$\mathrm{var}\left[\int_{v(r)} G(\xi(x))\,dx\right] = \left(\frac{C_m}{m!}\right)^2 c_7(n,m,\alpha)\ r^{2n-m\alpha} L^m(r)\ (1+o(1)). \qquad (2.2.16)$$

Proof

For any $\varepsilon > 0$

$$\begin{aligned}\sigma_m^2(r) &= k(n,m)\ r^n \left[\int_0^{\varepsilon} + \int_{\varepsilon}^{2r}\right] B^m(z)\ z^{n-1} I_{1-(z/2r)^2}\left(\frac{n+1}{2},\frac{1}{2}\right) dz\\ &= T_1(r) + T_2(r),\end{aligned}$$

where

$$k(n,m) = \frac{4m!\ \pi^n}{n\Gamma^2(n/2)}.$$

For $0 \leq z \leq \varepsilon,\ r > 0$

$$\left| B^m(z)\ z^{n-1} I_{1-(z/2r)^2}\left(\frac{n+1}{2}, \frac{1}{2}\right) \right| \leq k_1 z^{n-1},$$

and by the dominated convergence theorem as $r \to \infty$,

$$T_1(r) = k_1 r^n \varepsilon^n (1 + o(1)).$$

Transform the variable $u = (z/2r)^2$ in (2.2.13) we obtain

$$\begin{aligned} T_2(r) &= k(n,m)\ 2^{n-1-m\alpha} r^{2n-m\alpha} L^m(r) \\ &\quad \times \int_{\varepsilon/4r^2}^1 u^{(n-2-m\alpha)/2} I_{1-u}\left(\frac{n+1}{2}, \frac{1}{2}\right) \\ &\quad \times \left[\left(\frac{L^m(2r\sqrt{u}}{L^m(r)} - 1 \right) + 1 \right] du \\ &= S_1(r) + S_2(r), \end{aligned}$$

where

$$S_2(r) = c_7(n,m,\alpha)\ r^{2n-m\alpha} L^m(r)\ (1 + o(1)),$$

and

$$\begin{aligned} S_2(r) &= k(n,m)\ 2^{n-1-m\alpha} r^{2n-m\alpha} L^m(r) \\ &\quad \times \int_{\varepsilon/4r^2}^1 u^{(n-2-m\alpha)/2} I_{1-u}\left(\frac{n+1}{2}, \frac{1}{2}\right) \\ &\quad \times \left(\frac{L^m(2r\sqrt{u}}{L^m(r)} - 1 \right) du \\ &\leq k_2\ r^{2n-m\alpha} L^m(r) \\ &\quad \times \sup_{0<\sqrt{\varepsilon}\leq s\leq 2r} \left| \frac{L^m(s)}{L^m(r)} - 1 \right| \\ &\quad \times \int_{\varepsilon/4r^2}^1 u^{(n-2-m\alpha)/2} I_{1-u}\left(\frac{n+1}{2}, \frac{1}{2}\right) du. \end{aligned}$$

Utilizing the statement 4) of Lemma 1.4.12 we obtain

$$S_2(r) = o(r^{2n-m\alpha} L^m(r)),\quad T_1(r) = o(T_2(r)), \qquad \text{as } r \to \infty.$$

Remark 2.2.2. Let D, E hold and $r \to \infty$. Then for $\alpha = n/m$,

$$\sigma_m^2(r) = c_8(n,m) r^n L^m(r) \ \log(2r)(1 + o(1)),$$

where

$$c_8(n,m) = \frac{4m! \ \pi^n}{n\Gamma(n/2)}.$$

For $\alpha > n/m$, we obtain

$$\int_0^\infty z^{n-1} \left| B(z) \right|^m dz < \infty,$$

and if

$$b = \int_0^\infty z^{n-1} B(z)^m dz \neq 0,$$

then

$$\sigma_m^2(r) = c_8(n,m) b r^n (1 + o(1).$$

Similar formulas can be found in a section 2.1. of a book Ivanov and Leonenko (1989) for more general functions

$$\tilde{\sigma}^2(r) = var \left[\int_{v(r)} G(\xi(x)) \ dx \right], \qquad \text{where} \quad EG^2(\xi(0)) < \infty.$$

Theorem 2.2.1. Let assumptions C", D hold with $\sigma_m^2(r)$ given in (2.2.13). Then the limit of finite-dimensional distributions of the processes

$$Z_r(t) = \frac{\int_{v(r)} G(\xi(x)) \ dx - C_0 \ t \ r^n \left| v(1) \right|}{\left| C_m \right| \sigma_m(r) / \ m!}, \quad t \in [0,1],$$

and the processes

$$Z_{m,r}(t) = \text{sgn} \ \{C_m\} \frac{\int_{v(rt^{1/2})} H_m(\xi(x)) \ dx}{\sigma_m(r)}, \quad t \in [0,1],$$

coincide, where $\sigma_m(r)$ is defined in (2.1.13), $H_m(u)$ is defined in (2.1.5) and

$$C_q = \int_{-\infty}^\infty G(u) \ H_q(u) \ \phi(u) \ du, \quad q = 0, 1, 2 \ldots, \tag{2.2.17}$$

where $\phi(u)$ is defined in (2.1.3a).

If assumptions D, E, F hold and $\alpha \in (0, n/m)$, $m = H\operatorname{rank} G$, then the assertion of theorem remains true.

Proof

The assertion of the theorem follows from Lemma 2.2.3 with $r_m(z) = B^m(z)$, $\psi(z) = B(z)$, and Lemma 1.1.3.

D'. Let $\xi_p(x)$, $x \in R^n$, be a χ-squared random field defined in (2.1.19) (see example 2.1.3) with

$$E\xi_p(x) = \frac{p}{2}$$

and

$$\operatorname{cov}(\xi_p(0), \xi_p(x)) = pB^2(\|x\|)/2, \quad p \geq 1.$$

Under assumptions D' we derive from (2.1.21)

$$\begin{aligned} \sigma_m^2(r) &= \operatorname{var}\left[\int_{v(r)} e_m^{p/2}(\xi_p(x))dx\right] \\ &= \frac{4\pi^n}{n\Gamma^2(n/2)} r^n \\ &\times \int_0^{2r} z^{n-1} B^{2m}(z)\, I_{1-(z/2r)^2}\left(\frac{n+1}{2}, \frac{1}{2}\right) dz, \qquad (2.2.18) \end{aligned}$$

where $e_m^{p/2}(u)$ is defined in (2.1.10) with $\beta = p/2$.

F'. Let $G \in \mathbf{L}_2((0,\infty), p_\beta(u)\, du)$ be non-random function, and an integer $m \geq 1$ exists such that $C_1 = \cdots = C_{m-1} = 0, \quad C_m \neq 0$, where

$$C_q = \int_0^\infty G(u) e_m^{p/2}(u) p_\beta(u) du, \qquad (2.2.19)$$

and $p_\beta(u)$ is defined in (2.1.9) with $\beta = p/2$.

If F' hold, m is called the Laguerre rank of G, denoted as $m = L\operatorname{rank} G$. For example, $L\operatorname{rank} e_m^{(\beta)}(u) = m$.

Lemma 2.2.5. Let D', E hold and $r \to \infty$. Then

1) for $\alpha \in (0, n/2m)$

$$\sigma_m^2(r) = \frac{c_7(n, 2m, \alpha)}{m!} r^{2(n-m\alpha)} L^{2m}(r)(1 + o(1)), \qquad (2.2.20)$$

where $\sigma_m^2(r)$ is defined in (2.2.18) and $c_7(n, 2m, \alpha)$ is defined in (2.2.15).

2) If F' holds and $\alpha \in (0, n/2m)$, $m = L\operatorname{rank} G$, then

$$var\left[\int_{v(r)} G(\xi_p(x))dx\right] = C_m^2 \frac{c_7(n, 2m, \alpha)}{m!} r^{2(n-m\alpha)} L^{2m}(r)(1 + o(1)), \quad (2.2.21)$$

as $r \to \infty$, where C_m is defined in (2.2.19).

The proof is analogous to the proof of Lemma 2.2.4.

Theorem 2.2.2. Let assumptions D', E, F' hold and $\alpha \in (0, n/2m)$, $m = L\operatorname{rank} G$. Then the limit distributions of finite-dimensional distributions of the processes

$$Z_r'(t) = \frac{\int_{v(rt^{\frac{1}{n}})} G(\xi_p(x))dx - C_0\, r^n t\ |v(1)|}{r^{n-m\alpha} L^m(r)}, \quad t \in [0, 1],$$

and the processes

$$Z_{m,r}'(t) = \frac{C_m \int_{v(rt^{\frac{1}{n}})} e_m^{p/2}(\xi_p(x))dx}{r^{n-m\alpha} L^m(r)}, \quad t \in [0, 1],$$

coincide as $r \to \infty$, where C_m is defined in (2.2.19), $\xi_p(x)$ is defined in (2.1.19) and $e_m^{p/2}(u)$ is defined in (2.1.10).

Proof

The assertion of the theorem follows from Lemma 2.2.3. with $r_m(z) = B^{2m}(z)$, $m \geq 1$, $\psi(z) = B^2(z)$ and Lemma 1.1.3.

All the facts presented above for averages over balls becomes simpler in the case of spherical averages. We state the analogues of Lemma 2.2.4., Theorem 2.2.1. and Theorem 2.2.2. only.

Let assumptions D holds. Using Lemma 1.4.2. similar to (1.4.8) we derive for $n \geq 2$

$$\begin{aligned} \tilde{\sigma}_m^2(r) &= var\left[\int_{s(r)} H_m(\xi(x))d\sigma(x)\right] \\ &= c_8(n, m)\, r^{n-1} \\ &\times \int_0^{2r} z^{n-2}(1 - (z/2r)^2)^{(n-3)/2} B^m(z)dz, \end{aligned} \quad (2.2.22)$$

where

$$c_8(n, m) = \frac{m! 2^n \pi^{n-1}}{(n-2)!},$$

where $d\sigma(x)$ is defined in (1.1.4) and $s(r)$ is a sphere in R^n (see section 1.1.).

Lemma 2.2.6. Let $n \geq 2$, $m \geq 1$, $r \to \infty$, and suppose that assumptions D, E hold. Then for $\alpha \in (0, n/2m)$

$$\tilde{\sigma}_m^2(r) = c_9(n,m,\alpha)r^{2n-2-m\alpha}L^m(r)\ (1+o(1)), \tag{2.2.23}$$

where

$$c_9(n,m,\alpha) = c_8(n,m)\ 2^{n-2-m\alpha}B\left(\frac{n-1-m\alpha}{2}, \frac{n-1}{2}\right). \tag{2.2.24}$$

For $\alpha = (n-1)/m$,

$$\tilde{\sigma}_m^2(r) = \frac{c_8(n,m)}{4}r^{n-1}L^m(r)\log(2r)(1+o(1)).$$

For $\alpha > (n-1)/m$,

$$\tilde{\sigma}_m^2(r) = c_8(n,m)\ a\ r^{n-1}(1+o(1)),$$

if

$$a = \int_0^\infty z^{n-2}B^m(z)dz \neq 0.$$

If assumptions F holds, then

$$\text{var}\left[\int_{s(r)} G(\xi(x))d\sigma(x)\right] = (C_m/m!)^2 c_9(n,m,\alpha)\ r^{2n-2-m\alpha}L^m(r)(1+o(1)).$$

The proof is analogous to that of Lemma 2.2.4.

Theorem 2.2.3. Let assumptions D, E, F hold and $\alpha \in (0,(n-1)/m)$, $n \geq 2$, $m = H\,\text{rank}\,G$. Then the limits of finite-dimensional distributions of the processes

$$\tilde{Z}_r(t) = \frac{\int_{s(rt^{\frac{1}{n-1}})} G(\xi(x))\ dx - C_0 t r^{n-1} \mid s(1)|}{|C_m|\ \tilde{\sigma}_m(r)/m!}, \qquad t \in [0,1],$$

and the processes

$$\tilde{Z}_{m,r}(t) = \text{sgn}\ \{C_m\}\frac{\int_{s(rt^{\frac{1}{n-1}})} H_m(\xi(x))dx}{\tilde{\sigma}_m(r)}, \qquad t \in [0,1],$$

coincide as $r \to \infty$.

Let $\xi_p(x), x \in R^n$ be a χ-squared random field defined in (2.1.19), and

$$\tau_m^2(r) = (c_8(n,m)/m!) r^{n-1} \int_0^{2r} z^{n-2} B^{2m}(z) \left(1 - \left(\frac{z}{2r}\right)^2\right)^{(n-3)/2} dz.$$

Theorem 2.2.4. Let assumptions D', E, F' hold and $\alpha \in (0, (n-1)/2m)$, $n \geq 2$, $m = L \operatorname{rank} G$. Then

$$\tau_m^2(r) = (c_9(n,m,\alpha)/m!)\ r^{2(n-1-m\alpha)} L^{2m}(r)\ (1 + o(1))$$

as $r \to \infty$.

The limits of finite-dimensional distributions of the processes

$$U_r(t) = \frac{\int_{s(rt^{\frac{1}{n-1}})} G(\xi_p(x))\, d\sigma x - C_0\, t\, r^{n-1}\, |s(1)|}{r^{n-1-m\alpha} L^m(r)}, \quad t \in [0,1],$$

and the processes

$$U_{m,r}(t) = \frac{C_m \int_{s(rt^{\frac{1}{n-1}})} e_m^{p/2}(\xi_p(x)) d\sigma x}{r^{n-1-m\alpha} L^m(r)}, \quad t \in [0,1],$$

consider as $r \to \infty$, where $e_m^{p/2}(u)$ is defined in (2.1.10).

The proof of Theorems 2.2.3. and 2.2.4. are analogous to that of Theorems 2.2.1. and 2.2.2. (see Ivanov and Leonenko (1989) for details).

2.3 Multiple Stochastic Integrals

Multiple stochastic integrals were introduced by Wiener (1941), and their study was continued by Itô (1951 ,1952).

This section presents a brief account of multiple stochastic integrals with respect to Gaussian random measures.

Let $\mathbf{L}_2(\Omega)$ be a Hilbert space of complex r.v.s finite second moments, and $W(\cdot)$ be a complex Gaussian white noise random measure on $(R^n, \mathcal{B}(R^n))$, that is, a collection of jointly Gaussian r.v.s such that for any disjoint sets $\Delta_j \in \mathcal{B}(R^n)$

$$W\left(\bigcup_{j=1}^{\infty} \Delta_j\right) = \sum_{j=1}^{\infty} W(\Delta_j) \text{ a.s.},$$

$$\begin{aligned} W(\Delta) &= \overline{W(-\Delta)}, \\ -\Delta &= \{\lambda \in R^n : -\lambda \in \Delta\}, \\ EW(\Delta_1)\overline{W(\Delta_2)} &= |\Delta_1 \cap \Delta_2| \end{aligned}$$

for Borel sets of $R^n \setminus \{0\}$ with finite Lebesgue measure.

The r.v.s Re $W(\Delta)$ and Im $W(\Delta)$ are then independent and have the distribution $\mathcal{N}(0, |\Delta|/2)$.

Denote by $\mathbf{L}_2(R^{nm})$ the Hilbert space of the equivalence classes of complex-valued symmetric functions $f(\lambda_1, \ldots, \lambda_m)$, $\lambda_j \in R^n$, $1 \leq j \leq m$, satisfying the conditions :

$$\begin{aligned} f(-\lambda_1, \ldots, -\lambda_m) &= \overline{f(\lambda_1, \ldots, \lambda_m)}, \\ \|f\|^2 &= \int_{R^{nm}} |f(\lambda_1, \ldots, \lambda_m)|^2 \, d\lambda_1, \ldots, d\lambda_m < \infty. \end{aligned}$$

Let

$$\{\Delta\}_r = \{\Delta_j, \quad j = 0, \pm 1, \pm 2, \ldots\}$$

be measurable sets ($\Delta_j = -\Delta_{-j}, \quad j = 1, 2, \ldots$) disjoint for each r and forming a monotonic sequence in r of countable partitions of R^n such that

$$\lim_{r \to \infty} \sup_{\Delta \in \{\Delta\}_r} \operatorname{diam}\{\Delta\} = 0$$

and

$$\{\Delta\}_r^m = \{\Delta^m = \Delta_{j_1} \times \ldots \times \Delta_{j_m}, \ \Delta_j \in \{\Delta\}_r, \quad j_1, \ldots, j_m = \pm 1, \pm 2, \ldots\},$$

$$r = 1, 2, \ldots$$

is the induced sequence of partitions of the set R^{nm}. The function $\bar{f}_m = \bar{f}(\lambda_1, \ldots, \lambda_m)$ is said to be elementary if for some $r \geq 1$ it assumes constant values on sets of the form

$$\Delta^m = \Delta_{j_1} \times \ldots \times \Delta_{j_m}, \quad j_1, \ldots, j_m = \pm 1, \pm 2, \ldots;$$

$\bar{f}_m$ is non-vanishing only for a finite number of such sets ;

$$\bar{f}_m = 0$$

if $(\lambda_1, \ldots, \lambda_m) \in \mathcal{N}_m = \{\Delta^m = \Delta_{j_1} \times \ldots \times \Delta_{j_m},\ j_r \neq j_{r'} \text{ for some } r \neq r'\}$. For elementary functions f_m the multiple stochastic integral is defined by the formula

$$S_m(\bar{f}_m) = \sum_{j_1, \ldots, j_m} \bar{f}(\lambda_{j_1}, \ldots, \lambda_{j_m}) W(\Delta_{j_1}) \ldots W(\Delta_{j_m}).$$

The set of elementary functions is dense in $\mathbf{L}_2(R^{nm})$; if $f_m \in \mathbf{L}_2(R^{nm})$, then $S_m(f_m)$ can be defined (by extension) to all $\mathbf{L}_2(R^{nm})$. The fact that $\bar{f}_m$ vanishes on the 'diagonals' is necessary; this property allows the extension to $\mathbf{L}_2(R^{nm})$.

Thus, for an arbitrary function $f_m \in \mathbf{L}_2(R^{nm})$ the multiple stochastic integral

$$S_m(f_m) = \int'_{R^{nm}} f(\lambda_1, \ldots, \lambda_m) \prod_{j=1}^{m} W(d\lambda_j)$$

is an isometry mapping of $\mathbf{L}_2(R^{nm})$ into $\mathbf{L}_2(\Omega)$. The integral is defined as the mean square limit $S_m(f_m) = \lim S_m\left(\bar{f}_m^{(r)}\right)$, where $\left(\bar{f}_m^{(r)}\right)$ is a sequence of elementary functions associated with the partitions system $\{\Delta\}_r, r = 1, 2, \ldots$. One can show that such a limit exists and is independent of both the partitioning made and the choice of the sequence of elementary functions.

We note the following well-known properties of multiple stochastic integrals (see Major (1980), Engel (1982), Kwapien' and Woyczyński (1992)).

Theorem 2.3.1. If $f_m \in \mathbf{L}_2(R^{nm})$, then

1) $E\ S_m(f) = 0$;
2) $S_m(f)$ is a real-valued r.v. belonging to $\mathbf{L}_2(\Omega)$;
3) $E\ S_m(f)\ S_r(g) = 0, \quad r \neq m, \quad f, g \in \mathbf{L}_2(R^{nm})$;
4) $E\,|S_m(f)|^2 = \|f\|^2$;
5) the Itô formula : let

$$\xi(x) = \int_{R^n} e^{i<\lambda, x>} \sqrt{f(\lambda)} W(d\lambda)$$

be a spectral decomposition of homogeneous Gaussian random field with mean zero and spectral density $f(\lambda) \in \mathbf{L}_1(R^n)$ (see (1.2.6)), and $H_m(u)$ be an mth Chebyshev–Hermite polynomial (see (2.1.5)), then

$$H_m(\xi(x)) = \int'_{R^{nm}} e^{i<\lambda_1 + \cdots + \lambda_m, x>} \left\{\prod_{j=1}^{m} \sqrt{f(\lambda_j)}\right\} W(d\lambda_1) \ldots W(d\lambda_m). \tag{2.3.1}$$

It is well known that probability distribution of the multiple stochastic integral $S_1(f_1)$, $f_1 \in \mathbf{L}_2(R^{nm})$, is Gaussian with zero mean and variance $\|f_1\|^2$. It is shown in McKean (1974), Varberg (1966) and Dobrushin and Major (1979) that if f_2 is such that the operator K_2 in $\mathbf{L}_2(R^n)$ defined by

$$(K_2 g)(\lambda_2) = \int_{R^n} f_2(\lambda_1, \lambda_2) g(\lambda_1) d\lambda_1, \quad g \in \mathbf{L}_2(R^n),$$

is Hilbert–Schmidt, then $S_2(f_2)$ has the distribution of the random variable

$$R_2 = \sum_{j=1}^{\infty} \lambda_j \, (\chi_j^2 - 1), \qquad \sum_{j=1}^{\infty} \lambda_j^2 < \infty, \tag{2.3.2}$$

where $\chi_1^2, \chi_2^2, \ldots$ are independent χ-squared r.v.s with one degree of freedom and $\lambda_1, \lambda_2, \ldots$ are the eigenvalues of K_2.

The characteristic function of the r.v. R_2 is given by

$$\psi(t) = Ee^{itR_2} = \prod_{j=1}^{\infty} (1 - 2i \, \lambda_j \, t)^{-1/2} e^{-it\lambda_j}, \quad t \in R^1. \tag{2.3.3}$$

However, explicit expressions for the distribution function or the density of R_2 are not know. The paper by Imkeller (1994) contains a result on the exact tail behavior of the distribution of R_2.

For a general $m \geq 3$, all the moments of the r.v. $S_m(f_m)$ are finite. In fact, McKean (1974) has pointed out that

$$E \left| S_m(f_m) \right|^{2r} \leq \left(\frac{(2r)!}{r! 2^r} \right)^m \left[E \, (S_m(f_m))^2 \right]^r, \tag{2.3.4}$$

while Nualart, Üstünel and Zakai (1988), using the Nelson hypercontractivity theorem, prove that

$$E \left| S_m(f_m) \right|^r \leq (r-1)^{rm/2} \left[E \, S_m(f_m)^2 \right]^{r/2}. \tag{2.3.5}$$

Furthermore,

$$E \left| S_m(f_m) \right|^{2r} \approx \exp \left\{ \frac{rm}{2} \, \log r + O(n) \right\}$$

expression which is useful to show that the characteristic function of $S_m(f_m)$ is not analytic.

Shigekawa (1980) proves that the distribution of every finite same of multiple stochastic integrals is absolutely continuous but it is still an open problem to find the explicit expression for this density.

Exponential tail estimates for the distribution of $S_m(f_m)$ have been studied by Borel (1982), McKean (1974) and Plikusas (1981) among others. Pérez–Abreu and Tudor (1994) have recently shown that for each $m \geq 1$, $f_m \in \mathbf{L}_2(R^{nm})$, $u > 0$, and $\gamma > 2$,

$$P\{|S_m(f_m)| > u\} \leq K_\gamma \exp\left\{-\frac{1}{\gamma}\left(\frac{u^2}{m!\ \|f_m\|^2}\right)^{1/m}\right\}, \tag{2.3.6}$$

where the constant

$$K_\gamma = \sum_{p=0}^{\infty} \frac{(2p)!}{(\ p!)^2(2\gamma)^p} < \infty$$

is independent of m.

Using (2.3.6) it is possible to prove a large deviation principle for the r.v.s

$$\{\varepsilon^{m/2} S_m(f_m)\ ;\quad \varepsilon > 0\} \quad \text{and} \quad \left\{\sum_{m=0}^{\infty} \varepsilon^{m/2} S_m(f_m), \quad \varepsilon > 0\right\},$$

(see Nualart, Mayer-Wolf and Pérez–Abreu (1992) for details).

Let $F(\cdot)$ be the spectral measure of a homogeneous (possibly generalized) random field, that is, a σ-finite measure on $(R^n, \mathcal{B}(R^n))$ such that

$$F(\Delta) = F(-\Delta), \quad \Delta \in \mathcal{B}(R^n),$$

and for some $p > 0$

$$\int_{R^n} (1 + \|\lambda\|)^{-p} F(d\lambda) < \infty.$$

Suppose that $F(\cdot)$ is non-atomic measure: $F(\{x\}) = 0$ for any $x \in R^n$. Consider a complex Gaussian orthogonal measure $Z(\cdot)$ having the structure function $F(\cdot)$, that is, a collection of jointly Gaussian r.v.s such that for any $\Delta_j \in \mathcal{B}(R^n), j = 1, \ldots, r$, having a finite $F(\cdot)$-measure, the properties 1)–3) of section 1.1 are valid. Note that these properties imply that the r.v.s $\operatorname{Re} Z_F(\Delta)$ and $\operatorname{Im} Z_F(\Delta)$ are independent and have the distribution $\mathcal{N}(0, F(\Delta)/2)$. If $F(\Delta) = |\Delta|$, $Z(\cdot) = W(\cdot)$ is a complex Gaussian white noise in R^n.

Denote by $\mathbf{L}_2(R^{nm}, F)$ the Hilbert space of complex-valued symmetric functions $f_m = f(\lambda_1, \ldots, \lambda_m)$, $\lambda_j \in R^n, j = 1, \ldots, m$, satisfying the conditions

$$\begin{aligned} \|f\|^2 &= \int_{R^{nm}} |(\lambda_1, \ldots, \lambda_m)|^2 \prod_{j=1}^{m} F(d\lambda_j) \\ &< \infty. \end{aligned}$$

The multiple stochastic integral

$$S_m(f_m) = \int_{R^{nm}}' f_m(\lambda_1, \ldots, \lambda_m) \prod_{j=1}^{m} Z_F(d\lambda_j) \tag{2.3.7}$$

is an isometry mapping of $\mathbf{L}_2(R^{nm})$ into $\mathbf{L}_2(\Omega)$ such that the hyperdiagonals $\lambda_i = \pm\lambda_j$, $i, j = 1, \ldots, m$, $i \neq j$, are excluded from the integration (the elementary functions $\bar{f}_m$ vanish on the 'diagonals' $\mathcal{N}_m$ and this property allows the extension to $\mathbf{L}_2(\Omega)$). The properties 1)–5) of Theorem 2.3.1. are valid for multiple stochastic integrals (2.3.7). In particular, the Itô formula:

$$H_m(\xi(x)) = \int_{R^{nm}}' e^{i<\lambda_1+\cdots+\lambda_m, x>} \prod_{j=1}^{m} Z_F(d\lambda_j),$$

holds, where $H_m(\cdot)$ is mth Chebyshev–Hermite polynomial (see (2.1.5)).

A measure F of $(R^n, \mathcal{B}(R^n))$ is called locally finite if $F(\Delta) < \infty$ for any $\Delta \in \mathcal{B}(R^n)$. A collection of of locally finite measures $\{F_r\}$, $r > 0$, is said to converg locally weakly to the locally finite measure F_0, if

$$\lim_{r\to\infty} \int_{R^n} f(u) F_\mu(du) = \int_{R^n} f(u) F_0(du)$$

for any continuous function with a bounded support.

The definition of the multiple stochastic integral have been extended to dependent integrators by Fox and Taqqu (1987), Ho and Sun (1990), Sanchez and Noranio (1990), Arcones (1994), Daukhan and Leon(1996).

Analogously to the work of Fox and Taqqu (1987) we can introduce the multiple stochastic integrals of the the form

$$S_m(f_m) = \int_{R^{nm}}' f_m(\lambda_1, \ldots, \lambda_m) Z_1(d\lambda_1) \ldots Z_m(d\lambda_m), \tag{2.3.8}$$

where $f_m \in \mathbf{L}_2(R^{nm}, F)$ and $(Z_1, \ldots, Z_m)'$ is a vector Gaussian random measure such that

$$E\, W_j(\Delta_1)\, W_k(\Delta_2) = r_{jk}\, F(\Delta_1 \cap \Delta_2), \quad 1 \le j, \quad k \le m,$$

where $R = (r_{jk})_{1 \le j,k \le m}$ is symmetric positive semi-definite matrix. The integral is interpreted as to exlude integration over the hyperdiagonals. Note that this definition includes the case when R is the matrix with entried all equal to 1, in which case $Z_1 = \cdots = Z_m = Z$ and the integral (2.3.8) reduces to the integral (2.3.7).

The definition also includes the case where the random measures $Z_1, \ldots, Z_m$ are independent.

One can show that the following Itô-formula holds for multiple stochastic integrals (2.3.8):

$$\prod_{k=1}^{m} H_{r_k}\Big(\int_{R^n} \phi_k(\lambda) Z(d\lambda)\Big) = \int_{R^{N_n}}' \prod_{k=1}^{m} \prod_{i=N_{k-1}+1}^{N_k} \phi_k(\lambda_i) \prod_{k=1}^{m} \prod_{i=N_{k-1}+1}^{N_k} Z_k(d\lambda_i), \tag{2.3.9}$$

where $\{\phi_1, \ldots, \phi_m\}$ is an orthonormal collection of functions in $\mathbf{L}_2(f)$, $N_k = r_1 + \ldots + r_k$, $k = 1, \ldots, m$, $N_0 = 0$, $N = N_m$ (see Fox and Taqqu (1987), where this formula is done in more general form).

One can prove that the properties analogous to 1)–4) hold for the multiple stochastic integrals (2.3.8).

2.4 Non-Central Limit Theorems for Local Functionals of Random Fields

We now describe limiting distributions of the local functionals of Gaussian and χ-squared random fields with singular spectrum.Following the terminology of Dobrushin and Major (1979) we will call these theorems the non-central limit theorems. Note that non-central limit theorems were first derived by Rosenblatt (1961), Ibragimov (1963), Taqqu (1975, 1979) and Dobrushin and Major (1979).

We shall assume:

G. The random field $\xi(x)$, $x \in R^n$, is real, mean-square continuous homogeneous and isotropic Gaussian with $E\xi(x) = 0$, $E\xi^2(x) = 1$ and covariance function

$$B(\|x\|) = \|x\|^{-\alpha} L(\|x\|), \quad 0 < \alpha < n, \qquad \text{as} \quad \|x\| \to \infty,$$

where $L \in \mathcal{L}$. Suppose that there exist a spectral density $f(\|\lambda\|)$, $\lambda \in R^n$, and $f(\|\lambda\|)$ is decreasing for $\|\lambda\| \geq \lambda_0 > 0$ and cotinuos for all $\lambda \neq 0$.

Then the spectral representation (1.2.6) becomes

$$\xi(x) = \int_{R^n} e^{i<\lambda,x>} \sqrt{f(\|\lambda\|)} W(d\lambda), \tag{2.4.1}$$

where $W(\cdot)$ is the complex Gaussian white noise random measure, and by Tauberian Theorem 1.4.4 as $\|\lambda\| \to 0$:

$$f(\|\lambda\|) \sim \|\lambda\|^{\alpha-n} L\left(\frac{1}{\|\lambda\|}\right) c_5(n,\alpha), \ 0 < \alpha < n, \tag{2.4.2}$$

where the Tauberian constant $c_5(n, \alpha)$ is given in (1.4.13). Consider the random processes

$$\begin{aligned} X_m(t) &= \frac{(2\pi)^{n/2} t^{1/2}}{[c_5(n,\alpha)]^{m/2}} \\ &\times \int_{R^{nm}}' \frac{J_{n/2}(\|\lambda_1 + \cdots + \lambda_m\| \, t^{1/n})}{\|\lambda_1 + \cdots + \lambda_m\|^{n/2}} \\ &\times \frac{W(d\lambda_1) \ldots W(d\lambda_m)}{\|\lambda_1\|^{(n-\alpha)/2} \cdots \|\lambda_m\|^{(n-\alpha)/2}}, \end{aligned} \tag{2.4.3}$$

where $t \in [0,1]$, $0 < \alpha < n/m$, $m \geq 1$ and a multiple stochastic integral with respect to a complex Gaussian white noise random measure W appears on the right hand side of (2.4.3) and the Bessel function J_ν is defined in (1.2.12).

We note that the processes $X_m(t)$, $t \in [0,1]$, are not Gaussian for $m \geq 2$. The process $X_1(t)$, $t \in [0,1]$ is Gaussian. One can verify that

$$EX_m^2(t) < \infty, \quad m \geq 1.$$

Some moment inequalities and exponential tail estimations for the distribution of $X_m(t)$ are presented in the formulas (2.3.2)-(2.3.6).

The process $X_m(t)$, $t \in [0,1]$, is self-similar with parameter

$$H = 1 - \frac{\alpha m}{2n} \in (\frac{1}{2}, 1)$$

(see section 1.3.).

Theorem 2.4.1. Let assumption G hold and $\alpha \in (0, n/m)$, then as $r \to \infty$, the finite-dimensional distributions of the processes

$$X_{m,r}(t) = r^{(m\alpha)/2-n} L^{-m/2}(r) \int_{v(rt^{\frac{1}{n}})} H_m(\xi(x))\, dx, \quad t \in [0,1],$$

converge weakly to the finite-dimensional distributions of the process $X_m(t)$, $t \in [0,1]$, where H_m is a Chebyshev–Hermite polynomial defined in (2.1.5).

Proof

Using (2.4.3) the self-similarity of Gaussian white noise (formally $W(d(a\lambda)) \stackrel{D}{=} a^{n/2} W(d\lambda)$), and the Itó formula (2.3.1), we obtain

$$X_{m,r}(t) \stackrel{D}{=} \int_{R^{nm}}' \frac{1}{r^{n-(m\alpha/2)} L^{m/2}(r)} \left[\int_{v(rt^{\frac{1}{n}})} e^{i<x,\lambda_1+\cdots+\lambda_m>} dx \right]$$

$$\times \left[\prod_{j=1}^{m} \sqrt{f(\|\lambda_j\|)} \right] W(d\lambda_1) \ldots W(d\lambda_m)$$

$$\stackrel{D}{=} \frac{(2\pi)^{n/2} t^{1/2}}{r^{m(n-\alpha)/2} L^{m/2}(r)} \int_{R^{nm}}' \frac{J_{n/2}(\|\lambda_1 + \cdots + \lambda_m\|\ t^{1/n})}{\|\lambda_1 + \cdots + \lambda_m\|^{n/2}}$$

$$\times \sqrt{\prod_{j=1}^{m} \|\lambda_j\|^{(n-\alpha)/2} f\left(\frac{\|\lambda_j\|}{r}\right)} \frac{W(d\lambda_1) \ldots W(d\lambda_m)}{\|\lambda_1\|^{(n-\alpha)/2} \cdots \|\lambda_m\|^{(n-\alpha)/2}}$$

$$\stackrel{D}{=} \frac{(2\pi)^{n/2} t^{1/2}}{[c_5(n,\alpha)]^{m/2}} \int_{R^m}' \frac{J_{n/2}(\|\lambda_1 + \cdots + \lambda_m\|\ t^{1/n})}{\|\lambda_1 + \cdots + \lambda_m\|^{n/2}}$$

$$\times Q_r(\lambda_1, \ldots, \lambda_r) \frac{W(d\lambda_1) \ldots W(d\lambda_m)}{\|\lambda_1\|^{(n-\alpha)/2} \cdots \|\lambda_m\|^{(n-\alpha)/2}},$$

where

$$Q_r(\lambda_1, \ldots, \lambda_r) = r^{m(n-\alpha)/2} L^{-m/2}(r)\ [c_5(n,\alpha)]^{-m/2} \times \left[\prod_{j=1}^{m} \|\lambda_j\|^{(n-\alpha)/2} f\left(\frac{\|\lambda_j\|}{r}\right) \right]^{1/2}$$

It follows from (1.4.27), that the function

$$K(\lambda_1, \cdots, \lambda_m) = J_{n/2}^2 \left(\|\lambda_1 + \cdots + \lambda_m\|\ t^{1/n} \right) \|\lambda_1 + \cdots + \lambda_m\|^{-n} \times \prod_{j=1}^{m} \|\lambda_j\|^{(\alpha-n)},$$

where $0 < \alpha < n/m$ is absolutely integrable on R^{nm}.

From (2.4.2) and Theorem 1.4.12 we obtain

$$\lim_{r\to\infty} |Q_r(\lambda_1,\ldots,\lambda_m) - 1|^2 = 0.$$

So that we can pass to the limit under the integral by dominated convergence theorem in the assertion

$$\begin{aligned} E\,|X_{m,r}(t) - X_m(t)|^2 &= \int_{R^{nm}} K^2(\lambda_1,\ldots,\lambda_m)\,|Q_r(\lambda_1,\ldots,\lambda_m) - 1|^2 \times \\ d\lambda_1\ldots d\lambda_m &= \int_A \ldots + \int_{R^{nm}\setminus A} \ldots \end{aligned}$$

where $A = \{\lambda_1,...,\lambda_m : \|\lambda_j\| < r^{n-\alpha},\ \ j = 1,\ldots,m\}$. Using continuity and asymptotic behavior of the spectral density $f(\lambda)$ (see condition G and the formula (2.4.2)) we can make the first integral arbitrary small shoosing r large. As we assume that the spectral density is decreasing function after some λ_0 (see condition G), the form of $R^{nm}\setminus A$ assures that second integral tends to zero as $r \to \infty$. Note we have

$$\lim_{r\to\infty} |X_{m,r}(t) - X_m(t)|^2 = 0.$$

The proof of convergence of all finite-dimensional distributions $X_{m,r}(t)$, $t \in [0,1]$, to the finite-dimensional distributions of $X_m(t)$, $t \in [0,1]$, can be derive by Cramer–Wold method (see Lemma 1.1.3. and Ivanov and Leonenko (1989) for details).

From Theorems 2.2.1 and 4.1.1 we obtain.

Theorem 2.4.2. Let assumptions F, G hold for $\alpha \in (0, n/m)$, where $m = H\,\mathrm{rank}\,G$. Then the finite-dimensional distributions of the processes

$$X_r(t) = r^{(m\alpha/2)-n} L^{-m/2}(r)\left[\int_{v(rt^{\frac{1}{n}})} G(\xi(x))\,dx - C_0\,t\,r^n\,|v(1)|\right], \quad t \in [0,1],$$

converge weakly to the finite-dimensional distributions of the process $\frac{C_m}{m!} X_m(t)$, $t \in [0,1]$, where $X_m(t)$ is defined by (2.4.3) and

$$C_q = E\;G(\xi(0))\;H_q(\xi(0)), \quad q = 0,1,2,\ldots \tag{2.4.4}$$

We now provide the analogue of Theorem 2.4.1., and for spherical averages.

Theorem 2.4.3. Let assumptions F, G hold for $\alpha \in (0,(n-1)/m)$, $n \geq 2$, where $m = H\operatorname{rank} G$. Then the finite-dimensional distributions of the processes

$$\begin{aligned} Y_r(t) &= r^{(m\alpha/2)-n+1} L^{-m/2}(r) \\ &\times \left[\int_{s(rt^{\frac{1}{n-1}})} G(\xi(x))d\sigma(x) - C_0\, t\, r^{n-1}\, |s(1)| \right], \\ t &\in [0,1], \end{aligned}$$

converge weakly to the finite-dimensional distributions of the process

$$\begin{aligned} Y_m(t) &= \frac{C_m(2\pi)^{n/2} t^{n/2(n-1)}}{m!\, [c_5(n,\alpha)]^m} \\ &\times \int_{R^m}' \frac{J_{(n-2)/2}(\|\lambda_1 + \cdots + \lambda_m\|\ t^{1/(n-1)})}{\|\lambda_1 + \cdots + \lambda_m\|^{(n-2)/2}} \\ &\times \frac{W(d\lambda_1)\ldots W(d\lambda_m)}{\|\lambda_1\|^{(n-\alpha)/2} \cdots \|\lambda_m\|^{(n-\alpha)/2}} \end{aligned} \qquad (2.4.5)$$

where $t \in [0,1]$, C_q are defined in (2.4.4).

The proof of the Theorem 2.4.2 is analogous to that of Theorem 2.4.1.

The process $Y_m(t)$, $t \in [0,1]$, given in (2.4.4), is self-similar with parameter

$$H = 1 - \frac{\alpha m}{2(n-1)} \in (\frac{1}{2}, 1).$$

Note that $EY_m^2(t) < \infty$ for every $t \in [0,1]$, but the processes $Y_m(t)$ are non-Gaussian for $m \geq 2$.

E'. Suppose that assumption E hold and covariance function $B(\|x\|)$ such that there exist a spectral density $f(\|\lambda\|)$, $\lambda \in R^n$, and $f(\|\lambda\|)$ is decreasing for $\|\lambda\| \geq \lambda_0 > 0$ and continuous for all $\lambda \neq 0$.

We now describe limiting distributions of the spherical averages of χ-squared random fields with singular spectrum.

Theorem 2.4.4. Let assumptions D', E', F' hold and $L\operatorname{rank} G = 1$, $\alpha \in (0, n/2)$.

Then the finite-dimensional distributions of the random processes

$$Z_r'(t) = r^{\alpha-n} L^{-1}(r) \left[\int_{v(rt^{\frac{1}{n}})} G(\xi_p(x))\, dx - C_0\, t\, r^n\, |v(1)| \right], \quad t \in [0,1],$$

converge weakly to the finite-dimensional distributions of the non-Gaussian process $-(C_1/\sqrt{2p})\sum_{j=1}^{p} X_{i2}(t), \quad t \in [0,1]$ where $\xi_p(x)$ is χ-squared random field (2.1.19), $X_{i2}(t)$, $t \in [0,1]$, $i = 1,\dots,p$, are independent copies of the non-Gaussian random process $X_2(t)$, $t \in [0,1]$, defined in (2.4.3) with $m = 2$, and C_0, C_1 are defined in (2.2.19).

Proof

Put $m = 1$ in the Theorem 2.2.2., where $m = L\,\mathrm{rank}\,G$. Using (2.1.11) we have

$$e^{(p/2)}(u) = \sqrt{\frac{2}{p}}\left(\frac{p}{2} - u\right)$$

and the process $Z'_{1,p}(t)$ in the Theorem 2.2.2. becomes

$$\begin{aligned} Z'_{1,p}(t) &= C_1 \int_{v(rt^{\frac{1}{n}})} e^{(p/2)}(\xi_p(x))\,dx \;/\; r^{\alpha-n}L^{-1}(r) \\ &= -\frac{C_1}{\sqrt{2p}} \sum_{i=1}^{p} \frac{\int_{v(rt^{\frac{1}{n}})} H_2(\xi_i(x))\,dx}{r^{\alpha-n}L^{-1}(r)}, \qquad t \in [0,1], \end{aligned}$$

where $\xi_i(x), i = 1,\dots,p$ are independent copies of the Gaussian process $\xi(x)$, $x \in R^n$, with covariance function $B(\|x\|)$, $x \in R^n$, given in condition E (see formula (2.1.19)) and $H_2(u)$ is the second Chebyshev–Hermit polynomial, given in (2.1.6). Now we apply Theorem 2.4.1. or Theorem 2.4.2. with $m = H\,\mathrm{rank}\,G = 2$, if $G(u) = H_2(u)$.

Remark 2.4.2. Using (2.1.11) we may describe the limiting distributions of the spherical averages of χ-squared random fields with singular spectrum for $L\,\mathrm{rank}\,G \geq 2$ in the Theorem 2.2.4. They are non-Gaussian and can be given in terms of multiple stochastic integrals with dependent integrators (see Fox and Taqqu (1987)), Sanchez and Naranjo (1993) or Doukhan and León (1996)).

Thus the limiting distributions of the spectral averages of χ-squared random fields with singular spectrum are always non-Gaussian.

Let assumption G hold for $\alpha \in (0, n)$. Consider the spectral representations (1.2.4) and (1.2.6) where $F(\cdot)$ is the spectral measure of the field $\xi(x)$, $x \in R^n$, and $Z(\cdot)$ is an orthogonal Gaussian measure having the structure function $F(\cdot)$. Assume henceforth the $F(\cdot)$ is a non-atomic measure. We introduce the collection of locally finite measures

$$F_r(\Delta) = r^{\alpha}L^{-1}(r^{-1}\Delta), \quad \Delta \in \mathcal{B}(R^n).$$

Theorem 2.4.5. If assumptions G are valid with $\alpha \in (0, n)$, then the collection of locally finite measures $F_r(d\lambda)$ converges locally weakly to a focally finite measure $F_0(\Delta), \Delta \in \mathcal{B}(R^n)$, satisfying the self-similar condition: $F_0(a\Delta) = a^\alpha F_0(\Delta), \quad a > 0$, and the relation

$$\int_{R^n} \exp\{i < z, \lambda >\} J^2_{n/2}(\|\lambda\|) \|\lambda\|^{-n} F_0(d\lambda) = \int_{v(1)} \int_{v(1)} \|x - y + z\|^{-\alpha} \, dx \, dy.$$

Theorem 2.4.6. If assumptions G, F hold with $\alpha \in (0, n/m)$, where $m = H \operatorname{rank} G$, then the finite-dimensional distributions of the random processes

$$X_{m,r}(t) = r^{(m\alpha)/2-n} L^{-m/2}(r) \times \int_{v(rt^{\frac{1}{n}})} [G(\xi(x)) \, dx - C_0 \, t \, r^n \, |v(1)|] \, dx, \quad t \in [0,1],$$

converge to finite-dimensional distributions of the random processes

$$X'_m(t) = (2\pi)^{n/2} t^{1/2} (C_m/m! \,) \times \int'_{R_{nm}} \frac{J_{n/2}(\|\lambda_1 + \ldots + \lambda_m\| \; t^{1/n})}{\|\lambda_1 + \ldots + \lambda_m\|^{n/2}} \prod_{j=1}^{m} Z_{F_0}(d\lambda_j), \quad (2.4.6)$$

where $t \in [0,1]$, $C_q = E \; G(\xi(0)) H_q(\xi(0))$, $q = 0, \ldots, m$, and $Z_{F_0}(\cdot)$ is a orthogonal Gaussian random measure with the structure function $F_0(\cdot)$, defined in Theorem 2.4.5. The process $X'_m(t)$, $t \in [0,1]$, is self-similar with parameter $H = 1 - \alpha m/2n \in (1/2, 1)$.

Theorem 2.4.6. is continuous analogue of Theorem 1 from Dobrushin and Major (1979). Theorems 2.4.5–2.4.6 are proved in the book of Ivanov and Leonenko (1989), section 2.10.

Chapter 3

ASYMPTOTIC DISTRIBUTIONS OF GEOMETRIC FUNCTIONALS OF RANDOM FIELDS

3.1 Limit Distributions for Characteristics of the Excess above a Level for Gaussian Fields

We shall examine 'random volumes' contained by a Gaussian surface.

A. Let $\xi(x)$, $x \in R^n$, be a mean-square continuous homogeneous isotropic Gaussian random field with $E\xi(x) = 0$ and covariance function $B(\|x\|) \searrow 0$ as $\|x\| \to \infty$.

A'. For some $\alpha > 0$, $B(\|x\|) = \|x\|^{-\alpha} L(\|x\|)$,where $\|x\| > 0$, $\|x\| \to \infty$ and $L \in \mathcal{L}$.

$$V_1(r) = \int_{v(r)} max\{0, \xi(x)\}\ dx, \qquad V_2(r) = \int_{v(r)} |\xi(x)|\ dx$$

which have a clear geometric meaning for continuous random fields when

$n = 2$. Consider the functions (see (2.2.13))

$$\sigma_m^2(r) = c_{10}(n,m)\, r^n \int_0^{2r} z^{n-1} B^m(z)\, I_{1-(z/2r)^2}\left(\frac{n+1}{2}, \frac{1}{2}\right) dz, \qquad (3.1.1)$$

$m = 1, 2, \ldots$, where $I_\mu(p,q)$ is defined in (1.1.), and

$$c_{10}(n,m) = 4m!\ \pi^n /\ n\ \Gamma^2(n/2). \qquad (3.1.2)$$

Theorem 3.1.1. Let assumptions A hold and there exist $\delta \in (0,1)$ such that

$$\lim_{r\to\infty} \frac{\sigma_1^2(r)}{r^{n(1+\delta)}} = \infty, \qquad (3.1.3)$$

where $\sigma_1^2(r)$ is defined in (3.1.1). Then the r.v.

$$\frac{V_1(r) - (2\pi)^{-1/2} r^n\, |v(1)|}{\frac{1}{2}\sigma_1(r)} \xrightarrow{D} \mathcal{N}(0,1) \quad \text{as} \quad r \to \infty.$$

Proof

Using (2.1.6) and (2.2.14) we calculate the first two coefficients of the expansion (2.2.9) of the function $G(u) = \max\{0, u\}$ in a series in Chebyshev–Hermite polynomials $e_k(u) = H_k(u)/\sqrt{k!}$. These are respectively $C_0 = 1/\sqrt{2\pi}$, $C_1 = 1/2$. Thus, Theorem 3.1.1. is a corollary to Theorem 2.2.1. with $m = 1$.

Remark 3.1.1. If assumptions A, A' hold, then $\sigma_1^2(r)$ is defined in (2.2.14) with $m = 1$ and (3.1.3) satisfies for $\alpha \in (0, n)$.

A". The covariance function $B(\|x\|)$, $x \in R^n$, satisfies assumption A' and there exists a spectral density $f(\|\lambda\|)$, $\lambda \in R^n$, which is decreasing for $\|\lambda\| \geq \lambda_0 > 0$, and continuous for all $\lambda \neq 0$.

Theorem 3.1.2. Let assumptions A hold and there exist $\delta \in (0,1)$ such that

$$\lim_{r\to\infty} \frac{\sigma_2^2(r)}{r^{n(1+\delta)}} = \infty, \qquad (3.1.4)$$

where $\sigma_2^2(r)$ is defined in (3.1.1) with $m = 2$. Then the limiting distributions of the r.v.

$$\frac{V_2(r) - (2/\pi)^{1/2} r^n\, |v(1)|}{\sigma_2(r)\sqrt{2\pi}}$$

coincide with those of

$$\frac{1}{\sigma_2(r)} \int_{v(r)} H_2(\xi(x))\, dx \tag{3.1.5}$$

as $r \to \infty$.

In particular, if A', A" hold with $\alpha \in (0, n/2)$, then $\sigma_2^2(r)$ is denoted in (2.2.14) with $m = 1$, condition (3.1.4) satisfies and limiting distribution of the r.v. (3.1.5) is of the form $\{c_7(n, 2, \alpha)\}^{-1/2} X_2(1)$ (or $\{c_7(n, 2, \alpha)\}^{-1/2} X_2'(1)$, if A" is not hold necessarily), where $c_7(n, 2, \alpha)$ is given in (2.2.15) and r.v.s $X_2(1)$ and $X_2'(1)$ are defined in (2.4.3) and (2.4.6) respectively with $m = 2$ and $C_2/2 = 1$ in (2.4.6).

Proof

Using (2.1.6) and (2.2.14) we calculate the first three coefficients of the expansion (2.2.9) of the function $G(u) = |u|$ in a series in Chebyshev—Hermit polynomials $e_k(u) = H_k(u)/\sqrt{k!}$. These are respectively $C_0 = (2/\pi)^{1/2}$, $C_1 = 1$, $C_2 = (2/\pi)^{1/2}$. An application of Theorem 2.2.1. and 2.4.1. with $m = 2$ completes the proof.

Denote the density function and d.f. of the standard normal r.v. $\mathcal{N}(0, 1)$ by

$$\phi(a) = \frac{1}{\sqrt{2\pi}} e^{-a^2/2}, \qquad \Phi(a) = \int_{-\infty}^{a} \phi(b)\, db. \tag{3.1.6}$$

B. Let $a : (0, \infty) \to (0, \infty)$ be continuous monotonic function such that $a(r) \to \infty$ as $r \to \infty$.

Under assumptions A, B consider the functionals

$$W_1(r) = \int_{v(r)} max\{0, \xi(x) - a(r)\}\, dx, \quad W_2(r) = \int_{v(r)} max\{0, |\xi(x)| - a(r)\}\, dx, \tag{3.1.7}$$

which have a clear geometrical meaning for random fields with continuous realizations. For example, for $n = 2$ it is a part of the volume formed by a realization of the field $\xi(x)$, $x \in v(r)$, the plane $z = a(r)$ and a cylindrical surface passing through the boundary of the circle $v(r)$. Below we shall show that

$$EW_1(r) = A(r) = |v(1)|\ r^n \{\phi(a(r)) - a(r)\ [1 - \Phi(a(r))]\}.$$

Theorem 3.1.3. Let assumptions A, B be valid and there exists $\delta \in (0, 1)$, $\beta \in (0, \delta)$, such that for $r \to \infty$

$$\frac{\sigma_1^2(r)}{r^{n(1+\delta)} a^2(r)} \to \infty, \tag{3.1.8}$$

$$\frac{a^2(r)}{\log r} \to 0 : a^2(r)\ B(r^\beta) \to 0. \tag{3.1.9}$$

Then the r.v.

$$\frac{W_1(r) - A(r)}{\sigma_1(r)(1 - \Phi(a(r))}$$

has asymptotically $(0,1)$-normal distribution as $r \to \infty$, where $\sigma_1(r)$ is given in (3.1.1) with $m = 1$.

Proof

The function $G_r^{(1)}(u) = max\{0, u - a(r)\}$ satisfies assumptions B of chapter 2 with $p(u) = \phi(u)$, $u \in R^1$. The coefficients of the expansion of the function $G_r^{(1)}(u)$ in $\mathbf{L}_2(R^1, \phi(u)\ du)$ are

$$\begin{aligned} C_0(r) &= \int_{a(r)}^{\infty} (u - a(r))\ \phi(u)\ du \\ &= \phi(a(r)) - a(r)[1 - \Phi(a(r))], \end{aligned}$$

$$\begin{aligned} C_1(r) &= \int_{a(r)}^{\infty} u(u - a(r))\ \phi(u)\ du \\ &= 1 - \Phi(a(r)). \end{aligned}$$

Thus $EW_1(r) = \int_{v(r)} C_0(r)\ dx = A(r)$.

The following expansion of the function (2.1.7) is well-known (see Berman (1992), p. 31–34)

$$\phi(u, w, \rho) = \sum_{\nu=0}^{\infty} \frac{\rho^\nu}{\nu!} \Phi^{(\nu+1)}(u)\ \Phi^{(\nu+1)}(w). \tag{3.1.10}$$

Let

$$\begin{aligned} \eta_r(x) &= G_r^{(1)}(\xi(x)) \\ &= max\{0, \xi(x) - a(r)\}. \end{aligned}$$

Using (3.1.10) we integrate by parts and obtain

$$E\eta_r(x)\eta_r(y) = \int_{a(r)}^{\infty}\int_{a(r)}^{\infty} (u - a(r))(w - a(r))\phi(u, w, B(\|x - y\|))\ du\ dw$$

$$= \sum_{j=0}^{\infty} \frac{[B(\|x-y\|)]^j}{j!} \left[\int\limits_{a(r)}^{\infty} (u-a(r))\Phi^{(j+1)}(u)\, du \right]^2$$

$$= \left[\int\limits_{a(r)}^{\infty} (u-a(r))\phi(u)\, du \right]^2$$

$$+B(\|x-y\|) \left[\int\limits_{a(r)}^{\infty} \phi(u) du \right]^2$$

$$+\sum_{j=2}^{\infty} \frac{[B(\|x-y\|)]^j}{j!} \left[\Phi^{(j-1)}(a(r))\right]^2. \tag{3.1.11}$$

Similar to (2.4.7) we obtain

$$\begin{aligned}
\text{var } W_1(r) &= \int_{v(r)} \int_{v(r)} E\, \eta_r(x)\eta_r(y) dx\, dy - (EW_1(r))^2 \\
&= \int_{v(r)} \int_{v(r)} \Big\{ B(\|x-y\|) \left[1-\Phi(a(r))\right]^2 \\
&\quad + \sum_{j=2}^{\infty} \frac{[B(\|x-y\|)]^j}{j!} \left[\Phi^{(j-1)}(a(r))\right]^2 \Big\} dxdy \\
&= c_{10}(n,1) r^n \\
&\quad \times \left\{ \int_0^{2r} z^{n-1} B(z)\, I_{1-(z/2r)^2}\left(\frac{n+1}{2}, \frac{1}{2}\right) dz \right. \\
&\quad \times \left[1-\Phi(a(r))\right]^2 \\
&\quad + \sum_{j=2}^{\infty} \int_0^{2r} z^{n-1} B^j(z) I_{1-(z/2r)^2}\left(\frac{n+1}{2}, \frac{1}{2}\right) dz \\
&\quad \left. \times \left[\Phi^{(j-1)}(a(r))\right]^2 \frac{1}{j!} \right\}.
\end{aligned} \tag{3.1.12}$$

Using the relation (see relation (1.8.3) from Cramer and Leadbetter (1967))

$$\int_a^{\infty} \int_a^{\infty} \phi(u,w,\rho) dudw = \left(\int_0^{\infty} \phi(u) du \right)^2$$

$$+\frac{1}{2\pi}\int_0^\rho \exp\left\{-\frac{a^2}{1+v}\right\}\frac{dv}{\sqrt{1-v^2}} \quad (3.1.13)$$

we obtain

$$\begin{aligned}\sum_{j=2}^{\infty}\frac{\rho^j}{j!}\left[\Phi^{(j-1)}(a)\right]^2 &\le \frac{\rho}{2}\sum_{k=1}^{\infty}\frac{\rho^k}{k!}\left[\Phi^{(k)}(a)\right]^2 \\ &= \frac{\rho}{2\pi}\int_0^\rho \exp\left\{-\frac{a^2}{1+v}\right\}\frac{dv}{\sqrt{1-v^2}}. \quad (3.1.14)\end{aligned}$$

Formulas (3.1.12) and (3.1.14) yield:

$$\begin{aligned}\mathrm{var}W_1(r) \le c_{10}(n,1)\, r^n &\left\{\int_0^{2r} z^{n-1}B(z)I_{1-(z/2r)^2}\left(\frac{n+1}{2},\frac{1}{2}\right)dz\right.\\ &\times\left[1-\Phi(a(r))\right]^2\\ &+\int_0^{2r} z^{n-1}B(z)I_{1-(z/2r)^2}\left(\frac{n+1}{2},\frac{1}{2}\right)dz\\ &\left.\times\frac{1}{4\pi}\int_0^{B(z)}\exp\left\{-\frac{a^2(r)}{1+v}\right\}\frac{dv}{\sqrt{1-v^2}}\right\} \quad (3.1.15)\end{aligned}$$

Now we apply Lemma 2.2.1. in the case of Hilbert space $\mathbf{L}_2(R^1,\phi(u)du)$ to the functional

$$W_1(r)=\int_{v(r)} G_r^{(1)}(\xi(x)\, dx, \qquad G_r^{(1)}(u)=max\{0,\ u-a(r)\}.$$

Note that in our case (2.2.5) should be replaced by (3.1.1) and $e_m(u) = H_m(u)/\sqrt{m!}$.

Assumptions C (see(2.2.6)) of chapter 2 implies that if

$$\overline{\lim_{r\to\infty}}\frac{\mathrm{var}\ W_1(r)}{[1-\Phi(a(r))]^2\sigma_1^2(r)}\le 1, \quad (3.1.16)$$

then the limiting distribution of r.v.

$$\frac{W_1(r)-A(r)}{[1-\Phi(a(r))]\sigma_1(r)} \qquad \text{as} \quad r\to\infty$$

is the assume (in view of Lemma 2.2.1. with $m = 1$) as that of $\int_{v(r)} \xi(x)\, dx/\sigma_1(r)$, that is, it is $(0,1)$-normal.

Formula (3.1.15) yields

$$\overline{\lim_{r\to\infty}} \frac{\text{var } W_1(r)}{[1-\Phi(a(r))]^2\sigma_1^2(r)} = 1 + \overline{\lim_{r\to\infty}}\, Q_r, \tag{3.1.17}$$

where, taking into account the relation

$$\left(\frac{1}{a} - \frac{1}{a^3}\right)\phi(a) \le 1 - \Phi(a) \le \frac{\phi(a)}{a}, \tag{3.1.18}$$

we have

$$\begin{aligned} Q_r &= \left\{\frac{c_{10}(n,1)}{4\pi} r^n \int_0^{2r} z^{n-1} B(z) I_{1-(z/2r)^2}\left(\frac{n+1}{2}, \frac{1}{2}\right) dz \right.\\ &\quad \left. \times \int_0^{B(z)} \exp\left\{-\frac{a^2(r)}{1+v}\right\} \frac{dv}{\sqrt{1-v^2}}\right\} \left[(1-\Phi(a(r))^2\,\sigma_1^2(r)\right]^{-1} \\ &= \{S_1(r) + S_2(r)\} \left[\int_0^{2r} z^{n-1} B(z) I_{1-(z/2r)^2}\left(\frac{n+1}{2}, \frac{1}{2}\right) dz\right]^{-1} \\ &\quad \times (1 + o(1)), \end{aligned} \tag{3.1.19}$$

where

$$\begin{aligned} S_1(r) &= \int_0^{(2r)^\beta} U_r(z)\, dz, \\ S_2(r) &= \int_{(2r)^\beta}^{2r} U_r(z)\, dz, \\ Q_r(z) &= a^2(r)\, z^{n-1} B(z) I_{1-(z/2r)^2}\left(\frac{n+1}{2}, \frac{1}{2}\right) \\ &\quad \times \frac{1}{2} \int_0^{B(z)} \exp\left\{-a^2(r)\, \frac{v}{1+v}\right\} \frac{dv}{\sqrt{1-v^2}} \end{aligned}$$

and β is given in the conditions of the theorem.

For $(2r)^\beta \le z \le 2r$ and $r \to \infty$ we have (for $\frac{1}{1+v} \le 1$, $0 \le v \le B(z)$):

$$\begin{aligned} S_2(r) &\le a^2(r) \int_{(2r)^\beta}^{2r} z^{n-1} B(z)\, I_{1-(z/2r)^2}\left(\frac{n+1}{2}, \frac{1}{2}\right) dz \\ &\quad \times \frac{1}{2} \int_0^{B(z)} \exp\{a^2(r)\, v\}\, dv \end{aligned}$$

$$\leq \frac{1}{2}T(r)e^{T(r)}\int_{(2r)^\beta}^{2r} z^{n-1}B(z)\ I_{1-(z/2r)^2}\left(\frac{n+1}{2},\frac{1}{2}\right)dz, \tag{3.1.20}$$

where in view of (3.1.9), as $r \to \infty$

$$T(r) = a^2(r)\sup\{B(z),\ z \geq (2r)^\beta\} \to 0. \tag{3.1.21}$$

Formulas (3.1.20) and (3.1.21) imply that

$$\begin{aligned}
&\overline{\lim_{r\to\infty}}\frac{S_2(r)}{\int_0^{2r} z^{n-1}B(z)\ I_{1-(z/2r)^2}\left(\frac{n+1}{2},\frac{1}{2}\right)\ dz} \\
\leq\ & \overline{\lim_{r\to\infty}}\ \frac{1}{2}\ T(r)\ e^{T(r)}N(r) \\
=\ & 0,
\end{aligned} \tag{3.1.22}$$

since in view of assumption A

$$\begin{aligned}
N(r) = &\left[\int_{(2r)^\beta}^{2r} z^{n-1}B(z)I_{1-(z/2r)^2}\left(\frac{n+1}{2},\frac{1}{2}\right)dz\right]\times \\
&\left[\int_0^{2r} z^{n-1}B(z)I_{1-(z/2r)^2}\left(\frac{n+1}{2},\frac{1}{2}\right)dz\right]^{-1} \\
\leq\ & 1.
\end{aligned}$$

For $0 \leq z \leq (2r)^\beta$ we use the relation $|B(z)| \leq 1$, $\arcsin\{B(z)\} \leq \frac{\pi}{2}$, $\frac{v}{1+v} \leq 1$. The first of the relations (3.1.9) implies that for any $\varepsilon > 0$ as $r \to \infty$

$$\begin{aligned}
\overline{\lim_{r\to\infty}}\ S_1(r)\ \leq\ & \frac{1}{2}\ \overline{\lim_{r\to\infty}}\ a^2(r)\ e^{a^2(r)}\int_0^{(2r)^\beta} z^{n-1}B(z) \\
&\times I_{1-(z/2r)^2}\left(\frac{n+1}{2},\frac{1}{2}\right)\arcsin\ \{B(z)\}dz \\
\leq\ & \overline{\lim_{r\to\infty}}\ k_1(n,\beta)\ a^2(r)\ r^{n(\varepsilon+\beta)}.
\end{aligned} \tag{3.1.23}$$

Choosing $\varepsilon = \delta - \beta$ we obtain in view of (3.1.23) and (3.1.8) that

$$\begin{aligned}
&\overline{\lim_{r\to\infty}}\frac{S_1(r)}{\int_0^{2r} z^{n-1}B(z)I_{1-(z/2r)^2}\left(\frac{n+1}{2},\frac{1}{2}\right)dz} \\
\leq\ & \overline{\lim_{r\to\infty}}\left[\frac{k_2(n,\beta)}{a^2(r)\ r^{n(\varepsilon+\beta)}}.\int_0^{2r} z^{n-1}B(z)\ I_{1-(z/2r)^2}\left(\frac{n+1}{2},\frac{1}{2}\right)\ dz\right]^{-1} \\
=\ & 0.
\end{aligned} \tag{3.1.24}$$

Substituting (3.1.22) and (3.1.24) into (3.1.17), we obtain

$$\lim_{r\to\infty} Q_r = 0$$

and this together with relation (3.1.17) proves the relation (3.1.16). This completes the proof.

Theorem 3.1.4. Let assumptions A, B be valid and there exists $\delta \in (0,1)$, $\beta \in (0,\delta)$, such that

$$\frac{\sigma_2^2(r)}{r^{n(1+\delta)}} \to \infty, \tag{3.1.25}$$

$$\frac{a^2(r)}{\log\, r} \to 0\ ; \qquad a^2(r)\ B(r^{\beta}) \to 0. \tag{3.1.26}$$

Then as $r \to \infty$ the limiting distributions of the r.v.s

$$\zeta_r = \frac{W_2(r) - 2A(r)}{\phi(a(r))\ \sigma_2(r)}$$

and

$$\zeta_r' = \frac{\int_{v(r)} H_2(\xi(x))\ dx}{\sigma_2(r)}$$

consider (provided by one of this exists).

In particular, if A, A' hold with $\alpha \in (0, n/2)$ then $\sigma_2(r)$ is defined in (2.2.14) with $m = 2$, conditions (3.1.25), (3.1.26) satisfy and limiting distribution of r.v.s ζ_r and ζ_r', may be defined as $\{c_7(n,2,\alpha)\}^{-1/2}X_2(1)$ (or $\{c_7(n,2,\alpha)\}^{-1/2}X_2'(1)$ if A" is not hold necessarily), where $c_7(n,2,\alpha)$ is given in (2.2.15) and r.v.s $X_2(1)$ and $X_2'(1)$ are defined in (2.4.3) or (2.4.6) respectively with $m = 2$ and $C_2/2 = 1$ in (2.4.6).

Proof

The function $G_r^{(2)}(u) = \max\{0, |u| - a(r)\}$ satisfies assumptions B of chapter 2 with coefficients $C_q(r), q = 0,1,2$ computed by integration by parts:

$$\begin{aligned} C_0(r) &= 2\left\{\phi(a(r)) - a(r)\ (1 - \Phi(a(r)))\right\}, \\ C_1(r) &= 0, \\ C_2(r) &= 2\phi(a(r)). \end{aligned} \tag{3.1.27}$$

In view of (2.1.27)

$$EW_2(r) = \int_{v(r)} C_0(r)dx = 2A(R).$$

Let

$$\eta_r(x) = G_r^{(2)}(\xi(x)).$$

Then

$$\begin{aligned} E\eta_r(x)\eta_r(y) &= \int\limits_{|u|>a(r)} \int\limits_{|w|>a(r)} (|u|-a(r))(|w|-a(r)) \\ &\quad \times \phi(u,w,B(\|x-y\|))dudw \\ &= 2\int_{a(r)}^{\infty}\int_{a(r)}^{\infty} (u-a(r))(w-a(r)) \\ &\quad \times [\phi(u,w,B(\|x-y\|) \\ &\quad - \phi(u,w,-B(\|x-y\|)]\, dudv \end{aligned} \tag{3.1.28}$$

where $\phi(u,w,B)$ is given in (2.1.7).

Form (1.4.7), (3.1.10) (3.1.27) and (3.1.28) we obtain

$$\begin{aligned} \mathrm{var} W_2(r) &= \int_{v(r)}\int_{v(r)} E\ \eta_r(x)\ \eta_r(y)\ dx\ dy \\ &\quad -[E\ W_2(r)]^2 \\ &= c_{10}(n,2)r^n \int_0^{2r} z^{n-1} I_{1-(z/2r)^2}\left(\frac{n+1}{2},\frac{1}{2}\right) \\ &\quad \times \left\{\sum_{j=2}^{\infty}\frac{[B(z)]^j}{j!}\left[\Phi^{(j-1)}(a(r))\right]^2\right. \\ &\quad \left. + \sum_{j=2}^{\infty}\frac{[-B(z)]^j}{j!}\left[\Phi^{(j-1)}(a(r))\right]^2\right\} dz. \end{aligned} \tag{3.1.29}$$

Using inequalities of the type (3.1.14) we obtain from (3.1.29) that

$$\begin{aligned} \mathrm{var} W_2(r) &\le \frac{c_{10}(n,2)}{4\pi} r^n \int_0^{2r} z^{n-1} I_{1-(z/2r)^2}\left(\frac{n+1}{2},\frac{1}{2}\right) \\ &\quad \times \left\{\int_0^{B(z)} \left(\exp\left\{\frac{a^2(r)}{1+v}\right\}\right.\right. \\ &\quad \left.\left. + \exp\left\{-\frac{a^2(r)}{1+v}\right\}\right)\frac{dv}{\sqrt{1-v^2}}\right\} dz. \end{aligned} \tag{3.1.30}$$

Now we apply Lemma 2.2.1 in the case of Hilbert space $\mathbf{L}_2\ (R^1, \phi(u)\ du)$ to the functional $W_2(r) = \int_{v(r)} G_r^{(2)}(\xi(x))$ with $m = 2$. From (3.1.27) if follows that if

$$\overline{\lim_{r\to\infty}} \frac{\text{var } W_2(r)}{\phi^2(a(r))\ \sigma_2^2(r)} \leq 1, \tag{3.1.31}$$

then Theorem 3.1.4. is corollary to Lemma 2.6.1. and Theorem 2.4.1. with $m = 2$.

Utilizing (3.1.30), we have

$$\frac{\text{var } W_2(r)}{\phi^2(a(r))\ \sigma_2^2(r)} \leq \frac{S_1(r) + S_2(r)}{\int_0^{2r} z^{n-1} I_{1-(z/2r)^2}\left(\frac{n+1}{2}, \frac{1}{2}\right)\ dz}, \tag{3.1.32}$$

where for β given in the condition of the theorem and

$$\begin{aligned}
S_1(r) &= \int_0^{(2r)^\beta} Q_r(z)\ dz, \\
S_{21}(r) &= \int_{(2r)^\beta}^{2r} Q_r(z)\ dz, \\
Q_r(z) &= z^{n-1} I_{1-(z/2r)^2}\left(\frac{n+1}{2}, \frac{1}{2}\right)\ B(z) \\
&\quad \times \frac{1}{2} \int_0^{B(z)} \left[\exp\left\{a^2(r)\ \frac{y}{1+y}\right\}\right. \\
&\quad \left. + \exp\left\{-a^2(r)\ \frac{y}{1+y}\right\}\right] \frac{dy}{\sqrt{1-y^2}}.
\end{aligned}$$

For $(2r)^\beta \leq z \leq 2r$ we use the relations

$$\begin{aligned}
\exp\left\{a^2 \frac{y}{1+y}\right\} + \exp\left\{-a^2 \frac{y}{1+y}\right\} &\sim 2, \\
\frac{1}{\sqrt{1-y^2}} &\sim 1, \\
y &\to 0
\end{aligned}$$

or its refinements. Then as $r \to \infty$

$$\overline{\lim} \frac{S_2(r)}{\int_0^{2r} z^{n-1} B^2(z)\ I_{1-(z/2r)^2}\left(\frac{n+1}{2}, \frac{1}{2}\right)\ dz}$$

$$\leq \overline{\lim} \frac{\int_{(2r)^\beta}^{2r} z^{n-1} B^2(z)\ I_{1-(z/2r)^2}\left(\frac{n+1}{2}, \frac{1}{2}\right) dz}{\int_0^{2r} z^{n-1} B^2(z)\ I_{1-(z/2r)^2}\left(\frac{n+1}{2}, \frac{1}{2}\right) dz}$$
$$\leq 1. \tag{3.1.33}$$

For $0 \leq z \leq (2r)^\beta$ we obtain in the same manner as in the derivation of (3.1.24) that as $r \to \infty$

$$\overline{\lim}\, S_1(r) \leq \overline{\lim}\, \frac{1}{2} e^{a^2(r)} \left[\int_0^{2r} z^{n-1} B^2(z) \arcsin\{B(z)\} I_{1-(z/2r)^2}\left(\frac{n+1}{2}, \frac{1}{2}\right) dz \right]. \tag{3.1.34}$$

Formulas (3.1.25), (3.1.34) imply that for any $\varepsilon > 0$ and $r \to \infty$,

$$\overline{\lim}\ S_1(r) \leq \overline{\lim}\ k_3(n, \beta)\ r^{n(\varepsilon+\beta)}.$$

Choosing $\varepsilon = \delta - \beta$ and using (3.1.25), we obtain as $r \to \infty$ that

$$\begin{aligned}
\overline{\lim} \frac{S_1(r)}{\int_0^{2r} z^{n-1} B^2(z)\ I_{1-(z/2r)^2}\left(\frac{n+1}{2}, \frac{1}{2}\right) dz} &\leq k_4(n, \beta) \\
&\quad \times \overline{\lim} \frac{1}{r^{n(\varepsilon+\beta)}} \int_0^{2r} z^{n-1} B^2(z) \\
&\quad \times I_{1-(z/2r)^2}\left(\frac{n+1}{2}, \frac{1}{2}\right) dz \\
&= 0.
\end{aligned} \tag{3.1.35}$$

Substituting (3.1.35), (3.1.33) into (3.1.32) yields relation (3.1.31). An application of Lemma 2.2.1. and Theorem 2.4.1. (or Theorem 2.4.6.) with $m = 2$ completes the proof.

B'. Let $b : (0, \infty) \to (0, \infty)$ be continuous monotonic function such that $b(r) \to 0$ as $r \to \infty$.

Consider the functionals

$$W_3(r) = \int_{v(r)} \max\ \{0, b(r) - \xi(x)\}\ dx,$$
$$W_4(r) = \int_{v(r)} \max\ \{0, b(r) - |\xi(x)|\}\ dx.$$

Note that

$$A_2(r) = E\, W_3(r)$$

$$
\begin{aligned}
&= r^n |v(1)| \; [b(r)\; \Phi(b(r)) + \phi(b(r))] , \\
A_3(r) &= E\, W_4(r) \\
&= r^n |v(1)| \; \{b(r)\, [\Phi(b(r)) - \Phi(0)] \\
&\quad - [\phi(0) - \phi(b(r))]\} .
\end{aligned}
$$

Theorem 3.1.5. Let assumptions A, B' be valid and there exists $\delta \in (0,1)$ such that

$$\lim_{r\to\infty} \sigma_1^2(r)/r^{n(1+\delta)} = \infty.$$

Then the r.v. $[W_3(r) - A_2(r)] \;/\; [\Phi(b(r))\; \sigma_1(r)]$ has asymptotically $(0,1)$-normal distribution as $r \to \infty$.

Theorem 3.1.6. Let assumptions A, B' be valid and there exists $\delta \in (0,1)$ and $\beta \in (0,\delta)$ such that

$$b^2(r) \geq r^{-\beta},$$

$$\lim_{r\to\infty} \frac{\sigma_2^2(r)}{r^{n(1+\delta)}} = \infty.$$

Then as $r \to \infty$ the limiting distributions of the r.v.s

$$\eta_r = \frac{W_4(r) - A_3(r)}{\sigma_2(r)\; (\phi(0) - \phi(b(r)))}$$

and

$$\eta_r' = -\frac{\int_{v(r)} H_2(\xi(x))dx}{\sigma_2(r)}$$

consider.

In particular, if A, A' hold with $\alpha \in (0, n/2)$ then limiting distribution of r.v.s η_r and η_r' is of the form

$$-\{c_7(n,2,\alpha\}^{-1/2} X_2(1)$$

(or $-\{c_7(n,2,\alpha\}^{-1/2} X_2'(1)$ if A" is not hold necessarily), where $X_2(1)$ and $X_2'(1)$ are defined in (2.4.3) and (2.4.6) respectively with $m = 2$ and $C_2/2 = 1$ in (2.4.6).

The ideas of proofs of Theorems 3.1.5. and 3.1.6. are the same as in Theorems 3.1.3. and 3.1.4. (see Leonenko and Sabirov (1992) for details).

3.2 Limiting Distributions for the Excess Over a Radial Surface of χ-Squared Random Fields

We shall investigate measure of excess over a radial surface for χ-squared random fields with strong dependence.

C. Let $\xi(x)$, $x \in R^n$, be a homogeneous isotropic Gamma-correlated random field with

$$E\xi(x) = \beta > 0, \quad \text{var}\ \xi(x) = \beta$$

and

$$\text{cov}(\xi(x), \xi(0)) = \beta\gamma(\|x\|) > 0$$

(see example 2.1.2).

C'. Let

$$\xi_p(x) = \frac{1}{2}(\eta_1^2(x) + \cdots + \eta_p^2(x)), \quad x \in R^n,$$

be a homogeneous isotropic χ-squared random field, where $\eta_1(x), \ldots, \eta_p(x)$ are independent copies of a Gaussian field $\eta(x)$, $x \in R^n$, with

$$\begin{aligned} E\eta(x) &= 0, \quad E\eta^2(x) = 1, \\ \text{cov}\ \{\eta(0), \eta(x)\} &= B(\|x\|) \searrow 0 \quad \text{as} \ \|x\| \to \infty \end{aligned}$$

(see example, 2.1.3).

Note that $\xi_p(x)$, $x \in R^n$, is a Gamma-correlated random field with $\beta = 1/2$ and $\gamma(\|x\|) = B^2(\|x\|)$ (see section 2.1.).

D. Let $a(x) = a(\|x\|) : (0, \infty) \to [1, \infty)$ be a continuous radial monotonic function of $\|x\| > 0$, such that

$$\lim_{\|x\| \to \infty} a(\|x\|) = \infty.$$

Under the assumptions C, D consider the functionals

$$\begin{aligned} M_1(r) &= |\{x \in v(r) : \xi(x) > a(x)\}| \\ &= \int_{v(r)} \chi\left(\xi(x) > a(x)\right)\, dx, \end{aligned}$$

where $\chi(\cdot)$ is indicator and $|\cdot|$ is a Lebesgue measure in R^n.

Denote the incomplete Gamma-function by

$$\Gamma(\beta, u) = \int_u^{\infty} e^{-t} t^{\beta-1} dt, \quad \beta > 0, \quad u \geq 0. \tag{3.2.1}$$

Using (2.1.9) we obtain

$$\begin{aligned} E\ M_1(r) &= F_1(r) \\ &= |v(1)|\,\Gamma^{-1}(\beta) \int_0^r \Gamma(\beta, \phi(z))\ z^{n-1} dz, \end{aligned}$$

where

$$a(z) = a(\|x\|), \qquad \text{if } \|x\| = z.$$

We introduce the function

$$\tau^2(r) = c_{10}(n,1) r^n \int_0^{2r} z^{n-1} \gamma(z)\ I_{1-(z/2r)^2}\left(\frac{n+1}{2}, \frac{1}{2}\right)\ dz, \tag{3.2.2}$$

where $c_{10}(n,1)$ is defined in (3.1.2).

Theorem 3.2.1. Let assumptions C, D be valid and there exists $\delta \in (0,1)$, $\kappa \in (0, \delta)$ such that as $r \to \infty$

$$\lim \frac{\tau^2(r)}{r^{n(1+\delta)} a^{\kappa/2}(r)} = \infty, \tag{3.2.3}$$

$$a(r)\gamma^{1/2}(r^{\kappa}) \to 0, \qquad a(r) = o(\log r) \tag{3.2.4}$$

Then as $r \to \infty$ the limiting distributions of the r.v.s

$$\nu_r = \frac{M_1(r) - F_1(r)}{A_1(r)}$$

and r.v.s

$$\nu'_r = -\frac{\int_{v(r)} (\beta - \xi(x)) p_{\beta+1}(a(x)) dx}{A_1(r)}$$

coincide, provided one of them exists, where

$$A_1^2(r) = \beta \int_{v(r)} \int_{v(r)} p_{\beta+1}(a(x)) p_{\beta+1}(a(y)) \gamma(\|x-y\|) dx dy$$

and the function $p_\beta(u)$ is given in (2.1.9).

In particular, if A', A", C' hold with $\alpha \in (0, n/2)$ and

$$L_1(\|x\|) = p_{\beta+1}(a(\|x\|)) \in \mathcal{L}$$

with $\beta = p/2$, then as $r \to \infty$

$$A_1^2(r) = (p/2)^2 c_7(n, 2, \alpha) r^{2(n-\alpha)} L^2(r) L_1^2(r)(1 + o(1)),$$

where $c_7(n, 2, \alpha)$ is given in (2.2.15), and limiting distribution ν'_r and r.v.s

$$\nu''_r = \frac{1}{p\, c_7^{1/2}(n, 2, \alpha)} \sum_{i=1}^{p} \frac{\int_{v(r)} H_2(\eta_i(x))dx}{r^{n-\alpha} L(r)}$$

coincide, where $H_2(u)$ is given in (2.1.5). The limiting distribution of ν_r, ν'_r and ν''_r is defined by the sum of multiple stochastic integrals:

$$R_p = c_7^{-1/2}(n, 2, \alpha) \frac{1}{p} \sum_{i=1}^{p} X_{i2}(1),$$

where r.v.s $X_{i2}(1)$, $i = 1, \ldots, p$ are independent copies of r.v. $X_2(1)$ (or $X'_2(1)$, if A" is not hold necessarily), where $X_2(1)$ and $X'_2(1)$ are defined by (2.4.3) and (2.4.6) respectively with $m = 2$ or $C_2/2 = 1$ in (2.4.6).

Remark 3.2.1 . If assumption A' hold for $\alpha \in (0, n/2)$,then the conditions of the Theorem3.2.1 hold for the radial functional

$$a(\|x\|) = \log\log\|x\|, \quad \|x\| > e^e.$$

Proof

The function $G_x^{(1)}(u) = \chi(u > a(x))$ satisfies assumptions B of chapter 2 in the space $\mathbf{L}_2((0, \infty), p_\beta(u)\, du)$, where $p_\beta(u)$ is given in (2.1.9). We apply Lemma 2.2.2. in the Hilbert space $\mathbf{L}_2((0, \infty), p_\beta(u)\, du)$ to the functional $M_1(r) - F_1(r)$, where

$$M_1(r) = \int_{v(r)} G_x^{(1)}(\xi(x))dx.$$

Choose an integer $m = 1$ in the condition C' of chapter 2, and the functions $r_k(\|x\|) = \gamma^k(\|x\|)$ in the relations (2.1.2), (2.1.15).

From (2.1.11) we have $e_1^{(\beta)}(u) = (\beta - u)/\sqrt{\beta}$, and using the well-known property

$$\Gamma(\beta+1, u) = \beta\,\Gamma(\beta, u) + u^\beta e^{-u}, \tag{3.2.5}$$

we obtain

$$C_1(x) = \int_0^\infty G_x^{(1)}\, p_\beta(u)\, e_1^{(\beta)}(u)\, du \tag{3.1}$$

$$= \frac{\int_{a(x)}^\infty (\beta - u) p_\beta(u) du}{\sqrt{\beta}} \tag{3.2}$$

$$= -\sqrt{\beta} p_{\beta+1}(a(x)). \tag{3.2.6}$$

Using the partial differential equation (2.1.18) for a function (2.1.13) we obtain

$$\begin{aligned} var\ M_1(r) &= \int_{v(r)}\int_{v(r)} dxdy \int_{a(x)}^\infty \int_{a(y)}^\infty [p_\beta(u, w, \gamma(\|x-y\|)) \\ &\quad - p_\beta(u)p_\beta(w)dudw] \\ &= \int_{v(r)}\int_{v(r)} dxdy \int_{a(x)}^\infty \int_{a(y)}^\infty dudw \int_0^{\gamma(\|x-y\|)} \frac{\partial}{\partial t} p_\beta(u, w, t)dt \\ &= \beta \int_{v(r)}\int_{v(r)} dxdy \int_0^{\gamma(\|x-y\|)} dt \int_{a(x)}^\infty \int_{a(y)}^\infty \frac{\partial^2 p_\beta(u, w, t)}{\partial u\, \partial w} dudw \\ &= \beta \int_{v(r)}\int_{v(r)} dxdy \int_0^{\gamma(\|x-y\|)} p_{\beta+1}(a(x), a(y), t)dt, \end{aligned} \tag{3.2.7}$$

where $p_\beta(u, w, \gamma)$ is defined in(2.1.13).

From (3.2.6) and (2.1.17) we obtain

$$E\left[\int_{v(r)} e_1^{(\beta)}(\xi(x))C_1(x)dx\right]^2 = \beta \int_{v(r)}\int_{v(r)} p_{\beta+1}(a(x))p_{\beta+1}(a(y))\gamma(\|x-y\|)dxdy$$

$$= \quad A_1^2(r). \tag{3.2.8}$$

Assumption C' of chapter 2 (see (2.2.8)) implies that if

$$\overline{\lim_{r\to\infty}} \frac{\int\limits_{v(r)}\int\limits_{v(r)} dx\, dy \int\limits_0^{\gamma(\|x-y\|)} p_{\beta+1}(a(x), a(y), \rho) d\rho}{\int\limits_{v(r)}\int\limits_{v(r)} p_{\beta+1}(a(x)) p_{\beta+1}(a(y)) \gamma(\|x-y\|) dx dy} \leq 1 \tag{3.2.9}$$

then the limiting distributions of the r.v. ν_r and ν_r' are coincide by Lemma 2.2.2. with $m = L\,\mathrm{rank}\, G_x^{(1)} = 1$. Then, similarly to the proof of Theorem 2.5.2. of a book Ivanov and Leonenko (1989), we obtain that under A', A", C'

$$\lim_{r\to\infty} E(\nu_r' - \nu"_r)^2 = 0,$$

and application of Theorem 2.4.1. (or Theorem 2.4.6.) with

$$m = H\,\mathrm{rank}\, G = 2, \qquad G(u) = H_2(u)$$

completes the proof.

Thus, it suffices to verify (3.2.9) using the conditions (3.2.3) and (3.2.4) of the theorem.

We subdivide the set $V = v(r) \times v(r)$ into two subsets :

$$\begin{aligned} \Delta_1 &= \{(x,y) \in V : \|x-y\| \leq (2r)^{\kappa}\}, \\ \Delta_2 &= \{(x,y) \in V : \|x-y\| > (2r)^{\kappa}\}. \end{aligned}$$

Consider

$$Q(x,y) = p_{\beta+1}(a(x)) p_{\beta+1}(b(x)) \int_0^{\gamma(\|x-y\|)} \frac{p_{\alpha+1}(a(x), a(y), \rho)}{p_{\alpha+1}(a(x)) p_{\alpha+1}(a(y))} d\rho.$$

Then

$$\iint\limits_V Q(x,y) = \iint\limits_{\Delta_1} Q(x,y) dx dy + \iint\limits_{\Delta_2} Q(x,y) dx dy = S_1 + S_2. \tag{3.2.10}$$

From (2.1.12) we obtain the inequality

$$I_\nu(z) \leq \frac{(z/2)^\nu e^z}{\Gamma(\nu+1)}, \qquad z > 0. \tag{3.2.11}$$

From (2.1.13) and (3.2.11) we obtain for $(x, y) \in \Delta_2$:

$$\begin{aligned}\frac{p_{\beta+1}(a(x), a(y), \rho)}{p_{\beta+1}(a(x))\, p_{\beta+1}(a(y))} &\le (1-\rho)^{-1-\beta} \\ &\quad \times \exp\left\{(a(x)+a(y)) + \frac{\sqrt{\rho}}{1-\rho}\sqrt{a(x)a(y)} \right. \\ &\quad \left. - \frac{a(x)-a(y)}{1+\sqrt{\rho}}\right\} \\ &\le (1-\rho)^{-1-\beta}\exp\left\{\frac{2\sqrt{\rho}}{1+\sqrt{\rho}}(a(x)+a(y))\right\} \\ &\le (1-\rho)^{-1-\beta}\exp\left\{\frac{2\sqrt{\rho}}{1+\sqrt{\rho}}a(r)\right\}, \qquad (3.2.12)\end{aligned}$$

where $a(x) = a(\|x\|)$ with $\|x\| = r$.

Using the inequality $(1+\sqrt{\rho})^{-1} \le 1$ for $0 \le \rho \le \gamma(\|x-y\|)$, $\|x-y\| > (2r)^\kappa$, we obtain from (3.2.12)

$$\begin{aligned} S_2 &\le \exp\{-2a(r)\}\sup\left\{\sqrt{\gamma(z)},\ z \ge (2r)^\kappa\right\} \\ &\quad \times \iint_{\Delta_2} p_{\beta+1}(a(x))\, p_{\beta+1}(a(y)) \int_0^{\gamma(\|x-y\|)} (1-\rho)^{-1-\beta} d\rho \\ &\le T_r \iint_{\Delta_2} p_{\beta+1}(a(x)) p_{\beta+1}(a(y)) \gamma(\|x-y\|) dx dy, \qquad (3.2.13)\end{aligned}$$

where in view of (3.2.4)

$$\begin{aligned} T_r &= \exp\{2a(r)\}\sup\left\{\sqrt{\gamma(z)},\ z \ge (2r)^\kappa\right\} \\ \times &\quad [1-\sup\{\gamma(z),\ z \ge (2r)^\kappa\}]^{-\beta-1} \\ &\to 1. \qquad (3.2.14)\end{aligned}$$

Therefore it follows from (3.2.13) and (3.2.14) that

$$\overline{\lim_{r\to\infty}} \frac{\iint_{\Delta_2} Q(x,y) dx dy}{\iint_V p_{\beta+1}(a(x)) p_{\beta+1}(a(y)) \gamma(\|x-y\|) dx dy} \le 1. \qquad (3.2.15)$$

Thus it suffices to verify that as $r \to \infty$

$$\begin{aligned}\overline{\lim}\, S_1 &= \overline{\lim} \frac{\iint_{\Delta_2} Q(x,y) dx dy}{\iint_V p_{\beta+1}(a(x)) p_{\beta+1}(a(y)) \gamma(\|x-y\|) dx dy} \\ &= 0.\end{aligned}$$

From (2.1.3) we obtain that the function $p_\beta(u, w, \rho)$ is increasing in $\rho \in (0,1)$. Thus, for (3.2.16) it suffices to verify that for every $s \in (0,1)$

$$\begin{aligned}
&\overline{\lim_{r\to\infty}} \left[\iint_{\Delta_2} p_{\beta+1}(a(x))\, p_{\beta+1}(a(y))dxdy \int_s^1 p_{\beta+1}(a(x), a(y), \rho)\right.\\
&\left.\times\, p_{\beta+1}^{-1}(a(x))\, p_{\beta+1}^{-1}(a(y))\; d\rho\right]\\
&\times \left[\iint_V p_{\beta+1}(a(x))\, p_{\beta+1}(a(y))\; \gamma(\|x-y\|)dxdy\right]^{-1}\\
= &\overline{\lim_{r\to\infty}}\; U_r = 0. \qquad (3.2.16)
\end{aligned}$$

It is known, that

$$I_\nu(z) \le \frac{k_1\, e^z}{\sqrt{z}}, \quad z \ge 1, \quad k_1 > 0. \qquad (3.2.17)$$

Using (2.1.3) and (3.2.17) we have

$$\begin{aligned}
\frac{p_{\beta+1}(a(x), a(y), \rho)}{p_{\beta+1}(a(x))p_{\beta+1}(a(y))} \le\; & k_2 \exp\left\{2\frac{\sqrt{\rho\, a(x)\, a(y)}}{1-\rho}\right.\\
&+ (a(x) + a(y))\\
&\left.-\frac{a\,(x) + a\,(y)}{1-\rho}\right\}\\
&\times \left\{[a(x), a(y), \rho]^{\beta/2+1/4}\,(1-\rho)^{1/2}\right\}^{-1}\\
\le\; & k_3 \exp\left\{(a(x) + a(y))\,\frac{\sqrt{\rho}}{1+\sqrt{\rho}}\right\}\\
&\times \left\{[a(x), a(y), \rho]^{\beta/2+1/4}\,(1-\rho)^{1/2}]\right\}^{-1},\\
&k_2, k_3 > 0. \qquad (3.1.18)
\end{aligned}$$

Form (3.2.18) we obtain that the numerator of (3.2.16) does not exceed for a large $r > 0$

$$k_4 \iint_{\Delta_2} [a(x)\; a(y)]^{\beta/4-1/4}\; e^{-(a(x)+a(y))}dxdy$$

$$\times \int_s^1 \exp\left\{\frac{\sqrt{\rho}}{1+\sqrt{\rho}}\left(a(x)+a(y)\right)\rho^{-\beta/2-1/4}(1-\rho)^{-1/2}\right\} d\rho$$

$$\leq \quad k_4 \exp\left\{2a(r)\right\}^{\beta/2}(r) \iint_{\Delta_1} dx dy \int_s^1 (1-\rho)^{-1/2} d\rho$$

$$\leq \quad k_5 \; \exp\{2a(r)\} r^{n(1+\beta)} [a(r)]^{\beta/2}. \tag{3.2.19}$$

Similar (2.4.7) we obtain

$$\iint_V \gamma(\|x-y\|) dx dy \;=\; c_{10}(n,1) r^n \times \int_0^{2r} z^{n-1}\gamma(z) \times I_{1-(z/2r)^2}\left(\frac{n+1}{2},\frac{1}{2}\right) dz, \tag{3.2.20}$$

and

$$\iint_V p_{\beta+1}(a(x)) p_{\beta+1}(a(y)) \gamma(\|x-y\|) dx dy$$

$$\geq \quad \Gamma^{-2}(1+\beta)\; e^{-2a(r)} \iint_V [a(x)\; a(y)]^\beta \,\gamma(\|x-y\|) dx dy$$

$$\geq \quad k_6 \; e^{-2a(r)} \iint_V \gamma(\|x-y\|) dx dy$$

$$= \quad k_7 r^n \int_0^{2r} z^{n-1}\gamma(z) I_{1-(z/2r)^2}\left(\frac{n+1}{2},\frac{1}{2}\right) dz. \tag{3.2.21}$$

From (3.2.16),(3.2.19) and (3.2.21) we have

$$U_r \leq \frac{k_8 \; \exp\left\{4a(r)\; r^{n\kappa}\,[a(r)]^{\beta/2}\right\}}{r^n \int_0^{2r} z^{n-1}\gamma(z)\; I_{1-(z/2r)^2}\left(\frac{n+1}{2},\frac{1}{2}\right) dz}. \tag{3.2.22}$$

In view of (3.2.4), for any $\varepsilon > 0$

$$\exp\{4a(r)\} \leq k_9 r^{n\varepsilon}.$$

Then the expression (3.2.22) does not exceed

$$\frac{k_{10}}{[a(r)]^{-\beta/2}\, r^{-n(\kappa+\varepsilon)} \int\limits_0^{2r} z^{n-1}\gamma(z)\; I_{1-(z/2r)^2}\left(\frac{n+1}{2},\frac{1}{2}\right) dz}.$$

Choosing $\varepsilon = \delta - \kappa > 0$ and using (3.2.3) we obtain that $\lim U_r = 0$ as $r \to \infty$. Substituting (3.2.16), (3.2.15) into (3.2.10) we have (3.2.9).

Under the assumption A, D, consider the functional

$$\begin{aligned} M_2(r) &= \left|\left\{x \in v(r) : \xi(x) < \frac{1}{a(x)}\right\}\right| \\ &= \int_{v(r)} \chi\left(\xi(x) < \frac{1}{a(x)}\right) dx. \end{aligned}$$

Using (3.2.1) we obtain

$$\begin{aligned} EM_2(r) &= F_2(r) \\ &= \int_{v(r)} dx \int_0^{1/a(x)} p_\beta(u)du \\ &= |v(1)|\; \Gamma^{-1}(\beta) \int_0^r z^{n-1}\left[\Gamma(\beta) - \Gamma\left(\beta, \frac{1}{a(z)}\right)\right] dz. \end{aligned}$$

Theorem 3.2.2. Let assumptions C, D valid and there exists $\delta \in (0,1)$, $t \in (0, \delta n/2\beta)$ such that as $r \to \infty$

$$\frac{\tau^2(r)}{r^{n(1+\delta)}} \to \infty, \qquad a(r) \le k_{10} r^t, \tag{3.2.23}$$

where k_{10} and $\tau^2(r)$ is defined in (3.2.2). Then as $r \to \infty$ the limiting distributions of the r.v.s

$$\mu_r = \frac{M_2(r) - F_2(r)}{A_2(r)}$$

and r.v.s

$$\mu_r' = \frac{\int_{v(r)} (\beta - \xi(x)) p_{\beta+1}(\frac{1}{a(x)})dx}{A_2(r)}$$

coincide, where

$$A_2^2(r) = \beta \int_{v(r)} \int_{v(r)} p_{\beta+1}\left(\frac{1}{a(x)}\right) p_{\beta+1}\left(\frac{1}{a(y)}\right) \gamma(\|x-y\|)dxdy. \quad (3.2.24)$$

In particular, A', A", C' hold with $\alpha \in (0, n/2)$, and

$$L_2(\|x\|) = p_{\beta+1}(\frac{1}{a(\|x\|)}) \in \mathcal{L}$$

with $\beta = p/2$, then as $r \to \infty$

$$A_2^2(r) = \left(\frac{p}{2}\right)^2 c_7(n,2,\alpha) r^{2(n-\alpha)} L^2(r) L_2^2(t)(1+o(1)),$$

where $c_7(n,2,\alpha)$ is given in (2.2.15), and limiting distribution r.v.s μ_r' and r.v.s

$$\mu_r'' = -\frac{1}{p\, c_7^{-1/2}(n,2,\alpha)} \sum_{i=1}^{p} \frac{\int_{v(r)} H_2(\eta_i(x))dx}{r^{n-\alpha}L(r)}$$

coincide, where $\eta_1(x), \dots \eta_p(x)$ are independent copies of Gaussian field $\eta(x)$ (see condition C'). The limiting distribution of μ_r, μ_r' and μ_r'' is defined by the sum of multiple stochastic integrals

$$-c_7^{-1/2}(n,2,\alpha)\frac{1}{p}\sum_{i=1}^{p} X_{i2}(1),$$

where the r.v.s $X_{i2}(1), i = 1, \dots, p$ are independent copies of r.v. $X_2(1)$ (or $X_2'(1)$, if A" is not hold necessarily), where $X_2(1)$ and $X_2'(1)$ are given in (2.4.3) and (2.4.6) respectively with $m = 2$ and $C_2/2 = 1$, in (2.4.6).

Proof

The function

$$G_x^{(2)}(u) = \chi\left(u < \frac{1}{a(x)}\right)$$

satisfies assumptions B of chapter 2 in the Hilbert space $\mathbf{L}_2((0,\infty), p_\beta(u)\, du)$, where $p_\beta(u)$ is given in (2.1.9). We apply Lemma 2.2.2 in this space to the functional $M_2(r) - F_2(r)$, where

$$M_2(r) = \int_{v(r)} G_x^{(2)}(\xi(x))dx.$$

Chose an integer $m = 1$ in the condition C' of chapter 2 and the functions $r_k(\|x\|) = \gamma^k(\|x\|)$ is the relations (2.1.2) and (2.1.15). From (2.1.11) (3.1.2), we obtain

$$\begin{aligned} C_1(x) &= \frac{1}{\sqrt{\beta}} \int_0^{1/a(x)} (\beta - u) p_\beta(u) du \\ &= \sqrt{\beta}\, p_{\beta+1}\left(\frac{1}{a(x)}\right), \end{aligned}$$

and

$$E\left[\int\limits_{v(r)} e_1^{(\beta)}(\xi(x)) C_1(x) dx\right]^2 = A_2^2(r),$$

where $A_2^2(r)$ is given in (3.2.24).

Similar to (3.2.7) we have

$$var\ M_2(r) = \beta \int\limits_{v(r)} \int\limits_{v(r)} dxdy \int\limits_0^{\gamma(\|x-y\|)} p_{\beta+1}\left(\frac{1}{a(x)}, \frac{1}{a(y)}, \rho\right) d\rho.$$

Assumption C' of chapter 2 (see (2.2.8)) implies that if

$$\overline{\lim_{r\to\infty}} \frac{\int\limits_{v(r)} \int\limits_{v(r)} dxdy \int\limits_0^{\gamma(\|x-y\|)} p_{\beta+1}\left(\frac{1}{a(x)}, \frac{1}{a(y)}, \rho\right) d\rho}{\int\limits_{v(r)} \int\limits_{v(r)} p_{\beta+1}\left(\frac{1}{a(x)}\right) p_{\beta+1}\left(\frac{1}{a(y)}\right) \gamma(\|x-y\|) dxdy} \leq 1 \qquad (3.2.25)$$

then the limiting distributions of the r.v.s μ_r and μ_r' coincide by Lemma 2.2.2. with

$$m = L\,\mathrm{rank}\, G_x^{(2)}(u) = 1.$$

Then, using Lemma 1.4.12. we can prove that under A', A", C'

$$\lim_{r\to\infty} (\mu_r' - \mu_r'')^2 = 0$$

and application of Theorem 2.4.1. with $m = H\,\mathrm{rank}\, G = 2$, $G(u) = H_2(u)$, completes the proof. Thus, it suffices to verify (3.2.25) using the conditions (3.2.23). Let

$$Q(x,y) = p_{\beta+1}\left(\frac{1}{a(x)}\right) p_{\beta+1}\left(\frac{1}{a(y)}\right) \int\limits_0^{\gamma(\|x-y\|)} \frac{p_{\beta+1}(\frac{1}{a(x)}, \frac{1}{a(y)}, \rho) d\rho}{p_{\beta+1}(\frac{1}{a(x)}) p_{\beta+1}\left(\frac{1}{a(y)}\right)}.$$

Then

$$\iint\limits_V Q = \iint\limits_{\Delta_1} Q + \iint\limits_{\Delta_2} Q = S_1 + S_2, \tag{3.2.26}$$

where the sets V, Δ_1, Δ_2 are the same as in the proof of Theorem 3.1.1 with some $\kappa \in (0,1)$. Using (2.1.13) and (3.2.11) we have

$$\frac{p_{\beta+1}(\frac{1}{a(x)}, \frac{1}{a(y)}, \rho)}{p_{\beta+1}(\frac{1}{a(x)})\ p_{\beta+1}(\frac{1}{a(y)})} \leq (1-\rho)^{-\beta-1} \exp\left\{\left(\frac{1}{a(x)} + \frac{1}{a(y)}\right) \frac{\sqrt{\rho}}{1+\sqrt{\rho}}\right\}.$$

The we obtain

$$\begin{aligned} S_2 &\leq \iint_{\Delta_2} \Bigg[\Gamma^2(\beta+1)(a(x)a(y))^{-\beta} \exp\left\{-\left(\frac{1}{a(x)} + \frac{1}{a(y)}\right)\right\} \times \\ &\qquad \int\limits_0^{\gamma(\|x-y\|)} (1-\rho)^{-\beta-1} \exp\left\{\frac{\sqrt{\rho}}{1+\sqrt{\rho}} \left(\frac{1}{a(x)} + \frac{1}{a(y)}\right)\right\} d\rho\Bigg]\, dxdy \\ &\leq e^2 \Gamma^2(\beta+1) N_r \iint_{\Delta_2} (a(x)\ a(y))^{-\beta} \gamma(\|x-y\|) dxdy, \end{aligned} \tag{3.2.27}$$

where

$$\begin{aligned} N_r &= \exp\left\{-\frac{2}{a(r)} + \sup\left\{\sqrt{\gamma(z)},\ z \geq (2r)^\kappa\right\}\right. \\ &\quad \times \left\{1 - \sup\{\sqrt{\gamma(z)},\ z \geq (2r)^\kappa\}\right\}^{\beta+1} \\ &\to 1 \end{aligned}$$

as $r \to \infty$ by (3.2.23).

In view of (3.2.20) we can show that

$$\frac{A_2^2(r)}{\beta} \geq e^2 \Gamma^2(\beta+1) \int\limits_{v(r)} \int\limits_{v(r)} (a(x)a(y))^{-\beta} \gamma(\|x-y\|) dxdy. \tag{3.2.28}$$

Using (3.2.27), (3.2.28) we obtain

$$\overline{\lim_{r\to\infty}} \frac{\iint_{\Delta_2} Q(x,y)dxdy}{A_2^2(r)\frac{1}{\beta}} \le 1. \tag{3.2.29}$$

We shall prove that

$$\overline{\lim_{r\to\infty}} \frac{\iint_{\Delta_1} Q(x,y)dxdy}{A_2^2(r)\frac{1}{\beta}} = 0. \tag{3.2.30}$$

It is suffices to verify (see proof of Theorem 3.2.1.) that for any $s \in (0,1)$ as $r \to \infty$

$$\overline{\lim_{r\to\infty}} \frac{\int\limits_{v(r)}\int\limits_{v(r)} dxdy \int\limits_{1-s}^{\gamma(\|x-y\|)} p_{\beta+1}\left(\frac{1}{a(x)}, \frac{1}{a(y)}, \rho\right) d\rho}{A_2^2(r)/\alpha} = 0. \tag{3.2.31}$$

Using (3.2.11), (3.2.27) we have

$$\begin{aligned}
\int\limits_{1-s}^{\gamma(\|x-y\|)} p_{\beta+1}\left(\frac{1}{a(x)}, \frac{1}{a(y)}, \rho\right) d\rho &\le \int\limits_{1-s}^{1} p_{\beta+1}\left(\frac{1}{a(x)}, \frac{1}{a(y)}, \rho\right) d\rho \\
&\le k_{11} \int\limits_{1-s}^{1} \rho^{-\beta/2}(1-\rho)^{-1} \\
&\quad \times \exp\left\{-\frac{1}{1-\rho}\left(\frac{1}{a(x)} + \frac{1}{a(y)}\right)\right\} \\
&\quad \times I_\beta\left(\frac{2\sqrt{\rho}}{1+\sqrt{\rho}}\sqrt{\frac{1}{a(x)a(y)}}\right) d\rho \\
&\le k_{12} \int\limits_{1-s}^{1} \rho^{-\beta/2}(1-\rho)^{-1} \\
&\quad \times \exp\left\{-\frac{2}{(1-\rho)a(r)}\right\} \\
&\quad \times I_\beta\left(\frac{\sqrt{\rho}}{a(r)(1+\rho)}\right) d\rho
\end{aligned}$$

$$\leq k_{13} \int_0^{f(r)s} I_\beta\left(\frac{2}{w}\right) e^{-2/w} \frac{dw}{w}$$

$$\leq k_{14} \left[\int_0^1 + \int_0^{f(r)s}\right] I_\beta(\frac{2}{w}) e^{-2/w} \frac{dw}{w}$$

$$\leq k_{15}. \tag{3.2.32}$$

Substituting (3.2.32) into (3.2.31) we obtain

$$\int_{v(r)}\int_{v(r)} dx\, dy \int_{1-s}^{\gamma(\|x-y\|)} p_{\beta+1}\left(\frac{1}{a(x)}, \frac{1}{a(y)}, \rho\right) d\rho \leq k_{16}\, r^{n(1+\kappa)}. \tag{3.2.33}$$

Similar to (3.2.21) we have

$$A_2^2(r) \geq k_{17}\, (f(r))^{-2\beta} r^n \int_0^{2r} z^{n-1}\gamma(z)\, I_{1-(z/2r)^2}\left(\frac{n+1}{2}, \frac{1}{2}\right) dz. \tag{3.2.34}$$

Substituting (3.2.33) and (3.2.34) into (3.2.30) we obtain

$$\frac{\iint_{\Delta_1} Q(x,y)dxdy}{A_2^2(r)/\beta} \leq \left\{\frac{k_{18}}{r^{n\kappa} f^{2\beta}(r)} \int_0^{2r} z^{n-1}\gamma(z)\, I_{1-(z/2r)^2}\left(\frac{n+1}{2}, \frac{1}{2}\right) dz\right\}^{-1}.$$

Choosing $\kappa = \delta - 2\beta t/n$ ($t \in (0, n\delta/2\beta)$) we obtain that last expression does not exceed

$$k_{19}\left\{\frac{r^{-(n\kappa+2\beta t)}\tau^2(r)}{r^n}\right\}^{-1} \leq k_{20}\, \left\{r^{-n(\delta+1)}\tau^2(r)\right\},$$

where $\tau^2(r)$ is defined in (3.2.2). The last expression tends to zero as $r \to \infty$ in view of assumptions (3.2.23). Thus we have proved (3.2.31) and (3.2.30).

All the facts presented above becomes sumpler for the functionals

$$N_1(r) = |\{x \in v(r) : \xi(x) > a(r)\}|,$$

where $\xi(x)$ is a Gamma correlated random field (see assumptions C) or χ-squared random field (see assumptions C') and $a(r)$ satisfies condition B.

Note that

$$E\ N_1(r) \quad = \quad |v(1)|\,\frac{r^n\Gamma(\beta, a(r))}{\Gamma(\beta)}.$$

Theorem 3.2.3 Let assumptions B, C, hold and there exists $\delta \in (0,1)$, $\kappa \in (0,\delta)$ such that as $r \to \infty$,

$$\tau^2(r)/r^{n(1+\delta)} \to \infty, \quad a(r)\sqrt{\gamma\,(r^\kappa)} \to 0, \quad a(r) = o(\log r).$$

Then as $r \to \infty$ the limiting distributions of the r.v.s

$$\zeta_r = \frac{N_1(r) - E\ N_1(r)}{p_{\beta+1}(a(r))\ \tau(r)\sqrt{\beta}}$$

and r.v.'s

$$\zeta_r' = \frac{\int_{v(r)}(\xi(x) - \beta)dx}{\tau(r)\sqrt{\beta}}$$

coincide. In particularly, if A', A", C' hold with $\alpha \in (0, n/2)$ then the limiting distribution of ζ_r and ζ_r' is equal R_p (see Theorem 3.2.1. for definition of R_p).

The idea of the proof of Theorem 3.2.3 based on the used Lemma 2.2.1. instead of Lemma 2.2.2. (see Leonenko (1987b) for detailes).

3.3 Spherical Measures of Excess over of Moving Level

We shall examine spherical measures of excess above 'high' and 'low' levels for realizations of homogeneous isotropic random fields.

Under assumptions A, B consider the functional

$$\begin{aligned} U_1(r) &= \sigma(\{x \in s(r) : \xi(x) > a(r)\}) \\ &= \int\limits_{s(r)} \chi(\xi(x) > a(r))d\sigma(x), \end{aligned}$$

where $s(r)$ is sphere in R^n of radius r and the center at the origin, and $d\sigma(x)$ is an element of the Lebesgue measure on the sphere $s(r) \subset R^n$, $n \geq 2$, defined in (1.1.4).

The functionals $U_i(r)$, $i = 1, 2$, have a clear geometrical meaning. Consider a functions

$$\begin{aligned}\tau_m^2(r) &= \operatorname{var}\left[\int_{s(r)} H_m(\xi(x))d\sigma(x)\right] \\ &= c_8(n,m)r^{n-1}\int_0^{2r} z^{n-2}\left(1-\left(\frac{z}{2r}\right)^2\right)^{(n-3)/2} B^m(z)dz, \quad (3.3.1)\end{aligned}$$

where the constant $c_8(n, m)$ is defined in (2.2.22) and $m \geq 1$, $n \geq 2$.

Theorem 3.3.1. Let assumptions A, B hold and let there exist $\delta \in (0, 1)$, $\beta \in (0, \delta)$ such that

$$\lim_{r\to\infty} \frac{\tau_1^2(r)}{r^{(n-1)(\delta+1)}} = \infty \tag{3.3.2}$$

and

$$a^2(r) = o(\log r), \quad a^2(r)\; B(r^\beta) \to 0 \quad \text{as } r \to \infty.$$

Then the distribution of the r.v.

$$\frac{U_1(r) - |s(1)|\, r^{n-1}[1 - \Phi(a(r)]}{\phi(a(r))\; \tau_1(r)} \tag{3.3.3}$$

is asymptotically $(0, 1)$-normal as $r \to \infty$, where $\Phi(a)$ and $\phi(a)$ are defined in (3.1.6).

Proof

The coefficients (2.2.1) of the expansion of the function $G(u) = \chi(u > a)$ in $\mathbf{L}_2(R^1, \phi(u)\; du)$ are

$$C_q(a) = \begin{cases} 1 - \Phi(a), & q = 0, \\ \phi(a)\; H_{q-1}(u), & q \geq 1. \end{cases} \tag{3.3.4}$$

Using (3.3.4) we obtain the following expansion for the function $U_1(r)$ in $\mathbf{L}_2(\Omega)$:

$$\begin{aligned}U_1(r) &= EU_1(r) + \phi(a(r))\int_{s(r)} H_1(\xi(x))d\sigma(x) \\ &\quad + \phi(a(r))\sum_{k=2}^{\infty}\frac{H_{k-1}(a(r))}{k!}\int_{s(r)} H_k(\xi(x))d\sigma(x) \\ &= \eta_r + T_r.\end{aligned}$$

Using (2.1.8) similar to (1.4.8) we obtain

$$E \int_{s(r)} H_m(\xi(x))d\sigma(x) \int_{s(r)} H_k(\xi(y))d\sigma(y) = \delta_m^k \tau_m^2(r),$$

and

$$\operatorname{var} U_1(r) = \phi^2(a(r)) \operatorname{var} \eta_r + \operatorname{var} T_r.$$

Similarly to proof of Lemma 2.2.1. we obtain that if

$$\lim_{r\to\infty} \frac{\operatorname{var}\ U_1(r)}{\phi^2(a(r))\tau_1^2(r)} \leq 1, \tag{3.3.5}$$

then the r.v. (3.3.3) has a limiting distribution identical with that of

$$\frac{\int_{s(r)} \xi(x)d\sigma(x)}{\tau_1(r)},$$

namely, a standard normal distribution.

We shall show that conditions of Theorem 3.3.1. imply (3.3.5). Note that

$$\begin{aligned}
\frac{\operatorname{var} U_1(r)}{\phi^2(a(r))\tau_1^2(r)} &= c_8(n,1)\, r^{n-1} \int_0^{2r} u^{n-2} \left(1 - (\frac{u}{2r})^2\right)^{(n-3)/2} \\
&\quad \times \frac{\left[\int_0^{B(u)} \exp\left\{-\frac{a^2(r)}{1+v}\right\} \frac{dv}{\sqrt{1-v^2}}\right] du}{\phi^2(a(r))\tau_1^2(r)} \\
&= \left[\int_0^{(2r)^\beta} + \int_{(2r)^\beta}^{2r}\right] u^{n-2} \left(1 - \left(\frac{u}{2r}\right)^2\right)^{(n-3)/2} du \\
&\quad \times \int_0^{B(u)} \exp\left\{-\frac{v\, a^2(r)}{1+v}\right\} \frac{dv}{\sqrt{1-v^2}} \\
&\quad \times \left[\int_0^{2r} u^{n-2} \left(1 - \left(\frac{u}{2r}\right)^2\right)^{(n-3)/2} B(u)\, du\right]^{-1} \\
&= A_1(r) + A_2(r).
\end{aligned} \tag{3.3.6}$$

The first term in the numerator (3.3.6) is most equal to

$$e^{a^2(r)} \int_0^{(2r)^\beta} u^{n-2} \left(1 - (\frac{u}{2r})^2\right)^{(n-3)/2} \arcsin\{B(u)\}\, du \leq$$

$$\frac{\pi}{2}\, e^{a^2(r)} (2r)^{(n-1)\beta} \int_0^1 v^{n-1} \left(1 - \frac{v^2}{(4r^2)^{1-\beta}}\right)^{(n-3)/2} dv = k_1 \frac{\pi}{2}\, e^{a^2(r)} (2r)^{(n-1)\beta} (1+o(1)).$$

For any $\varepsilon > 0$

$$e^{a^2(r)} = o(r^{(n-1)\varepsilon}).$$

Dividing this expression by the denominator of (3.3.6) for $\varepsilon = \delta - \beta$ yields

$$A_1(r) \leq \left(\frac{\tau_1^2(r)}{r^{(n-1)(\delta+1)}}\right)^{-1}.$$

The latter expression tends to zero in view of assumption (3.3.2). The second term in the numerator (3.3.6) isat most equal to

$$\int_{(2r)^\beta}^{2r} u^{n-2} \left(1 - \left(\frac{u}{2r}\right)^2\right)^{(n-3)/2}$$

$$\times \left[\int_0^{B(u)} \exp\{va^2(r)\} \frac{dv}{\sqrt{1-v^2}}\right] du$$

$$\leq \frac{1}{\sqrt{1 - B^2((2r)^\beta)}} \int_{(2r)^\beta}^{2r} u^{n-2} \left(1 - \left(\frac{u}{2r}\right)^2\right)^{(n-3)/2}$$

$$\times \left[\int_0^{B(u)} \exp\{B(u)\ a^2(r)\} dv\right] du$$

$$\leq \frac{\exp\{B((2r)^\beta)\ a^2(r)\}}{\sqrt{1 - B^2((2r)^\beta)}}$$

$$\times \int_{(2r)^\beta}^{2r} u^{n-2} \left(1 - \left(\frac{u}{2r}\right)^2\right)^{(n-3)/2} B(u) du$$

$$\leq \exp\left\{B((2r)^{\beta})\, a^2(r)\right\}$$
$$\times \int\limits_{(2r)^{\beta}}^{2r} u^{n-2}\left(1-\left(\frac{u}{2r}\right)^2\right)^{(n-3)/2} B(u)du \qquad (3.3.7)$$

as $r \to \infty$.

Note that

$$\lim_{r\to\infty} \exp\{B((2r)^{\beta})a^2(r)\} = 1.$$

Therefor for τ large, expression (3.3.7) does not exceed

$$\int\limits_{(2r)^{\beta}}^{2r} u^{n-2}\left(1-\left(\frac{u}{2r}\right)^2\right)^{(n-3)/2} B(u)du$$

which, clearly, does not exceed the denominator of (3.3.6). Hence (3.3.5) hold.

Under assumptions A, B consider the functional

$$\begin{aligned} U_2(r) &= \sigma(\{x \in s(r) : |\xi(x)| > a(r)\}) \\ &= \int_{s(r)} \chi\left(|\xi(x)| > a(r)\right) d\sigma(x), \end{aligned}$$

which may be expanded in $\mathbf{L}_2(\Omega)$ as follows:

$$\begin{aligned} U_2(r) &= 2(1-\Phi(a(r))\; r^{n-1}\,|s(1)| \\ &\quad +2\phi(a(r)) \sum_{q=1}^{\infty} \frac{H_{2q}(a(r))}{(2q)!} \int_{s(r)} H_{2q}(\xi(x))d\sigma(x). \end{aligned}$$

Direct computations allows us to obtain the variance of this functional as follows

$$\begin{aligned} \operatorname{var} U_2(r) &= c_8(n,1)\,\frac{r^{n-1}}{2\pi} \int_0^{2r} z^{n-2}\left(1-(\frac{z}{2r})^2\right)^{(n-3)/2} dz \\ &\quad \times \int_0^{B(z)} \left[\exp\left\{-\frac{a^2(r)}{1+y}\right\} - \exp\left\{-\frac{a^2(r)}{1-y}\right\}\right] \frac{dy}{\sqrt{1-y^2}}. \end{aligned}$$

Theorem 3.3.2. If assumptions A, B hold and

$$\overline{\lim_{r\to\infty}} \frac{\operatorname{var} U_2(r)}{a^2(r)\phi^2(a(r))\tau_2^2(r)} \leq 1, \qquad (3.3.8)$$

then the r.v.

$$\nu_r = \frac{U_2(r) - 2r^{n-1}\,|s(1)|\,(1 - \Phi(a(r)))}{a(r)\phi(a(r))\tau_2(r)}$$

has the same distribution as the r.v.

$$\nu'_r = \frac{\int_{s(r)} H_2(\xi(x))d\sigma(x)}{\tau_2(r)}.$$

In particularly, if A, A' hold with $\alpha \in (0,(n-1)/2)$, $n \geq 2$, then limiting distribution of r.v.s ν_r and ν'_r is of the form

$$[c_9(n,2,\alpha)]^{-1/2}Y_2(1),$$

where $c_9(n,2,\alpha)$ is given in (2.2.23) and $Y_2(1)$ is given in (2.4.5) with $m=2$ and $C_m = 1$.

Theorem 3.3.3. If assumptions A, B hold and there exists $\delta \in (0,1)$, $\beta \in (0,\delta)$ such that

$$\lim_{r\to\infty} \frac{\tau_2^2(r)}{r^{(n-1)(1+\delta)}} = 0, \quad a^2(r)\;B(r^\beta) \to 0, \quad a^2(r) = o(\log r).$$

Then (3.3.8) holds.

The proof of Theorem 3.3.3. is similar to the proof of Theorem 3.3.1. (see also the proof of the Theorem 2.2.3. from Ivanov and Leonenko (1989) or Rybasov (1977a)).

Under assumptions A, B' consider the functional

$$U_3(r) = \sigma(\{x \in s(r) : |\xi(x)| < b(r)\})$$

which is expanded in the space $\mathbf{L}_2(\Omega)$ in the form

$$\begin{aligned} U_3(r) &= 2r^{n-1}\,|s(1)|\left(\Phi(b(r)) - \frac{1}{2}\right) \\ &\quad -2\phi(b(r))\sum_{q=1}^{\infty}\frac{H_{2q-1}(b(r))}{(2q)!}\int_{s(r)} H_{2q}(\xi(x))b(x)d\sigma(x). \end{aligned}$$

Theorem 3.3.4. Let assumptions A, B' hold and there exists $\delta \in (0,1)$, $\gamma \in (0,(n-1)\delta)$ such that

$$\lim_{r\to\infty} \frac{\tau_2^2(r)}{r^{(n-1)(1+\delta)}} = \infty, \quad b^2(r) \geq kr^{-\gamma}, \quad k > 0.$$

Then the limiting distributions of the r.v.

$$\mu_r = \frac{U_3(r) - r^{n-1}\,|s(1)|\,(2\Phi(b(r)) - 1/2)}{b(r)\phi(b(r))\tau_2(r)}$$

and r.v.

$$\mu_r' = -\frac{\int_{s(r)} H_2(\xi(x))d\sigma(x)}{\tau_2(r)}$$

coincide. In particular, if A, A' hold with $\alpha \in (0, (n-1)/2)$, then the limiting distributions of r.v.s μ_r and μ_r' is of the form

$$-[c_9(n, 2, \alpha)]^{-1/2} Y_2(1),$$

where r.v. $Y_2(1)$ is given in (2.4.5) with $m = 2$ and $C_m = 1$.

The proof of Theorem 3.3.4. is similar to the proof of Theorem 3.3.1. (see also the proof of Theorem 2.2.5.) in Ivanov and Leonenko (1989) and Rybasov (1977a)).

3.4 Sojourns of Multi-Dimensional Gaussian Fields with Dependent Components

We now describe limiting distributions of the sojourn measures of vector Gaussian fields. The statements of this section extend the results of Maejima (1985, 1986a, b).

1. Reduction theorem for vector fields

E. Let $\xi(x) = [\xi_1(x), \ldots, \xi_p(x)]'$, $x \in R^n$, be a vector mean-square continuous homogeneous isotropic Gaussian random field with $E\xi(x) = 0$ and covariance matrix

$$E\xi(0)\xi(x)' = (B_{ij}(\|x\|))_{1 \le i,j \le p},$$

where

$$\begin{aligned} B_{ii}(\|x\|) &= a(\|x\|), \quad i = 1, \ldots, p, \\ B_{ij}(\|x\|) &= b(\|x\|), \quad i \neq j, \quad i, j = 1, \ldots, p, \end{aligned}$$

where

$$a(0) = 1, \quad b(0) = \rho_0 \in [0, 1),$$

$$\begin{aligned} a(\|x\|) &= \frac{L(\|x\|)}{\|x\|^\alpha}, \quad b(\|x\|) = \frac{\rho_\infty L(\|x\|)}{\|x\|^\alpha} \quad \text{as } \|x\| \to \infty, \\ L &\in \mathcal{L}, \quad \rho_\infty \in [0,1), \alpha > 0. \end{aligned}$$

Consider the following matrix

$$T = (t_{ij})_{1 \le i,j \le p},$$

where

$$\begin{aligned} t_{1j} &= \{p(1+(p-1)\rho_0\}^{-1/2}, \quad 1 \le j \le p, \\ t_{ij} &= \{i(i-1)(1-\rho_0)\}^{-1/2}, \quad 2 \le i \le p, \quad 1 \le j \le i, \\ t_{ii} &= -(i-1)^{1/2}\{i(1-\rho_0)\}^{-1/2}, \quad 2 \le i \le p, \quad t_{ij} = 0, \text{ otherwise,} \end{aligned}$$

and diagonal matrix

$$\mathcal{D} = \operatorname{diag}\{d_1, \ldots d_p\},$$

where

$$d_1 = \frac{1+(p-1)\rho_\infty}{1+(p-1)\rho_0}, \quad d_2 = \cdots = d_p = \frac{1-\rho_\infty}{1-\rho_0}. \tag{3.4.1}$$

Remark 3.4.1 . Note that

$$T^{-1} = (s_{ij})_{1 \le i,j \le p},$$

where

$$\begin{aligned} s_{i1} &= \sqrt{\frac{1+(p-1)\,\rho_0}{p}}, \quad 1 \le i \le p, \\ s_{ij} &= \sqrt{\frac{(1-\rho_0)}{i\,(i-1)}}, \quad 2 \le i \le p, \quad 1 \le i \le j, \\ s_{ii} &= -(i-1)\sqrt{\frac{1-\rho_0}{i\,(i-1)}}, \quad 2 \le i \le p, \\ s_{ij} &= 0, \quad 2 \le i \le p-1, \quad j < i < p. \end{aligned}$$

Consider the numbers

$$\mu_1 = [1+((p-1)\,\rho_0]^{-1/2}, \quad \mu_2 = \cdots = \mu_p = [1-\rho_0]^{-1/2}. \tag{3.4.2}$$

Define vector random field by

$$\eta(x) = [\eta_1(x), \ldots, \eta_p(x)]' = T\ \xi(x). \tag{3.4.3}$$

Then $\eta(x)$, $x \in R^n$, is homogeneous isotropic Gaussian random field with $E\eta(x) = 0$ and covariance matrix

$$E\eta(0)\eta(x)' = (\tilde{B}_{ij}(\|x\|)_{1\leq i,j\leq p},$$

where

$$\begin{aligned}\tilde{B}_{11}(\|x\|) &= [a(\|x\| + (p-1)\ b(\|x\|)]\ \mu_1^2,\\ \tilde{B}_{ii}(\|x\|) &= [a(\|x\| - b(\|x\|)]\ \mu_i^2,\end{aligned} \tag{3.4.4.}$$

and $\tilde{B}_{ij}(\|x\|) = 0, \quad i \neq j.$ Therefore $\eta_i(x)$, $i = 1, \ldots, p$ are independent Gaussian fields. Note that $\tilde{B}_{ii}(0) = 1$, $i = 1, \ldots, p$, and

$$E\eta(0)\ \eta(x)' = \mathcal{D} \cdot \|x\|^{-\alpha} L(\|x\|), \quad \alpha > 0 \quad \text{as} \quad \|x\| \to \infty. \tag{3.4.5}$$

Let $H_k(w)$, $k \geq 0$, $w \in R^1$, be a Chebyshev–Hermite polynomials defined in (2.1.5).

Lemma 3.4.1. Let $(\xi_1, \ldots, \xi_{2p})$ be $2p$-dimensional Gaussian vector with

$$\begin{aligned}E\xi_j &= 0, \quad E\xi_j^2 = 1, \quad E\xi_j\xi_{j+p} = r_j, \quad 1 \leq j \leq p,\\ E\xi_j\xi_k &= 0, \quad if \quad (i,j) \notin \{(i,j) : (j+p, j+p),\ (j, j+p), \quad 1 \leq j \leq p\}.\end{aligned}$$

Then

$$E\prod_{j=1}^{p} H_{k_j}(\xi_j) H_{m_j}(\xi_{j+p}) = \prod_{j=1}^{p} \delta_{k_j}^{m_j}\ k_j!\ r_j^{k_j}. \tag{3.4.6}$$

Proof (see Taqqu (1977)). Note that (3.4.6) is a special case of diagram formula (see Malyshev and Minlos (1985)).

Denote by

$$\begin{aligned}e_\nu(u) &= \prod_{j=1}^{p} H_{k_j}(u_j),\\ u &= (u_1, \ldots, u_p)' \in R^p, \quad \nu = (k_1, \ldots, k_p), \quad k_j \geq 0.\end{aligned} \tag{3.4.7}$$

The polynomials $\{e_\nu(u)\}_\nu$ form a complete orthogonal system in the Hilbert space

$$\mathbf{L}_2(R^p, \phi(\|u\|)\, du) = \{G : \int_{R^p} G^2(u)\, \phi(\|u\|)\, du < \infty\},$$
$$\phi(\|u\|) = \prod_{j=1}^{p} \phi(u_j).$$

The function $G_r(u) \in \mathbf{L}_2(R^p, \phi(\|u\|)\, du)$ admits expansion in mean-square convergent series

$$G_r(u) = \sum_{k=0}^{\infty} \sum_{\nu \in S_k} \frac{C_\nu(r) e_\nu(u)}{\nu!}, \qquad C_\nu(r) = \int_{R^p} G_r(u) e_\nu(u) \phi(\|u\|) du, \tag{3.4.8}$$

where

$$S_k = \{\nu = (k_1, \ldots, k_p) : \sum_{j=1}^{p} k_j = k,\ k_j \geq 0\},$$
$$\nu! = k_1!, \ldots, k_p!$$

and by Parseval equality

$$\sum_{k \geq 0} \sum_{\nu \in S_k} \frac{C_\nu^2(r)}{\nu!} = \int_{R^p} G_r(u)\, \phi(\|u\|)\, du. \tag{3.4.9}$$

F. Let $G_r(u) \in \mathbf{L}_2(R^p, \phi(\|u\|)\, du)$, and there exists an integer $m \geq 1$ such that $C_\nu(r) = 0$, $\nu \in S_k$, $1 \leq k \leq m-1$, but $C_\nu(r) \neq 0$ for at least one tuple $\nu = (k_1, \ldots, k_p) \in S_m$.

If F holds, m is called the Hermitian rank of $G_r(u)$, denoted as

$$m = H \operatorname{rank} G_r(u).$$

We define the random processes

$$K_r(t) = \int_{v(rt^{\frac{1}{n}})} G_r(\eta(x)) dx,$$
$$K_{r,m}(t) = \sum_{\nu \in S_k} \frac{C_\nu(r)}{\nu!} \int_{v(rt^{\frac{1}{n}})} e_\nu(\eta(x)) dx,$$

where $t \in [0,1]$.

Theorem 3.4.1. Suppose that assumptions E, F hold for $\alpha \in (0, n/m)$, $m = H \operatorname{rank} G_r(u)$, and as $r \to \infty$

$$\zeta_r = \left\{ \sum_{\nu \in S_k} \frac{C_\nu^2(r)}{\nu!} \right\}^{-1} \int_{R^p} G_r^2(u)\ \phi(\|u\|)\ du = o(r^\gamma) \tag{3.4.10}$$

for some $\gamma \in (0, \min\{\alpha, n - m\alpha\})$.

Then for $r \to \infty$ the finite-dimensional distributions of the random processes

$$X_r(t) = \frac{K_r(t) - EK_r(t)}{\sqrt{\operatorname{var} K_r(1)}}, \quad t \in [0,1],$$

and the processes

$$X_{r,m}(t) = \frac{K_{r,m}(t)}{\sqrt{\operatorname{var}\ K_{r,m}(1)}}, \quad t \in [0,1],$$

have the same limits.

Proof

It is easy to verify that

$$EK_r(t) = tr^n\, |v(1)|\, C_r(0, \ldots, 0).$$

Let

$$Q_r = \sum_{k \geq 0} \sum_{\nu \in S_k} \frac{C_\nu(r)}{\nu!} \int_{v(rt^{\frac{1}{n}})} e_\nu(\eta(x))dx,$$

then by Lemma 3.4.1.

$$\operatorname{var} K_r(t) = \operatorname{var} K_{r,m}(t) + \operatorname{var} Q_r,$$

similar to (1.4.7) we obtain from (3.4.4):

$$\begin{aligned} \operatorname{var}\ K_{r,m}(t) \;&=\; \sum_{\nu \in S_k} \frac{C_\nu^2(r)}{\nu!} \int\limits_{v(rt^{\frac{1}{n}})} \int\limits_{v(rt^{\frac{1}{n}})} (a(\|x-y\|) + (p-1)\ b(\|x-y\|)\ \mu_1^2)^{m_1} \\ &\times \prod_{2 \leq j \leq p} (a(\|x-y\|) - b(\|x-y\|)\ \mu_j^2)^{m_j} dx\ dy \end{aligned}$$

$$
\begin{aligned}
&= c_{10}(n,1)tr^n \sum_{\nu \in S_k} \frac{C_\nu^2(r)}{\nu!} \int_0^{(2rt^{\frac{1}{n}})} z^{n-1} I_{1-(z/2r)^2}\left(\frac{n+1}{2},\frac{1}{2}\right) \\
&\quad \times \left([a(z)+(p-1)\,b(z)]\,\mu_1^2\right)^{m_1} \prod_{j=2}^{\infty} \left([a(z)-b(z)]\,\mu_j^2\right)^{m_j} dz \\
&= c_7(n,m,\alpha)\left[\sum_{\nu\in S_k} \frac{C_\nu^2(r)}{\nu!} \prod_{j=2}^{\infty} d_j^{m_j}\right] \\
&\quad \times t^{2-(m\alpha)/n} r^{2n-m\alpha} L^m(r\,t^{1/n})(1+o(1))
\end{aligned}
$$

as $r \to \infty$.

Similar

$$
\begin{aligned}
\operatorname{var} Q_r &= c_{10}(n,1)\;t\;r^n \sum_{l\geq m+1} \sum_{\nu\in S_l} \frac{C_\nu^2(r)}{\nu!} \\
&\quad \times \int_0^{(2rt^{\frac{1}{n}})} z^{n-1} I_{1-(z/2r)^2}\left(\frac{n+1}{2},\frac{1}{2}\right) \\
&\quad \times \left(\frac{a(z)+(p-1)b(z)}{\mu_1^{-2}}\right)^{k_1} \prod_{j=2}^{p} \left(\frac{a(z)-b(z)}{\mu_j^{-2}}\right)^{k_j} dz.
\end{aligned}
$$

Note that

$$
\left|\frac{a(z)+(p-1)b(z)}{\mu_1^{-2}}\right| \leq 1, \qquad \left|\frac{a(z)-b(z)}{\mu_j^{-2}}\right| \leq 1.
$$

Hence for any $k_1, \ldots, k_p$ with $k_1 + \cdots + k_p \geq 1$ there exist $q_1, \ldots, q_p$ such that

$$
\begin{aligned}
& \left|\frac{a(z)+(p-1)b(z)}{\mu_1^{-2}}\right|^{k_1} \prod_{j=2}^{p} \left(\frac{a(z)-b(z)}{\mu_j^{-2}}\right)^{k_j} \\
\leq\; & \left|\frac{a(z)+(p-1)b(z)}{\mu_1^{-2}}\right|^{q_1} \prod_{j=2}^{p} \left(\frac{a(z)-b(z)}{\mu_j^{-2}}\right)^{q_j}
\end{aligned}
$$

and

$$
\begin{aligned}
q_1 &\leq k_1, \ldots, q_p \leq k_p, \\
q_1 + \cdots + q_p &= m+1.
\end{aligned}
$$

Note that $(q_1, \ldots, q_p)$ depends on $(k_1, \ldots, k_p)$. Similarly to Ivanov and Leonenko (1989), p.59, we obtain from (3.4.9):

$$\begin{aligned}
\operatorname{var} Q_r &\leq c_{10}(n,1)\, t\, r^n \sum_{l\geq m+1} \sum_{\nu\in S_l} \frac{C_\nu^2(r)}{\nu!} \prod_{j=2}^{\infty} d_j^{q_j} \\
&\quad \times \int_0^{(2rt^{\frac{1}{n}})} z^{n-1} I_{1-(z/2r)^2}\left(\frac{n+1}{2}, \frac{1}{2}\right) \frac{L(z)}{z^{(m+1)\alpha}} dz \\
&\leq k_1 \left[\int_{R^p} G_r^2(u)\phi(\|u\|)du\right] \\
&\quad \times t^{2-(m+1)\alpha/n} r^{2n-(m+1)\alpha} L^{m+1}(r\, t^{1/2}) + O(r^n)].
\end{aligned}$$

Hence by condition (3.4.10)

$$\begin{aligned}
\operatorname{var} Q_r &= o(\operatorname{var} K_{r,m}(t)), \\
\operatorname{var} K_r(t) &\sim \operatorname{var} K_{r,m}(t) \quad \text{as } r \to \infty.
\end{aligned}$$

Thus

$$\lim_{r\to\infty} E(X_r(t) - X_{r,m}(t))^2 = 0 \quad \text{and} \quad \lim_{r\to\infty} E\left[\sum_{j=1}^{s} a_j\, (X_r(t) - X_{r,m}(t)\right]^2 = 0$$

completing the proof.

Remark 3.4.2. If $G(u) \in \mathbf{L}_2(R^p, \phi(\|u\|)\, du)$ does not depend on $r > 0$ and has the Hermitian rank m, then (3.4.10) is satisfied.

2. Sojourn measures

Suppose that assumptions E hold. Define

$$\begin{aligned}
T(r) &= |\{x \in v_n(r) : \xi(x) \in E\}| \\
&= \int_{v_n(r)} \chi\, (\xi(x) \in E)\, dx,
\end{aligned}$$

where $\chi\,(\cdot)$ is the indicator function and $E \subset R^p$. In this section we write $v_n(r)$ specifying the dimension of the ball. The functionals $T(r)$ have a geometrical meaning of sojourn measures.

Let $\eta(x)$, $x \in R^n$, be a homogeneous isotropic Gaussian fields with independent components defined in (3.4.3). We thus have

$$T(r) = \int_{v_n(r)} \chi\left(\eta(x) \in \tilde{E}\right) dx,$$

where

$$\tilde{E} = \{y \in R^p : T^{-1}y \in E\}.$$

We shall study the limiting distributions of sojourn functionals

$$T_i(r) = |\{x \in v_n(r) : \xi(x) \in E_i\}|, \quad i = 1,2,3,$$

where

$$\begin{aligned} E_1 &= \{y \in R^p : y_1 \geq 0, -\infty < y_j < \infty, \ j = 2,\dots,p\}, \\ E_2 &= v_p(1) = \{y \in R^p : \|y\| < 1\}, \\ E_3 &= E_3(r) = \left|\{x \in v_n(r) : \xi(x) \notin v_p(\sqrt{a(r)})\}\right|, \end{aligned}$$

where $a(r) \to \infty$ as $r \to \infty$.

Note that

$$\tilde{E}_1 = \{y \in R^p :< w, y > \geq 0\},$$

where

$$\begin{aligned} w &= (w_1,\dots,w_p), \quad w_1 = \left\{\frac{1+(p-1)\rho_0}{p}\right\}^{1/2}, \\ w_j &= \left\{\frac{1-\rho_0}{j\,(j-1)}\right\}^{1/2}, \quad j = 2,\dots,p. \end{aligned}$$

We obtain

$$ET_1(r) = |v(1)| \ r^n \int_{\tilde{E}_1} \phi(\|y\|) \, dy,$$

and

$$\text{var } T_1(r) = c_7(n,1,\alpha) \, c_{11}(p) \, r^{2n-\alpha} L(r) \, (1+o(1)),$$

as $r \to \infty$, where $c_7(n,1,\alpha)$ is defined in (2.2.5) for $0 < \alpha < n$, and

$$c_{11}(p) = \sum_{j=1}^{p} \kappa_j^2 \, d_j, \quad \kappa_j = \int_{\tilde{E}_1} y_j \, \phi(\|y\|) \, dy, \quad j = 1,\dots,p,$$

where d_j are defined in (3.4.1). Note that $\kappa_1 \neq \kappa_2$, because of

$$\begin{aligned}\kappa_1 &= \int_{-\infty}^{\infty} y_1\phi(y_1)\left[1-\Phi\left(-y_1\left[\frac{1+\rho_0}{1-\rho_0}\right]^{1/2}\right)\right]dy_1 \\ &> 0, \\ \kappa_2 &= \int_{-\infty}^{\infty} \phi(y_2)\phi\left(-y_2\left[\frac{1+\rho_0}{1-\rho_0}\right]^{1/2}\right)dy_2 \\ &> 0.\end{aligned}$$

Theorem 3.4.2. Let assumptions E hold for $\alpha \in (0, n)$. Then the distribution of r.v.

$$\frac{T_1(r) - ET_1(r)}{r^{n-\alpha/2}L^{1/2}(r)\sqrt{c_7(n,1,\alpha)c_{11}(p)}}$$

is asymptotically $(0,1)$-normal as $r \to \infty$.

Proof

The function $G_1(u) = \chi(u \in \tilde{E}_1)$ satisfies condition F with coefficients $C_\nu = \kappa_j > 0$, for $\nu = (k_1, \ldots, k_p)$ such that $k_j = 1$ and $k_i = 0$ for $i \neq j$. Therefore

$$H \operatorname{rank} G_1(u) = 1.$$

Now we apply Theorem 3.4.1. with $m = 1$ to the functional

$$T_1(r) = \int_{v_n(r)} \chi\left(\eta(x) \in \tilde{E}_1\right) dx.$$

If $0 < \alpha < n$ then the limiting distribution of the r.v.

$$\frac{T_1(r) - ET_1(r)}{\sqrt{\operatorname{var} T_1(r)}}$$

is the same as that of

$$\frac{\kappa_1\eta_1 + \ldots + \kappa_p\eta_p}{\sqrt{\operatorname{var}(\kappa_1\eta_1 + \ldots + \kappa_p\eta_p)}},$$

where

$$\eta_j = \frac{\int_{v_n(r)} H_1(\eta_j(x))dx}{\sqrt{\text{var} \int_{v_n(r)} H_1(\eta_j(x))dx}},$$

that is, it is $(0,1)$-normal, because of

$$\text{var} \int_{v_n(r)} H_1(\eta_j(x))dx = d_j c_7(n,1,\alpha) r^{2n-\alpha} L(r)(1+o(1))$$

as $r \to \infty$. Thus, by Theorem 3.4.1., we conclude the statement.

For the functional $T_2(r) = |\{x \in v_n(r) : \xi(x) \in v_p(1)\}|$ we obtain

$$\begin{aligned} \tilde{E}_2 &= \{y \in R^p : T^{-1}y \in v_p(1)\} \\ &= \left\{ y \in R^p : \sum_{j=1}^{p} \left(\frac{y_j}{\mu_j}\right)^2 < 1 \right\}, \end{aligned}$$

where μ_j are defined in (3.4.2). We have for $0 < \alpha < n/2$

$$ET_2(r) = |v(1)| \ r^n \int_{\tilde{E}_2} \phi(\|u\|)du,$$

and

$$\text{var } T_2(r) = c_7(n,2,\alpha) c_{12}(p) r^{2(n-\alpha)} L^2(r)(1+o(1)),$$

as $r \to \infty$, where $c_7(n,2,\alpha)$ is defined in (2.2.5), and

$$c_{12}(p) = \frac{1}{4} \sum_{j=1}^{p} d_j^2 s_j^2, \tag{3.4.11}$$

where

$$\begin{aligned} s_1 &= \int_{\tilde{E}_2} H_2(y_1)\phi(\|y\|)dy \\ &= -2 \int_{U_1} \psi_1 \phi(\psi_1) \prod_{j=2}^{p} \phi(y_j)dy_j \\ &< 0, \end{aligned}$$

$$U_1 = \left\{ y \in R^p : \sum_{j=1}^{p} \left(\frac{y_j}{\mu_j} \right)^2 < 1 \right\}, \qquad \psi_1 = \mu_1 \left[1 - \sum_{j=1}^{p} \left(\frac{y_j}{\mu_j} \right)^2 \right]^{1/2},$$

and

$$\begin{aligned} s_2 &= \cdots = s_p \\ &= \int_{\tilde{E}_2} H_2(y_2)\phi(\|y\|)dy \\ &= -\int_{U_1} \psi_2 \phi(\psi_2) \prod_{j=1, j\neq 2}^{p} \phi(y_j)dy_j \\ &< 0, \end{aligned}$$

$$\begin{aligned} U_2 &= \left\{ (y_1, y_3, \ldots, y_p) : \sum_{j=1, j\neq 2}^{p} \left(\frac{y_j}{\mu_j} \right)^2 < 1 \right\}, \\ \psi_2 &= \left\{ \mu_2^2 \left[1 - \left(\frac{y_j}{\mu_j} \right)^2 \right] - \sum_{j=3}^{p} y_j^2 \right\}^{1/2}. \end{aligned}$$

Theorem 3.4.3. Let assumptions E hold for $\alpha \in (0, n/2)$. Then as $r \to \infty$ the limiting distributions of r.v.

$$W_r = \frac{T_2(r) - ET_2(r)}{\sqrt{\operatorname{var} T_2(r)}}$$

and r.v.

$$W_r' = \frac{1}{2} \sum_{j=1}^{p} s_j \frac{\int_{v_n(r)} H_2(\eta_j(x))dx}{r^{n-\alpha} L(r) \sqrt{c_7(n, 2, \alpha) c_{12}(p)}}$$

coincide (provided one of them exists), where $c_{12}(p)$ is depended in (3.4.11). In particular, if A" holds every covariance function $B_{ij}(\|x\|), i, j \in \{1, \ldots, p\}$ then the limiting distribution of W_r and W_r' is of the form

$$\{c_7(n, 2, \alpha)\ c_{12}(p)\}^{-1/2} \frac{1}{2} \sum_{i=1}^{p} d_i\ s_i\ X_{i2}(1),$$

where r.v.s $X_{i2}(1)$, $i = 1, \ldots, p$, are independent copies of the r.v. $X_2(1)$ (or $X_2'(1)$, if A" does not hold necessarily), where $X_2(1)$ and $X_2'(1)$ are given in (2.4.3) and (2.4.6) respectively with $m = 2$ and $C_2/2 = 1$ in (2.4.6).

Proof

The function $G_2(u) = \chi(u \in \tilde{E}_2)$ satisfies condition F with coefficients $C_\nu = 0$, if $\nu = (k_1, \dots, k_p) : k_1 + \cdots + k_p = 1$, $k_j \geq 0$, and $C_\nu = 0$, if $\nu = (k_1, \dots, k_p) : k_1 + \cdots + k_p = 2$, $k_i = k_j = 1$, for some $i \neq j$. Using the relation

$$\int_{-a}^{a} H_2(y)\phi(y)dy = -2a\phi(a),$$

we obtain $C_{2,0\dots,0} = s_1$, $C_{0,2,0\dots,0} = C_{0,\dots,0,2} = s_2$.

Therefore $H\operatorname{rank} G_2 = 2$. Thus by Theorem 3.4.1. with $m = 2$ we complete the proof of the first part of our statement. Application of Theorem 2.4.1. (or Theorem 2.4.6.) to r.v. W_r' completes the proof.

Consider now the functional

$$T_3(r) = \int_{v_n(r)} \chi\left(\xi(x) \notin v_p(\sqrt{a(r)})\right) dx,$$

where the function $a(r)$ satisfies condition B.

Note that

$$k_* = \frac{1}{2}(\mu_2^2 - \mu_1^2) \geq 0, \tag{3.4.12}$$

where μ_j are defined in (3.4.2). We remark that $k_* = 0$ if and only if $\rho_0 = 0$, that is the vector field $\xi(x)$ has independent components. We shall show that the limiting distributions of the functional $T_3(r)$ are different for $k_* = 0$ and $k_* > 0$.

Note that

$$\tilde{E}_3(r) = R^p \setminus \left\{ y \in R^p : \sum_{j=1}^{p} \left(\frac{y_j}{\mu_j}\right)^2 < a(r) \right\}$$

and

$$ET_3(r) = |v_n(1)|\, r^n \int_{\tilde{E}_3(r)} \phi(\|u\|)du.$$

We shall show that for $k_* = 0$, $\alpha \in (0, n/2)$ and $r \to \infty$

$$\operatorname{var} T_3(r) = c_{13}(n, \alpha, p) e^{-a(r)} [a(r)]^p r^{2(n-\alpha)} L^2(r)(1 + o(1)), \tag{3.4.13}$$

where

$$c_{13}(n, \alpha, p) = (2\pi)^{-p} |v_n(1)|^2 c_7(n, 2, \alpha) \sum_{j=1}^{p} \frac{d_j^2}{4}.$$

On the other hand if $k_* > 0$, $\alpha \in (0, n/2)$ and $r \to \infty$

$$\operatorname{var} T_3(r) = c_{14}(n,\alpha,p)\ e^{-\mu_1^2 a(r)} r^{2(n-\alpha)} L^2(r)(1+o(1)), \tag{3.4.14}$$

where

$$c_{14}(n,\alpha,p) = \frac{1}{4}\mu_1^2\ \mu_2^{2(p-1)}(2\pi)^{-p}\left|v_n(1)\right|^2 \Gamma^2\left(\frac{p-1}{2}\right) k_*^{1-p} d_1^2 c_7(n,2,\alpha), \tag{3.4.15}$$

and k_* is defined in (3.4.12).

Theorem 3.4.4. Let assumptions B, E hold for $\alpha \in (0, n/2)$ and $a(r) = o(\log r)$, $r \to \infty$. Then

1) for $k_* = 0$ the limiting distributions of r.v.s

$$V_r = \frac{T_3(r) - ET_3(r)}{e^{-a(r)/2}[a(r)]^{p/2} r^{n-\alpha} L(r)\sqrt{c_{13}(n,\alpha,p)}}$$

and r.v.s

$$V_r' = \sum_{j=1}^{p} \frac{\int_{v_n(r)} H_2(\eta_j(x))dx}{r^{n-\alpha}L(r)\sqrt{c_7(n,2,\alpha)\sum_{j=1}^p d_j^2}}$$

coincide.

In particular, if every covariance function $B_{ij}(\|x\|)$ from condition E satisfies condition A", then the limiting distribution of V_r and V_r' is of the form

$$\frac{\sum_{j=1}^p d_j X_{j2}(1)}{\sqrt{c_7(n,2,\alpha)\sum_{j=1}^p d_j^2}},$$

where the r.v.s $X_{j2}(1)$, $j = 1, \dots p$, are independent copies of the r.v. $X_2(1)$ (or $X_2'(1)$ if A" does not necessarily hold), where $X_2(1)$ and $X_2'(1)$ are given in (2.4.3) and (2.4.6) respectively with $m = 2$ and $C_2/2 = 1$ in (2.4.6).

2) for $k_* > 0$ the limiting distributions of the r.v.s

$$V_r''' = \frac{T_3(r) - ET_3(r)}{\sqrt{a(r)}\ e^{-\mu_1^2 a(r)/2} r^{n-\alpha} L(r)\sqrt{c_{14}(n,\alpha,p)}}$$

and r.v.

$$V_r'' = \frac{\int_{v_n(r)} H_2(\eta_1(x))dx}{r^{n-\alpha}L(r)\sqrt{c_7(n,2,\alpha)}d_1}$$

coincide.

In particular, if every covariance function $B_{ij}(\|x\|)$ from condition E satisfies condition A", then limiting distribution of r.v. V_r''' and V_r'' is of the form $X_2(1)/\sqrt{c_7(n,\alpha,p)}$ or $X_2'(1)/\sqrt{c_7(n,\alpha,p)}$ if A" does not hold necessarily, where $X_2(1)$ and $X_2'(1)$ are defined in (2.4.3) and (2.4.6) respectively with $m=2$ and $C_2/2=1$ in (2.4.6).

In the proof of Theorem 3.4.4. we need the so called Watson's Lemma which we recall for the sake of completeness (see for example, Zorich (1984), p. 609).

Lemma 3.4.2. Let $a>0$, $b>0$, $f\in C[0,a]$. Then as $r\to\infty$

$$\int_0^a x^{b-1} f(x)\, e^{-rx^a} dx = \frac{f(0)}{a}\Gamma\left(\frac{b}{a}\right) r^{-b/a} + O(r^{-(b+1)/a})$$

if

$$f(x) = f(0) + O(x) \qquad \text{as} \quad x\to 0.$$

Proof

Consider the function

$$G_r(u) = \chi\left(\|u\|^2 > a(r)\right) \in \mathbf{L}_2(R^p, \phi(\|u\|)\, du), \quad u\in R^p$$

for every $r>0$. This function admits the expansion in the series (3.4.8) with coefficients

$$C_\nu(r) = \int_{\tilde{E}_3(r)} e_\nu(u) G_r(u)\phi(\|u\|)du.$$

Note that in the case $k_*=0$ we have $C_\nu(r)=0$, if $\nu=(k_1,\ldots,k_p)\in S_1$, that is, $k_1+\cdots+k_p=1$ or $\nu\in S_2$, but $k_i=k_j=1$ for some $i\neq j$. We obtain

$$\begin{aligned} C_{2,0,\ldots,0}(r) &= \int_{\tilde{E}_3(r)} H_2(u_1)\phi(\|u\|)du \\ &= |v_n(1)|\,(2\pi)^{-p/2} e^{-a(r)/2}[a(r)]^{p/2} \end{aligned}$$

and

$$\begin{aligned} C_{0,2,0,\ldots,0}(r) &= \cdots = C_{0,\ldots,0,2}(r) \\ &= C_{2,0,\ldots,0}(r). \end{aligned}$$

Thus $H\,\mathrm{rank}\, G_r(u) = 2$.

On the other hand, as $r \to \infty$

$$\int_{R^p} G_r^2(u)\phi(\|u\|)du \sim k_2 \, [a(r)]^{(p-1)/2} e^{-a(r)/2}.$$

Hence in (3.4.10)

$$\begin{aligned} \zeta_r &\sim k_2 \frac{[a(r)]^{p-1} e^{-a(r)/2}}{[a(r)]^p e^{-a(r)}} \\ &= k_2 \frac{e^{a(r)/2}}{a(r)} \\ &\leq k_3 \, r^{\varepsilon} \end{aligned}$$

for any $\varepsilon > 0$, by the assumption that $a(r) = o(\log r)$. Hence (3.4.10) is fulfilled. Consequently by Theorem 3.4.1., the r.v.

$$\frac{T_3(r) - ET_3(r)}{\sqrt{\operatorname{var} T_3(r)}}$$

has the same limiting distribution as

$$\frac{1}{2} C_{2,0,\ldots,0}(r) \frac{\sum_{j=1}^p \int_{v_n(r)} H_2(\eta_j(x))dx}{\sqrt{\operatorname{var} T_3(r)}},$$

where

$$\begin{aligned} \operatorname{var} T_3(r) &= \frac{1}{4} C_{2,0,\ldots,0}^2(r) \operatorname{var} \left[\sum_{j=1}^p \int_{v_n(r)} H_2(\eta_j(x)) \, dx \right] \\ &= c_7(n,2,\alpha) \left(\sum_{j=1}^p d_j^2 \right) \left(\frac{1}{4} C_{2,0,\ldots,0}^2(r) \right) r^{2(n-\alpha)} L^2(r)(1 + o(1)). \end{aligned}$$

Thus we have proved (3.4.13).

In particular, if A" holds for every covariance function $B_{ij}(\|x\|)$, then by Theorem 2.4.1. (or Theorem 2.4.6.) the limiting distribution of V, V' is of the form

$$\frac{\sum_{j=1}^p d_j X_{i2}(1)}{\left\{ c_7(n,2,\alpha) \left(\sum_{j=1}^p d_j^2 \right) \right\}^{1/2}},$$

where r.v.s $X_{i2}(1)$, $i = 1, \ldots, p$, are independent copies of r.v. $X_2(1)$ given in (2.4.3) with $m = 2$.

In the case $k_* > 0$ we have $C_\nu(r) = 0$, if $\nu \in S_1$ or $\nu \in S_2$, but $k_i = k_j = 1$ for some $i \neq j$, where $\nu = (k_1, \ldots, k_p)$. For $\nu = (2, 0, \ldots, 0)$ we obtain

$$\begin{aligned} C_{2,0,\ldots,0}(r) &= \int_{\tilde{E}_3(r)} H_2(u_1)\phi(\|u\|)du \\ &= -2\mu_1\mu_2^{p-1} \int\limits_{y_2^2+\cdots+y_p^2 \leq a(r)} \prod_{j=2}^{p} \phi(\mu_2 y_j)dy_j \\ &\quad \times \int^{\sqrt{a(r)-(y_2^2+\cdots+y_p^2)}} H_2(\mu_1 y_1)\phi(\mu_1 y_1)dy_1 \\ &= 2(2\pi)^{-p/2}\mu_1\ \mu_2^{p-1} \\ &\quad \times \int\limits_{y_2^2+\cdots+y_p^2 \leq a(r)} \exp\left\{-\mu_2^2\, \frac{y_2^2+\cdots+y_p^2}{2}\right\} \\ &\quad \times \exp\left\{-\mu_1^2\, \frac{a(r)-(y_2^2+\cdots+y_p^2)}{2}\right\} \\ &\quad \times \sqrt{a(r)-(y_2^2+\cdots+y_p^2)}\ dy_2\ldots dy_p \\ &= 2(2\pi)^{-p/2}\mu_1\mu_2^{p-1} \\ &\quad \times \exp\left\{-\mu_1^2\ a(r)\ /\ 2\right\} |v_p(1)|\ [a(r)]^{p/2} \\ &\quad \times \int_0^1 s^{p-2}e^{-s^2 k_* a(r)}\sqrt{1-s^2}ds \\ &\sim (2\pi)^{-p/2}\mu_1\ \mu_2^{p-1}k_*^{(1-p)/2} \\ &\quad \times \sqrt{a(r)}e^{-\mu_1^2 a(r)/2}\, |v_p(1)|\, \Gamma\left(\frac{p-1}{2}\right) \end{aligned}$$

by Lemma 3.4.2 as $r \to \infty$.

Similarly

$$\begin{aligned} C_{0,2,0\ldots,0}(r) &= \cdots = C_{0,\ldots,02}(r) \\ &\sim (2\pi)^{-p/2}\mu_1\mu_2^{p-1}k_*^{(1-p)/2} \\ &\quad \times \sqrt{a(r)}e^{-\mu_2^2 a(r)/2}\, |v_p(1)|\, \Gamma\left(\frac{p-1}{2}\right) \end{aligned}$$

as $r \to \infty$.

Note that for $k_* > 0$,

$$\begin{aligned}\frac{C_{0,2,0\ldots,0}(r)}{C_{2,0,\ldots,0}(r)} &= e^{-k_* a(r)} \\ &\to 0 \qquad \text{as } r \to \infty.\end{aligned}$$

Therefore

$$\begin{aligned}\operatorname{var} T_3(r) &= \frac{1}{4}\left\{C^2_{2,0,\ldots,0}(r) \operatorname{var}\left[\int_{v_n(r)} H_2(\eta_1(x))dx\right]\right. \\ &\quad \left. + C^2_{0,2,0,\ldots,0}(r) \operatorname{var}\left[\sum_{j=2}^{p}\int_{v_n(r)} H_2(\eta_j(x))dx\right]\right\} \\ &= \frac{1}{4}\, C^2_{2,0,\ldots,0}(r) c_7(n,2,\alpha) d_1^2\, r^{2(n-\alpha)} L^2(r)(1+o(1))\end{aligned}$$

as $r \to \infty$. Thus we have proved (3.4.14).

On the other hand for (3.4.10) we have

$$\begin{aligned}\zeta_r &\sim \frac{k_4\,[a(r)]^{p-1} e^{-a(r)/2}}{a(r)\; e^{-\mu_1^2 a(r)}} \\ &= k_5 a(r) e^{a(r)[1/2-\mu_1^2]]} \\ &\leq k_6 r^{-q}\end{aligned}$$

for any $\varepsilon > 0$. Hence, (3.4.10) is fulfilled.

Consequently by Theorem 3.4.1. r.v.s V_r''' have the same limiting distributions as

$$\frac{\int_{v_n(r)} H_2(\eta_1(x))dx}{r^{n-\alpha} L(r)\sqrt{c_7(n,2,\alpha)}d_1} \tag{3.4.16}$$

as $r \to \infty$. If every covariance function $B_{ij}(\|x\|)$ satisfies condition A", then by Theorem 2.4.1. (or Theorem 2.4.6.) the r.v.s (3.4.16) has limiting distribution of the form $X_2(1)$ (or $X_2'(1)$).

Remark 3.4.3. All non-Gaussian limiting distributions in Theorems 3.4.3.–3.4.4. can be given also in term of a multiple stochastic integral with dependent integrator (see Fox and Taqqu (1987) and Sanchez (1993). In the case $n = 1$ some of these representations can be found in Taqqu (1986) and Maejima (1985,1986a,b)).

3. Spherical sojorn measures

Under assumptions B, E consider the functionals

$$\begin{aligned} A_i(r) &= \sigma\{x \in s_{n-1}(r) : \xi(x) \in E_i\} \\ &= \int_{s_{n-1}(r)} \chi\left(\xi(x) \in E_i\right) d\sigma(x), \end{aligned}$$

$i = 1, 2, 3$, where $s_{n-1}(r) \subset R^n$, $n \geq 2$, is a sphere in R^n, $d\sigma(x)$ is an element of the Lebesgue measure on the sphere $s_{n-1}(r)$ defined in (1.1.4), and

$$\begin{aligned} E_1 &= \{x \in R^p : a_i \leq y_i < b_i, \ i = 1, \dots, p\}, \\ E_2 &= \{x \in R^p : \|y\| < a\}, \\ E_3 &= E_3(r) = R^p \backslash v_p(\sqrt{a(r)}). \end{aligned}$$

Consider the sets

$$\tilde{E}_i = \{y \in R^p : T^{-1}y \in E_i\}, \qquad i = 1, 2, 3,$$

where the matrix T^{-1} is defined in remark 3.4.1., and the constants

$$\kappa_{ij} = \int_{\tilde{E}_i} H_i(y_j)\phi(\|y\|)dy, \quad i = 1, 2; \quad j = 1, \dots, p.$$

Theorem 3.4.5. Let assumptions E hold for $\alpha \in (0, n-1)$, $n \geq 2$. Then the r.v.s

$$\frac{A_1(r) - EA_1(r)}{\sqrt{\operatorname{var} A_1(r)}}$$

have asymptotically $(0, 1)$-normal distribution, and

$$\begin{aligned} EA_1(r) &= |s_{n-1}(1)| \, r^{n-1} \int_{\tilde{E}_1} \phi(\|u\|)du, \\ \operatorname{var} A_1(r) &= c_9(n, 1, \alpha) \left[\sum_{j=1}^{p} d_j \, (\kappa_{1j})^2 \right] r^{2n-2-\alpha} L^{1/2}(r)(1 + o(1)), \end{aligned}$$

where the constant $c_9(n, 1, \alpha)$ is defined in (2.2.24).

Theorem 3.4.6. If condition E is satisfied and $\alpha \in (0, n-1)$, $n \geq 2$, then the limiting distributions of r.v.s

$$\frac{A_2(r) - EA_2(r)}{\sqrt{\operatorname{var} A_2(r)}}$$

and r.v.s

$$\frac{\sum_{j=1}^{p} \kappa_{2j} \int_{s_{n-1}(r)} H_2(\eta_j(x))d\sigma(x)}{r^{n-1-\alpha}L(r)\sqrt{c_9(n,2,\alpha)\sum_{j=1}^{p} d_j^2\kappa_{2j}^2}} \tag{3.4.17}$$

coincide, where

$$\begin{aligned} EA_2(r) &= |s_{n-1}(1)|\ r^{n-1} \int_{\tilde{E}_2} \phi(\|u\|)du, \\ \operatorname{var} A_2(r) &= c_9(n,2,\alpha)\left[\sum_{j=1}^{p} d_j^2\kappa_{2j}^2\right] r^{2n-2-2\alpha}L^2(r)(1+o(1)). \end{aligned}$$

In particular, if A" holds for every covariance function $B_{ij}(\|x\|)$ from condition E, then the limiting distributions of (3.4.17) are of the form

$$\frac{\sum_{j=1}^{p} d_j\kappa_{2j}Y_{j2}(1)}{\sqrt{c_9(n,2,\alpha)\sum_{j=1}^{p} d_j^2\kappa_{2j}^2}},$$

where the r.v.s $Y_{j2}(1)$, $j = 1, \ldots, p$, are independent copies of r.v. $Y_2(1)$ given in (2.4.5) with $m = 2$ and $C_m = 1$, and $c_9(n,2,\alpha)$ defined in (2.2.24).

Theorem 3.4.7. Let assumptions B, E hold for $\alpha \in (0,(n-1)/2)$, $n \geq 2$, and $a(r) = o(\log r)$ as $r \to \infty$. Then

1) for $k_* = 0$ the limiting distributions of r.v.s

$$\frac{A_3(r) - EA_3(r)}{e^{-a(r)/2}[a(r)]^{p/2}r^{n-1-\alpha}L(r)\sqrt{c_{15}(n,\alpha,p)}} \tag{3.4.18}$$

and r.v.s

$$\frac{\sum_{j=1}^{p} \int_{s_{n-1}(r)} H_2(\eta_j(x))d\sigma(x)}{r^{n-1-\alpha}L(r)\sqrt{c_9(n,2,\alpha)\sum_{j=1}^{p} d_j^2}} \tag{3.4.19}$$

coincide, where

$$c_{15}(n,\alpha,p) = \frac{1}{4}(2\pi)^{-p}\,|v_p(1)|^2\,c_9(n,2,\alpha)\sum_{j=1}^{p} d_j^2.$$

In particular, if A" holds for every $B_{ij}(\|x\|)$, then the limiting distribution of r.v.s (3.4.18) and (3.4.19) is of the form

$$\frac{\sum_{j=1}^{p} d_j\ Y_{2j}(1)}{\sqrt{c_9(n,2,\alpha)\sum_{j=1}^{p} d_j^2}},$$

where the r.v.s $Y_{2j}(1)$, $j = 1, \ldots, p$, are independent copies of the r.v. $Y_2(1)$ defined in (2.4.5) with $m = 2$ and $C_m = 1$.

2) for $k_* > 0$ the limiting distributions of the r.v.s

$$\frac{A_3(r) - EA_3(r)}{e^{-\mu_1^2 a(r)/2}\sqrt{a(r)}r^{n-1-\alpha}L(r)\sqrt{c_{16}(n,\alpha,p)}} \tag{3.4.20}$$

and r.v.s

$$\frac{\int_{s_{n-1}(r)} H_2(\eta_1(x))d\sigma(x)}{r^{n-1-\alpha}L(r)\sqrt{c_9(n,2,\alpha)}d_1} \tag{3.4.21}$$

coincide, where

$$c_{16}(n,\alpha,p) = \frac{1}{4}\mu_1^2\mu_2^{2(p-1)}(2\pi)^{-p}\left|v_p(1)\right|^2\Gamma^2\left(\frac{p-1}{2}\right)k_*^{1-p}d_1^2 c_9(n,2,\alpha).$$

In particular, if A" holds for every $B_{ij}(\|x\|)$, then the limiting distribution of the r.v.s (3.4.20) and (3.4.21) is of the form $Y_2(1)\sqrt{c_9(n,2,\alpha)}$, where the r.v. $Y_2(1)$ is defined in (2.4.5) with $m = 2$ and $C_m = 1$.

The proofs of Theorems 3.4.5. -3.4.7. are analogous to that of Theorems 3.4.1–3.4.4. (see Leonenko and Parkhomenko (1992)).

3.5 Asymptotic Normality of Random 'Area of Surface' of Planar Gaussian Field

We shall examine the limiting distribution of the 'area of surface' of a planar Gaussian field. The conditions of asymptotic normality of the 'length of realization' of the stationary Gaussian process presented in a book of Ivanov and Leonenko (1989), p. 82–83.

K. Let $\xi(x)$, $x = (x_1, x_2) \in R^2$, be a homogeneous isotropic mean-square differentiable Gaussian random field with

$$E\xi(x) = 0, \quad E\xi^2(x) = 1$$

and covariance function

$$B(\|x\|) = \int_0^\infty J_0(\rho\,\|x\|)\; G(d\rho),$$

where $G(u)$ is isotropic spectral measure (see (1.2.15)) and $J_\nu(z)$ is a Bessel function defined in (1.2.12).

Define

$$S(r) = \int_{v(r)} \sqrt{1 + \dot{\xi}^2_{x_1}(x) + \dot{\xi}^2_{x_2}(x) dx_1 dx_2},$$

where

$$\dot{\xi}_{x_i}(x) = \frac{\partial}{\partial x_i}\xi(x), \quad i = 1, 2.$$

Remark 3.5.1 . If $\xi(x)$ have differentiable sample paths, then $S(r)$ is a 'area of surface' of a realization of $\xi(x)$, $x \in R^2$, over a circle $v(r) = \{x \in R^2 : x_1^2 + x_2^2 < r\}$.

Note that random fields $\dot{\xi}_{x_1}(x)$, $x \in R^2$, and $\dot{\xi}_{x_2}(x)$, $x \in R^2$, are homogeneous (see section 1.2.) with covariance functions of the form

$$\begin{aligned} E\,\dot{\xi}_{x_1}(0)\,\dot{\xi}_{x_1}(x) &= -\ddot{B}_{x_1x_1}(x) \\ &= \|x\|^{-1} \int_0^\infty \rho\, J_1(\rho\,\|x\|)\, G(d\rho) \\ &\quad - \|x\|^{-2} x_1^2 \int_0^\infty \rho^2 J_2(\rho\,\|x\|)\, G(d\rho), \end{aligned}$$

$$\begin{aligned} E\,\dot{\xi}_{x_2}(0)\,\dot{\xi}_{x_2}(x) &= -\ddot{B}_{x_2x_2}(x) \\ &= \|x\|^{-1} \int_0^\infty \rho\, J_1(\rho\,\|x\|)\, G(d\rho) \\ &\quad - \|x\|^{-2} x_2^2 \int_0^\infty \rho^2 J_2(\rho\,\|x\|)\, G(d\rho), \end{aligned}$$

$$\begin{aligned} E\dot{\xi}_{x_1}(0)\,\dot{\xi}_{x_2}(x) &= -\ddot{B}_{x_1x_2}(x) \\ &= E\dot{\xi}_{x_2}(0)\,\dot{\xi}_{x_1}(x) \\ &= -\ddot{B}_{x_2x_1}(x) \\ &= -x_1\, x_2\, \|x\|^{-2} \int_0^\infty \rho^2 J_2(\rho\,\|x\|)\, G(d\rho). \end{aligned}$$

Let

$$\sigma^2 = -\ddot{B}_{x_1x_1}(0) = -\ddot{B}_{x_2x_2}(0).$$

The random field

$$\nu(x) = \dot{\xi}^2_{x_1}(x) + \dot{\xi}^2_{x_2}(x), \ x \in R^2,$$

is homogeneous and isotropic with covariance function

$$\begin{aligned} B_\nu(\|x\|) &= E\ \nu(0)\ \nu(x) \\ &= 2\left(\left\{\int_0^\infty \|x\|^{-1}\rho J_1(\rho\|x\|)G(d\rho)\right\}^2\right. \\ &\quad \left.+\left\{\int_0^\infty \left[\rho^2 J_0(\rho\|x\|) - \rho\|x\|^{-1}\rho J_1(\rho\|x\|)\right]G(d\rho)\right\}^2\right). \end{aligned}$$

L. Suppose that

$$\int_{R^2} |B_\nu(\|x\|)|\, dx < \infty. \tag{3.5.1}$$

Remark . 3.5.2. Let condition K hold and

$$B(\|x\|) = \frac{L(\|x\|)}{\|x\|^\alpha}, \quad \alpha > 0, \quad L \in \mathcal{L} \qquad \text{for} \quad \|x\| \to \infty.$$

Suppose that (1.2.24) holds with $n = 2$, that is, there exists an isotropic spectral density

$$g(\rho) : G'(p) = 2\pi g(\rho)\rho, \quad \rho g(\rho) \in \mathbf{L}_1([0,\infty)),$$

such that

$$\rho(\rho) = \rho^{\alpha-2} f(\rho), \quad \alpha > 0,$$

where

a) $f(\rho)$ is bounded at zero ;
b) $f^{(m)}(\rho)$ is continuous on $(0,\infty)$ for $m = [\alpha] + 3$;
c) $(\rho^{i+\alpha} f(\rho))^{(j)} \rho^{-1/2} \to 0$, as $\rho \to 0$ for $j = 0, 1, \ldots, m-1, \quad i = 0, 1$;
d) $\int_0^\infty J_{i+m}(u\rho)\rho^{2-i}(\rho^{\alpha-1}f(\rho))^{(m)}(d\rho) < \infty$, $i = 1, 2$, and the convergence of integrals are uniform for $u > u_0 > 0_0$.

Then (see Wong (1986)) the covariance function

$$B_\nu(\|x\|) = O(\|x\|^{-2\alpha-4}) + o(\|x\|^{-2\alpha-4}) \quad \text{as} \quad \|x\| \to \infty,$$

and condition (3.5.1) is satisfied. Note that for $0 < \alpha < 2$ the random field $\xi(x)$, $x \in R^n$, has singular spectrum (see Theorem 1.4.4.).

It is easy to see that

$$ES(r) = \pi r^2 \sigma e^{(1/2\sigma^2)} \sqrt{2\pi} \int_{1/\sigma}^{\infty} \frac{4}{\sqrt{2\pi}} e^{-(u^2/2)} du.$$

Let

$$\begin{aligned} \lambda_1 &= B_{11}(\|x\|) \\ &= \|x\|^{-1} \int_0^{\infty} \rho J_1(\rho \|x\|) G(d\rho), \end{aligned} \tag{3.5.2}$$

$$\begin{aligned} \lambda_2 &= B_{22}(\|x\|) \\ &= \int_0^{\infty} \left[\rho^2 J_0(\rho \|x\|) - \rho \|x\|^{-1} \rho J_1(\rho \|x\|)\right] G(d\rho), \end{aligned} \tag{3.5.3}$$

be a roots of the equation

$$\det |B_\eta - \lambda\, I| = 0,$$

where I is a unique matrix, and

$$B_\eta = \begin{pmatrix} -\ddot{B}_{x_1 x_1} & -\ddot{B}_{x_1 x_2} \\ -\ddot{B}_{x_2 x_1} & -\ddot{B}_{x_2 x_2} \end{pmatrix}$$

is a covariance matrix of a vector homogeneous isotropic Gaussian field

$$\eta(x) = \nabla \xi(x) = (\dot{\xi}_{x_1}(x),\ \dot{\xi}_{x_2}(x))',$$

where ∇ is the gradient.

Let us denote

$$B_i(\|x\|) = \frac{B_{ii}(\|x\|)}{\sigma^2}, \quad i = 1, 2,$$

where $B_{ii}(\|x\|)$, $i = 1, 2$, are defined in (3.5.2) and (3.5.3).

M. Suppose that

$$0 < \int_0^{\infty} \rho^2 G(d\rho) < \infty, \qquad \int_0^{\infty} u\, B_i^2(u) du < \infty, \quad i = 1, 2,$$

where G is a spectral measure of a field, and B_i, $i = 1, 2$, are defined by (3.5.2) and (3.5.3).

Consider the coefficients (3.4.2) for a function

$$\bar{G}(u) = \sigma\sqrt{\sigma^{-2} + u_1^2 + u_2^2}, \quad u = (u_1, u_2) \in R^2,$$

that is,

$$\bar{C}_{k,l} = \int_{R^2} \bar{G}(u)\, e_{k,l}(u)\phi(\|u\|)du, \quad k \geq 0, \quad l \geq 0$$

where

$$e_{k,l}(u) = H_k(u_1)H_l(u_2),$$

and the functions

$$\begin{aligned} \sigma_{k,l}^2(r) &= 2k!l!\pi^2 r^2 \int_0^{2r} u I_{1-(u/2r)^2}\left(\frac{3}{2}, \frac{1}{2}\right) B_1^k(u) B_1^l(u) du, \\ k, l &\geq 1, \quad k + l \geq 2, \end{aligned}$$

where (see remark 1.4.3.)

$$I_{1-(u/2r)^2}\left(\frac{n+1}{2}, \frac{1}{2}\right) = \frac{2}{\pi}\left\{\arccos\left(\frac{u}{2r}\right) - \left(\frac{u}{2r}\right)\sqrt{1 - \left(\frac{u}{2r}\right)^2}\right\}.$$

Under assumptions K, L, M

$$c_{17} = \sum_{k,l \geq 0, k+l \geq 2} \frac{\bar{C}_{k,l}^2\, \psi(k,l)}{k!l!} \in (0, \infty),$$

where

$$\psi(k,l) = \lim_{r \to \infty} \sigma_{k,l}^2(r)\, r^{-2}, \quad k, l \geq 2,$$

and we have for $r \to \infty$

$$\text{var}\, S(r) = c_{17}\, r^2(1 + o(1)).$$

Theorem 3.5.1. Let assumptions K, L, M hold. Then the random variable

$$\frac{S(r) - ES(r)}{r\sqrt{c_{17}}}$$

has asymptotically (0, 1)-normal distribution as $r \to \infty$.

The proof of Theorem 3.5.1. is based on the procedure of orthogonalization of a vector Gaussian field $\eta(x) = \nabla\xi(x)$ and diagrammatic formalism (see Ivanov and Leonenko (1989), section 2.3). Theorem 3.5.1. is proved in a paper of Leonenko and Rarkhomenko (1990).

Remark 3.5.3. Note that $\nu(x)$, $x \in R^2$, is a χ-squared random field (see example 2.1.3.)). From (2.1.13) we obtain

$$\begin{aligned} c_{17} &= 2\pi^2\sigma^2 \int_0^\infty z \left[\int_0^\infty \int_0^\infty \sqrt{\sigma^{-2}+2u}\sqrt{\sigma^{-2}+2v}\,(1-B_\nu(z))^{-1} \right. \\ &\qquad \left. \times \exp\left\{-\frac{u+v}{1-B_\nu(z)}\right\} I_0\left(\frac{2\sqrt{uvB_\nu(z)}}{1-B_\nu(z)}\right) du dv \right] dz, \end{aligned}$$

where

$$B_\nu(\|x\|) = \operatorname{cov}(\nu(0),\ \nu(x)), \quad x \in R^2,$$

and $I_0(z)$ is modified Bessel function defined in (2.1.12).

3.6 Asymptotics for Occupation Densities of Gaussian and χ-Squared Random Fields

We generalize to random fields the results on limit distributions for local times of Gaussian stochastic processes with long-range dependence obtained by Berman (1982). We also present limit distributions for occupation densites of χ-squared random fields.

The occupation density of a real measurable random field $\xi(x)$, $x \in R^n$, is defined (see, for example, Geman and Horowitz (1980), Berman (1982)) as the Radon–Nikodym derivative of the random measure

$$\begin{aligned} \nu_T(A) &= |\{x \in T : \xi(x) \in A\}| \\ &= \int_T \chi\left(\xi(x) \in A\right) dx, \end{aligned}$$

where $T \in \mathcal{B}(R^n)$, $A \in \mathcal{B}(R^1)$, $|\cdot|$ is the Lebesgue measure on the measurable space $(R^n, \mathcal{B}(R^n))$, $\chi(\cdot)$ is the indicator function (under the condition that the derivative exists).

The occupation density with respect to a set $T \in \mathcal{B}(R^n)$ will be denoted by $l_T(t)$. According to the definition

$$\nu_T(A) = \int_A l_T(t)\,dt \qquad \text{a.s.}$$

We invesigate the limit distributions of the occupation density $l_{v(r)}$ (with respect to a ball $v(r) \subset R^n$) as $r \to \infty$ for some classes of random fields.

Introduce the following conditions.

V . There exists $\delta \in (0,1)$ and an interger $m \geq 1$ such that

$$\lim_{r\to\infty} r^{-n(1+\delta)}\sigma_m^2(r) = \infty,$$

where

$$\begin{aligned}\sigma_m^2(r) &= \int_{v(r)}\int_{v(r)} B^m(\|x-y\|)dxdy \\ & c_{10}(n,m)r^n \int_0^{2r} z^{n-1}B^m(z)I_{1-(z/2r)^2}\left(\frac{n+1}{2},\frac{1}{2}\right)dz,\end{aligned}$$

where $c_{10}(n,m)$ is defined in (3.1.2).

W . For any $r > 0$

$$\int_{v(r)}\int_{v(r)} (1-B(\|x-y\|))^{-1/2}\,dxdy < \infty.$$

Under the condition W the local time $l_{v(r)}(t)$ of the Gaussian field $\xi(x)$, $x \in R^n$, described in the condition A exists and is integrable in $L_2(\lambda_1 \times P)$ (see, for example, Geman and Horowitz(1980)), where $L_2(\lambda_1 \times P)$ is a Hilbert space of integrable with square functions on the measure $\lambda_1 \times P$ (λ_1 is one dimensional Lebesgue measure).

The extension of results in Berman(1982) yields that under the condition W the occupation density $l_{v(r)}(t)$ of the field $\xi(x)$ has the following expansion in $L_2(\Omega)$:

$$l_{v(r)}(t) = \phi(t)\sum_{k=0}^{\infty}(k!)^{-1}H_k(t)\int_{v(r)} H_k(\xi(x))dx, \tag{3.6.1}$$

where $H_k(t)$, $k = 0,1,\ldots$, are Chebyshev–Hermite polynomials defined in (2.1.5).

Asymptotic analysis of the expansion (3.6.1) leads to the following results.

Theorem 3.6.1. Let $\xi(x)$, $x \in R^n$, be a Gaussian random field satisfying the conditions A, V under $m = 1$ and W.

Then the finite-dimensional distributions of the stochastic processes

$$\frac{l_{v(r)}(t) - \phi(t)|\ v(r)|}{\sigma_1(r)}, \quad t \neq 0,$$

converge weakly, as $r \to \infty$, to those of the process

$$t\phi(t)\mathcal{N}, \quad t \neq 0,$$

where $\mathcal{N}$ is a standard Gaussian r.v., and $\phi(t)$ is defined by (3.1.6).

Theorem 3.6.2. Let under the conditions of Theorem 3.6.1. condition V holds with $m = 2$.

Then the limit distributions of the r.v.

$$\frac{l_{v(r)}(0) - \phi(0)|v(r)|}{\sigma_2(r)}$$

and

$$-\frac{\phi(0)\int_{v(r)} H_2(\xi(x))dx}{\sigma_2(r)}$$

coincide as $r \to \infty$.

The proofs of these results one can obtain using the same scheme as in the work of Berman (1982).

Observe that in the expansion (3.6.1) random variables $\int_{v(r)} H_k(\xi(x))\,dx$, $k \geq 1$, have zero mean, for differnt k they are orthogonal in pairs, and

$$\mathrm{E}\left(\int_{v(r)} H_k(\xi(x))dx\right)^2 = k! \int_{v(r)} \int_{v(r)} B^k(\|x - y\|)dxdy. \tag{3.6.2}$$

From the expansion (3.6.1) one can obtain first two moments of the process $l_{v(r)}(t)$. They are of the form:

$$\mathrm{E}l_{v(r)}(t) = \phi(t)|v(r)|, \tag{3.6.3}$$

$$\mathrm{E}l_{v(r)}(t)l_{v(r)}(s) = \int_{v(r)} \int_{v(r)} \phi(s, t; x, y)\,dxdy, \tag{3.6.4}$$

where

$$\begin{aligned}\phi(s,t;x,y) &= \phi(s,t;B(\|x-y\|)) \\ &= \phi(s)\phi(t)\sum_{k=0}^{\infty}\frac{B^k(\|x-y\|)H_k(t)H_k(s)}{k!}\end{aligned}$$

is the two-dimensional Gaussian density (2.1.7).

For $t \neq 0$ represent the expansion (3.6.1) in the form

$$\begin{aligned}l_{v(r)}(t)-\phi(t)|v(r)| &= \phi(t)t\int_{v(r)}\xi(x)dx \\ &\quad +\phi(t)\sum_{k=2}^{\infty}(k!)^{-1}H_k(t)\int_{v(r)}H_k(\xi(x))dx \\ &= S_1(r)+S_2(r). \end{aligned} \tag{3.6.5}$$

To prove the assertion of Theorem 3.6.1. it is sufficient to show that

$$\begin{aligned}&\overline{\lim}_{r\to\infty}\left\{\int_{v(r)}\int_{v(r)}\left(\phi(t,t;B(\|x-y\|))-\phi^2(t)\right)\,dx\,dy\right\} \\ &\times\left\{\phi^2(t)\,t^2\int_{v(r)}\int_{v(r)}B(\|x-y\|)\,dx\,dy\right\}^{-1} \\ \leq\ & 1. \end{aligned} \tag{3.6.6}$$

Really, from (3.6.3) and (3.6.4) we have that $\operatorname{var} l_{v(r)}(t)$ is equal to the first factor in the expression under limit sign in (3.6.6). From the other hand, by (3.6.2) and (3.6.5) and orthogonality of the r.v. $\int_{v(r)} H_k(\xi(x))\,dx$ we conclude that

$$\operatorname{var} l_{v(r)}(t)=\phi^2(t)t^2\int_{v(r)}\int_{v(r)}B(\|x-y\|)\,dx\,dy+\mathrm{E}S_2^2(r).$$

From the last relation and (3.6.6) we derive:

$$\operatorname{var} l_{v(r)}(t)\left\{\phi^2(t)t^2\int_{v(r)}\int_{v(r)}B(\|x-y\|)dxdy\right\}^{-1}=1,$$

as $r\to\infty$, and, hence, $S_2(r)/\sigma(r)\to 0$ in probability.

Dividing now both sides of (3.6.5) by

$$\left\{\int_{v(r)}\int_{v(r)}B(\|x-y\|)\,dx\,dy\right\}^{1/2}$$

we obtain the assertion of Theorem 3.6.5.

Let us show that the relation (3.6.6) holds.

Using the formula (1.4.7) we can write

$$\begin{aligned}
&\int_{v(r)}\int_{v(r)} \left(\phi(t,t;B(\|x-y\|)) - \phi^2(t)\right)\,dxdy \\
&= c_{10}(n,1)\,r^n \int_0^{2r} z^{n-1}\left(\phi(t,t;B(z)) - \phi^2(t)\right) \\
&\quad \times I_{1-(z/2r)^2}\left(\frac{n+1}{2},\frac{1}{2}\right)\,dz.
\end{aligned}$$

Represent the last expression as a sum:

$$\begin{aligned}
I_1 + I_2 + I_3 &= (2\pi)^{-1} c_{10}(n,1) \\
&\quad \times r^n \left[\int_0^a + \int_a^{(2r)^\delta} + \int_{(2r)^\delta}^{2r}\right] z^{n-1} \\
&\quad \times \left\{\exp\left\{\frac{-t^2}{1+B(z)}\right\}(1-B^2(z))^{-1/2} - \exp\left\{-t^2\right\}\right\} \\
&\quad \times I_{1-(z/2r)^2}\left(\frac{n+1}{2},\frac{1}{2}\right)\,dz.
\end{aligned}$$

We have by condition W

$$\begin{aligned}
I_1 &\le (2\pi)^{-1} c_{10}(n,1)\,r^n \Big[\exp\left\{-t^2/2\right\} \\
&\quad \times \int_0^a z^{n-1}\left(1-B^2(z)\right)^{-1/2} I_{1-(z/2r)^2}\left(\frac{n+1}{2},\frac{1}{2}\right)\,dz \\
&\quad - \exp\left\{-t^2\right\}\,a^n/n\Big] \\
&= k_1\,r^n.
\end{aligned}$$

Consider now I_2. Let

$$B(z) \le B_0 \qquad \text{for } a \le z \le (2r)^\delta.$$

Using the relation

$$\int_0^{(2r)^\delta} z^{n-1} I_{1-(z/2r)^2}\left(\frac{n+1}{2},\frac{1}{2}\right)\,dz \le k_2\,r^{\delta n},$$

we obtain

$$I_2 \le k_3\ r^{n(1+\delta)}.$$

By condition V we conclude

$$\frac{I_1}{\sigma_1^2(r)} \to 0 \quad \text{and} \quad \frac{I_2}{\sigma_1^2(r)} \to 0 \quad \text{as} \quad r \to \infty.$$

In order to estimate I_3 apply the expansion

$$\exp\left\{-\frac{t^2}{1+z}\right\} = \exp\{-t^2\} + t^2 \exp\{-t^2\}\ z + o(z), \quad z \to 0$$

and the relation

$$(1 - B^2(z))^{-1/2} \sim 1 \qquad \text{as } z \to \infty.$$

We have

$$\begin{aligned} I_3 \ &\le\ (2\pi)^{-1} c_{10}(n,1)\ r^n t^2 \exp\{-t^2\} \\ &\times \int_{(2r)^\delta}^{2r} B(z) z^{n-1} I_{1-(z/2r)^2}\left(\frac{n+1}{2}, \frac{1}{2}\right) dz, \end{aligned}$$

and, hence,

$$\frac{I_3}{t^2\phi^2(t)\sigma_1^2(r)} \le 1, \quad r \to \infty.$$

Therefore, the relation (3.6.6) holds and Theorem 3.6.1. is proved. Theorem 3.6.2. can be obtained by the same reasoning.
For $t = 0$ the local time has the representation

$$\begin{aligned} l_{v(r)}(0) - \phi(0)|v(r)| \ &=\ -\phi(0) \int_{v(r)} H_2(\xi(x))dx \\ &+ \phi(0) \sum_{k=3}^{\infty} (k!)^{-1} H_k(t) \int_{v(r)} H_k(\xi(x))dx. \end{aligned}$$

We have to verify the relation

$$\overline{\lim_{r\to\infty}} \frac{\int_{v(r)}\int_{v(r)} \left((1 - B^2(\|x-y\|))^{-1/2} - 1\right) dxdy}{\frac{1}{2}\int_{v(r)}\int_{v(r)} B^2(\|x-y\|)dxdy} \le 1. \tag{3.6.7}$$

Analogously to the above we can write

$$\begin{aligned}\int\limits_{v(r)}\int\limits_{v(r)}(1-B^2(\|x-y\|))^{-1/2}-1)dxdy &= I_1+I_2+I_3\\ &= c_{10}(n,1)\,r^n\left[\int\limits_0^a+\int\limits_a^{(2r)^\delta}+\int\limits_{(2r)^\delta}^{2r}\right]z^{n-1}\\ &\quad\times((1-B^2(z))^{-1/2}-1)\\ &\quad\times I_{1-(z/2r)^2}\left(\frac{n+1}{2},\frac{1}{2}\right)dz.\end{aligned}$$

×

The integrals I_1 and I_2 can be estimated in the same manner as it was done before.

To estimate I_3 apply the relation

$$(1-x^2)^{-1/2}-1\sim\frac{x^2}{2},\quad x\to 0.$$

We have

$$(1-B^2(z))^{-1/2}-1\sim\frac{B^2(z)}{2},\quad z\to\infty,$$

and, hence, as $r\to\infty$

$$I_3\left\{\frac{1}{2}\int_{v(r)}\int_{v(r)}B^2(\|x-y\|)\,dx\,dy\right\}^{-1}\leq 1.$$

Therefore the relation (3.6.7) is satisfied and Theorem 3.6.2. is proved.

Applying the results of Section 2.4 we can describe the limiting distribution of $l_{v(r)}(0)$ more precisely.

Theorem 3.6.2′. Let $\xi(x)$, $x\in R^n$, be a Gaussian random field satisfying the conditions A", W and $\alpha\in(0,n/2)$.

Then the distributions of the random variables

$$r^{\alpha-n}L^{-1}(r)\left(l_{v(r)}(0)-\phi(0)\,|v(r)|\right)$$

converge to that of the random variable

$$-(2\pi)^{(n-1)/2}c_5^{-1}(n,\alpha)\int_{R^{2n}}'\frac{J_{n/2}(\|\lambda_1+\lambda_2\|)}{\|\lambda_1+\lambda_2\|^{n/2}}\frac{W(d\lambda_1)W(d\lambda_2)}{\|\lambda_1\|^{(n-\alpha)/2}\|\lambda_2\|^{(n-\alpha)/2}},$$

where the constant $c_5(n,\alpha)$ is given in (1.4.13) (see, also, (2.4.2)).

The study of the limiting distributions of occupation density of χ-squared random fields is based on the same technique, i.e., on the asymptotic analysis of orthogonal expansions of local times.

Suppose, that

C' . The field $\eta_d(x)$, $x \in R^n$, is a χ-squared random field with d degrees of freedom (see example 2.1.3.):

$$\eta_d(x) = \frac{1}{2}\sum_{i=1}^{d} \xi_i^2(x), \tag{3.6.8}$$

where $\xi_i(x)$, $x \in R^n$, $i = 1, \ldots, d$, are independent copies of a real homogeneous isotropic Gaussian random field $\xi(x)$, $x \in R^n$, with $\mathrm{E}\xi(x) = 0$, $\mathrm{E}\xi^2(x) = 1$, and continuous correlation function $B(\|x\|) = \mathrm{E}\xi(0)\xi(x) \downarrow 0$ as $\|x\| \to \infty$.

Recall that the one- and two-dimensional densities of the χ-squared random field $\eta_d(x)$ with d degrees of freedom, $p_\beta(t)$ and $p_\beta(t,s,\gamma)$ respectively, have been introduced in Section 2.1 (see formulas (2.1.9) and (2.1.13) with $p = d$),

$$\gamma = \gamma(\|x-y\|) = B^2(\|x-y\|).$$

In what follows we omit the indeces β and write simply $p(t)$ and $p(t,s,\gamma)$.

Introduce the condition

W' . For any $r > 0$

$$\int_{v(r)}\int_{v(r)} \left(1 - B^2(\|x-y\|\right)^{-1/2} dx dy < \infty.$$

Analogously to the case $n = 1$ (see Berman(1982)) one can show that under condition W' there exists the local time of the field (3.6.8) and for $d \geq 5$ the following expansion holds in $L_2(\Omega)$:

$$l_{v(r)}(t) = p(t)\sum_{k=0}^{\infty} e_k^{(d/2)}(t) \int_{v(r)} e_k^{(d/2)}(\eta_d(x))dx,$$

where the polynomials $\{e_k^{(d/2)}(t)\}_{k=0}^{\infty}$ are defined by formulas (2.1.11).

Theorem 3.6.3. Let $\eta_d(x)$, $x \in R^n$, be a χ-squared random field with $d \geq 5$ degrees of freedom described by condition C' and condition V under $m = 2$ and W' hold.

Then the finite-dimensional distributions of the stochastic processes

$$\frac{l_{v(r)}(t) - p(t)|v(r)|}{\sigma_2(r)}, \quad t \neq \frac{d}{2},$$

converge weakly, as $r \to \infty$, to those of the processes

$$\frac{p(t)d^{-1}(2t-d)\sum_{i=1}^{d}\int_{v(r)} H_2(\xi_i(x))dx}{\sigma_2(r)}, \quad t \neq \frac{d}{2}.$$

Theorem 3.6.4. Let under the conditions of Theorem 3.6.3. condition V holds with $m = 4$.

Then the random variables

$$\frac{l_{v(r)}\left(\frac{d}{2}\right) - p\left(\frac{d}{2}\right)|v(r)|}{\sigma_4(r)} \tag{3.6.9}$$

and

$$-\frac{p\left(\frac{d}{2}\right) d^{1/2}(2d+4)^{-1/2}\int_{v(r)} e_2^{(d/2)}(\eta(x))dx}{\sigma_4(r)}$$

have the same limit distribution as $r \to \infty$.

Proof

We have for $t \neq d/2$:

$$\begin{aligned} l_{v(r)}(t) - p(t)|v(r)| &= p(t)e_1^{(d/2)}(t)\int_{v(r)} e_1^{(d/2)}(\eta_d(x))dx \\ &\quad + p(t)\sum_{k=2}^{\infty} e_k^{(d/2)}(t)\int_{v(r)} e_k^{(d/2)}(\eta_d(x))dx \end{aligned}$$

and for $t = d/2$:

$$\begin{aligned} l_{v(r)}\left(\frac{d}{2}\right) - p\left(\frac{d}{2}\right)|v(r)| &= p\left(\frac{d}{2}\right)e_2^{(d/2)}\left(\frac{d}{2}\right) \\ &\quad \times \int_{v(r)} e_2^{(d/2)}(\eta_d(x))dx \\ &\quad + p\left(\frac{d}{2}\right)\sum_{k=3}^{\infty} e_k^{(d/2)}\left(\frac{d}{2}\right) \end{aligned}$$

$$\times \int_{v(r)} e_k^{(d/2)}(\eta_d(x))dx,$$

$$e_2^{(d/2)}\left(\frac{d}{2}\right) = -\left(\frac{d}{2d+4}\right)^{1/2}.$$

Analogously to the Gaussian case considered above one can see that to prove Theorem 3.6.3. it sufficient to show that

$$\begin{aligned}&\overline{\lim}_{r\to\infty}\left\{\int_{v(r)}\int_{v(r)}\left[p(t,t;\gamma(\|x-y\|))-p^2(t)\right]dxdy\right\}\\&\times\left\{p^2(t)e_1^{(d/2)}(t)\int_{v(r)}\int_{v(r)}\gamma(\|x-y\|)dxdy\right\}^{-1}\\&\leq 1.\end{aligned} \tag{3.6.10}$$

The expression under the limit sign can be written in the form

$$\begin{aligned}c_{10}(n,1) \quad & r^n\int_0^{2r}\left[p(t,t;\gamma(z))-p^2(t)\right]z^{n-1}I_{1-(z/2r)^2}\left(\frac{n+1}{2},\frac{1}{2}\right)dz\\&\times\left[p^2(t)(d/2-t)^2(d/2)^{-1}\sigma_2^2(r)\right]^{-1}.\end{aligned}$$

We break up the first factor in the last expression into the sum of integrals

$$\begin{aligned}I_1+I_2+I_3 &= c_{10}(n,1)r^n\left[\int_0^a+\int_a^{(2r)^\delta}+\int_{(2r)^\delta}^{2r}\right]z^{n-1}\\&\times\left[p(t,t;\gamma(z))-p^2(t)\right]I_{1-(z/2r)^2}\left(\frac{n+1}{2},\frac{1}{2}\right)dz.\end{aligned}$$

Using the relation

$$I_\nu(z)=\left(\frac{e^z}{\sqrt{2\pi z}}\right)(1+O(z)), \quad z\to\infty$$

and Condition C' one can see that

$$I_1 \leq k_1\, r^n$$

and, thus, $I_1/\sigma_2^2(r)\to 0$, as $r\to\infty$, by condition V.

To estimate I_2 we can use the inequalities

$$B(z) \leq B_0 < 1 \qquad \text{as} \quad z \in [a, (2r)^\delta],$$

and

$$I_\nu(x) \leq \left(\frac{x}{2}\right)^\nu e^x \Gamma^{-1}(\nu + 1).$$

We obtain

$$\begin{aligned} I_2 &\leq c_{10}(n,1) r^n k_2 \int_0^{(2r)^\delta} z^{n-1} I_{1-(z/2r)^2}\left(\frac{n+1}{2}, \frac{1}{2}\right) dz \\ &\leq k_3 r^{n(\delta+1)}, \end{aligned}$$

and, hence, by condition V,

$$I_2/\sigma_2^2(r) \to 0, \qquad \text{as } r \to \infty.$$

To estimate I_3 we can use the relation

$$I_\nu(x) \sim \left(\frac{x}{2}\right)^\nu e^x \Gamma^{-1}(\nu+1), \qquad x \to 0,$$

to obtain :

$$\begin{aligned} I_3 &\leq c_{10}(n,1)\, r^n \int_{(2r)^\delta}^{2r} \left(\exp\left\{-\frac{2t}{1 - B^2(z)}\right\} (1 - B^2(z))^{-d/2} - \exp\{-2t\}\right) \\ &\quad \times t^{d-2}\Gamma^{-2}(d/2)\, z^{n-1} I_{1-(z/2r)^2}\left(\frac{n+1}{2}, \frac{1}{2}\right) dz. \end{aligned}$$

Applying then the Taylor expansion for the expression

$$e^{-2t/(1-\gamma)}(1-\gamma)^{-d/2} - e^{-2t}$$

we come to the estimate

$$\frac{I_3}{p^2(t)(d/2 - t)^2 (d/2)^{-1} \sigma_2^2(r)} \leq 1$$

which completes the proof of the inequality (3.6.10), and, so, Theorem 3.6.3. holds.

To prove Theorem 3.6.4. we have to verify the relation

$$\overline{\lim}_{r\to\infty}\left\{\int_{v(r)}\int_{v(r)}\left[p(d/2,d/2;\gamma(\|x-y\|))-p^2\left(\frac{d}{2}\right)\right]dxdy\right\}$$
$$\times\left\{p^2\left(\frac{d}{2}\right)\left(e_2^{(d/2)}\left(\frac{d}{2}\right)\right)^2\int_{v(r)}\int_{v(r)}\gamma^2(\|x-y\|)dxdy\right\}^{-1}$$
$$\leq\ 1.$$

Following the scheme used above we write

$$\int_{v(r)}\int_{v(r)}\left[p(d/2,d/2;\gamma(\|x-y\|))-p^2(d/2)\right]\,dx\,dy=$$

$$=c_{10}(n,1)\,r^n\left[\int_0^a+\int_a^{(2r)^\delta}+\int_{(2r)^\delta}^{2r}\right]z^{n-1}\left[p(d/2,d/2;\gamma(z))-p^2(d/2)\right]\times$$
$$\times I_{1-(z/2r)^2}\left(\frac{n+1}{2},\frac{1}{2}\right)\,dz=I_1+I_2+I_3.$$

The estimating of I_1 and I_2 is performed as in the proof of Theorem 3.6.3. It remains to show that

$$\frac{I_3}{p^2(d/2)\,(e_2^{(d/2)}(d/2))^2\sigma_4^2(r)}\;=\;=\int\limits_{(2r)^\delta}^{2r}z^{n-1}\left[p(d/2,d/2;\gamma(z))-p^2(d/2)\right]$$
$$\times\ I_{1-(z/2r)^2}\left(\frac{n+1}{2},\frac{1}{2}\right)\,dz$$
$$\times\left\{p^2(d/2)\,(e_2^{(d/2)}(d/2))^2\right.$$
$$\left.\times\int_0^{2r}z^{n-1}\gamma^2(z)\,I_{1-(z/2r)^2}\left(\frac{n+1}{2},\frac{1}{2}\right)\,dz\right\}^{-1}$$
$$\leq\ 1 \tag{3.6.11}$$

as $r\to\infty$.

To verify (3.6.11) it suffices to show that

$$p\left(\frac{d}{2},\frac{d}{2};x\right)-p^2\left(\frac{d}{2}\right)\sim p^2\left(\frac{d}{2}\right)d(2d+4)^{-1}x^2,\quad x\to 0.$$

Consider the expression

$$\frac{p\left(\frac{d}{2},\frac{d}{2};x\right)-p^2\left(\frac{d}{2}\right)}{p^2\left(\frac{d}{2}\right)d(2d+4)^{-1}x^2}. \tag{3.6.12}$$

Using the relation

$$I_\nu(z)\sim(z/2)^\nu\Gamma^{-1}(\nu+1),\qquad \text{as } z\to\infty,$$

it is easy to conclude that (3.6.12) is an indeterminacy of the form 0/0 as $x\to 0$. Using the relation

$$\frac{d}{d\gamma}(f^{-\nu}(\gamma)I_\nu(f(\gamma)))=f^{-\nu}(\gamma)\ I_{\nu+1}(f(\gamma))f'(\gamma)$$

and applying twice the L'Hopital rule to the expression (3.6.12) we obtain that the expression (3.6.12) tends to one as $x\to 0$. Theorem 3.6.4. is proved.

Using the following expression for the second Laguerre polynomial (see (2.1.11))

$$e_2^{(d/2)}(u)=\left(\frac{1}{2}d\left(\frac{d}{2}-1-u\right)+(2-u)\left(\frac{d}{2}-u\right)\right)\left(d\left(\frac{d}{2}+1\right)\right)^{-1/2}$$

we can represent the variable $e_2^{(d/2)}(\eta_d(x))$ in the form

$$e_2^{(d/2)}(\eta_d(x))=\frac{1}{4}\left(d\left(\frac{d}{2}+1\right)\right)^{-1/2}\left[\sum_{\substack{i,j=1\\i\neq j}}^{d}H_2(\xi_i(x))H_2(\xi_j(x))-\sum_{i=1}^{d}H_4(\xi_i(x))\right]$$

Therefore, according to Theorem 3.6.4., the limit distribution of the random variable (3.6.9) coincides with that of the random variables

$$c_{18}\ (\sigma_4(r))^{-1}\left[\sum_{i=1}^{d}\int_{v(r)}H_4(\xi_i(x))dx-\sum_{\substack{i,j=1\\i\neq j}}^{d}\int_{v(r)}H_2(\xi_i(x))H_2(\xi_j(x))dx\right] \tag{3.6.13}$$

as $r\to\infty$, where $c_{18}=p(d/2)(4d+8)^{-1}$.

As an intermidiate consequence of Theorem 3.6.3. and Theorem 2.4.1. we have the following result.

Theorem 3.6.3'. Let $\eta_d(x)$, $x \in R^n$, be a χ-squared random field with $d \geq 5$ degrees of freedom described by condition C' and Conditions W' and A" with $\alpha \in (0, n/2)$ hold.

Then the finite-dimensional distributions of the stochastic processes

$$r^{\alpha-n}L^{-1}(r)\left(l_{v(r)}(t) - p(t)\,|v(r)|\right), \quad t \neq \frac{d}{2},$$

converge weakly, as $r \to \infty$, to those of the process

$$\begin{aligned}&(2\pi)^{n/2}p(t)d^{-1}(2t-d)c_5^{-2}(n,\alpha)\\ &\times \sum_{i=1}^{d} \int_{R^{2n}}' \frac{J_{n/2}(\|\lambda_1+\lambda_2\|)}{\|\lambda_1+\lambda_2\|^{n/2}} \frac{W_i(d\lambda_1)W_i(d\lambda_2)}{\|\lambda_1\|^{(n-\alpha)/2}\|\lambda_2\|^{(n-\alpha)/2}},\\ t &\neq \frac{d}{2},\end{aligned}$$

where $W_i(\cdot)$, $i = 1, \ldots, d$, are independent copies of a complex Gaussian white noise on R^n.

As a corollary of Theorem 3.6.4. we can obtain the following

Theorem 3.6.4'. Let under the conditions of Theorem 3.6.3'. condition A" holds with $\alpha \in (0, n/4)$.

Then the distributions of the random variables

$$r^{2\alpha-n}L^{-2}(r)\left(l_{v(r)}\left(\frac{d}{2}\right) - p\left(\frac{d}{2}\right)|v(r)|\right) \tag{3.6.14}$$

converge, as $r \to \infty$, to that of the random variable

$$\begin{aligned}&(2\pi)^{n/2}p\left(\frac{d}{2}\right)(4d+8)^{-1}c_5^{-2}(n,\alpha)\\ &\times\left\{\sum_{i=1}^{d}\int_{R^{4n}}' \frac{J_{n/2}(\|\lambda_1+\cdots+\lambda_4\|)}{\|\lambda_1+\cdots+\lambda_4\|^{n/2}} \prod_{k=1}^{4}\|\lambda_k\|^{(\alpha-n)/2}\prod_{k=1}^{4}W_i(d\lambda_k)\right.\\ &-2\sum_{\substack{i,j=1\\ i<j}}^{d}\int_{R^{4n}}' \frac{J_{n/2}(\|\lambda_1+\cdots+\lambda_4\|)}{\|\lambda_1+\cdots+\lambda_4\|^{n/2}}\\ &\left.\times\prod_{k=1}^{4}\|\lambda_k\|^{(\alpha-n)/2}W_i(d\lambda_1)W_i(d\lambda_2)W_j(d\lambda_3)W_j(d\lambda_4)\right\},\end{aligned} \tag{3.6.15}$$

where $W_i(\cdot)$, $i = 1, \ldots, d$, are independent copies of a complex Gaussian white noise on R^n and multiple stochastic integrals in (3.6.15) are defined in section 2.3.

Proof of Theorem 3.6.4'. According to Theorem 3.6.4. and formula (3.6.13) to obtain the limit distribution of the variables (3.6.14) we have to find those for the functionals

$$X_r = r^{2\alpha-n} L^{-2}(r) \int_{v(r)} H_4(\xi_i(x))dx$$

and

$$Y_r = r^{2\alpha-n} L^{-2}(r) \int_{v(r)} H_2(\xi_i(x))H_2(\xi_j(x))dx.$$

From Theorem 2.4.1. we have that the distributions of the random variables X_r converge weakly to those of the random variable

$$X = (2\pi)^{n/2} c_5^{-2}(n,\alpha) \int'_{R^{4n}} \frac{J_{n/2}(\|\lambda_1 + \cdots + \lambda_4\|)}{\|\lambda_1 + \cdots + \lambda_4\|^{n/2}} \prod_{k=1}^{4} \|\lambda_k\|^{(\alpha-n)/2} \prod_{k=1}^{4} W_i(d\lambda_k).$$

Consider the functional Y_r.

By the representation

$$\xi_i(x) = \int_{R^n} e^{i(\lambda,x)} \sqrt{f(\|\lambda\|)} W_i(d\lambda)$$

from the formula (2.3.9) we obtain

$$\begin{aligned} H_2(\xi_i(x))\ H_2(\xi_j(x)) &= \int'_{R^{4n}} e^{i(x,\lambda_1+\cdots+\lambda_4)} \\ &\times \prod_{j=1}^{4} \sqrt{f(\|\lambda_j\|)} W_i(d\lambda_1)W_i(d\lambda_2)W_j(d\lambda_3)W_j(d\lambda_4) \end{aligned}$$

and next, reasoning exactly as in the proof of Theorem 2.4.1. we come to the equality

$$\begin{aligned} Y_r &= (2\pi)^{n/2} r^{2\alpha-n} L^{-2}(r) \\ &\times \int'_{R^{4n}} \frac{J_{n/2}(\|\lambda_1 + \cdots + \lambda_4\|)}{\|\lambda_1 + \cdots + \lambda_4\|^{n/2}} r^{-n} \\ &\times \prod_{j=1}^{4} \sqrt{f\left(\frac{\|\lambda_j\|}{r}\right)} W_i(d\lambda_1)W_i(d\lambda_2)W_j(d\lambda_3)W_j(d\lambda_4). \end{aligned}$$

Denote

$$\begin{aligned} Y &= (2\pi)^{n/2} c_5^{-2}(n,\alpha) \\ &\quad \times \int_{R^{4n}}' \frac{J_{n/2}(\|\lambda_1 + \cdots + \lambda_4\|)}{\|\lambda_1 + \cdots + \lambda_4\|^{n/2}} \\ &\quad \times \prod_{k=1}^{4} \|\lambda_k\|^{(\alpha-n)/2} W_i(d\lambda_1) W_i(d\lambda_2) W_j(d\lambda_3) W_j(d\lambda_4). \end{aligned}$$

Applying Theorem 2.4.1. and relation (2.3.10) we can see that

$$\mathrm{E}|Y_r - Y|^2 = \mathrm{E}|X_r - X|^2 \to 0 \qquad \text{as } r \to \infty.$$

Theorem 3.6.4'. is proved.

The results analogous to Theorem 3.6.1., 3.6.2'. and 3.6.3'., 3.6.4'. can also be obtained for occupation densities of vector fields.

Chapter 4

LIMIT THEOREMS FOR SOLUTIONS OF THE BURGERS' EQUATION WITH RANDOM DATA

4.1 Physical Motivation and Recent History

A nonlinear diffusion equation known as the Burgers' equation describes various physical phenomena, such as nonlinear acoustic waves, kinematic waves, growth of molecular interfaces, and the formation of large scale quasi-Voronoi tesselation structures of self-gravitating matter in the late stages of the Universe (see, for example, Burgers (1974),Whitam (1974), Chorin (1975), Gurbator, Malachov and Saichev (1991), Kardar, Parisi and Zhang (1986), Shandarin and Zeldovich (1989), Kofman, Pogosyan, Shandarin and Melott (1992), Weinberg and Gunn (1990), Vergassola, Dubrulle, Frisch and Noullez (1994), Albeverio, Molchanov and Surgailis (1994), Molchanov, Surgailis and Woyczynski (1997)) and other types of irrotational flows.

The Burgers' equation can be viewed as a simplified version of the Navier–Stokes equation and other types of nonlinear equations of mathematical physics (see, for example, Monin and Yaglom (1980), Vishik and Fursikov (1988), Avellaneda and Majda (1990,1991,1992a, b), Kelbert and Suhov (1995), Klyatskin, Woyczynski and Gurarie (1996) Kotelenez (1995b), Fannjiang and Papanicolau (1996), E. Rykov and Sinai (1996), Kuksin (1996),

Kelbert and Sazonov (1996), Farge, Kavlahan, Perrier and Schneider (1997a, b, c).

Equations related to the one-dimensional Burgers' equation have been proposed in models of financial markets (option pricing (see Hodges and Carverhill(1993)).

Note that another viewpoint states that the differences between the Burgers' equation and the Navier–Stokes equation are more interesting than their similarities.

Rosenblatt (1968, 1987) was one of the first to have considered the Burgers' equation with random initial data from the rigorous perspective of probability theory and, more recently, numerous researchers studied solutions of the Burgers' equation depending on different types of random initial conditions. In particular, Bulinski and Molchanov (1991), Bulinski (1992), Giraitis, Molchanov and Surgailis (1993), Albeverio, Molchanov and Surgailis (1994), Funaki, Surgailis, Woyczynski (1995), Leonenko and Orsingher (1995), and Leonenko, Parkhomenko and Woyczynski (1996), Ryan (1998) studied solutions of the Burgers' equation when the initial condition was either a Gaussian random field or short-noise (or Gibbs–Cox) random fields with weak and strong dependence. They obtain Gaussian and non-Gaussian distributions of parabolic scaling limits of distributions of the solution random fields.

Leonenko and Li Zhanbing (1994), Leonenko, Orsingher and Rybasov (1994), Leonenko, Li Zhanbing and Rybasov (1995), Leonenko and Orsingher (1995), Leonenko, Orsingher and Parkhomenko (1996a) also obtain non-Gaussian limit distributions when the initial condition is a χ-squared field with singular spectrum. In the Gaussian model with nonintegrable oscillating correlations, the limit solution is also non-Gaussian (see Surgailis and Woyczynski (1994a, b). Analogous results under suitable non-Gaussian initial conditions with weak dependence can be found in Surgailis and Woyczynski (1993, 1994b), Hu and Woyczynski (1994), Leonenko and Deriev (1994), Surgailis (1997) and Deriev and Leonenko (1997). For the other results concerning limiting distributions of average solutions of Burgers' equation see Rosenblatt (1987) and Hu and Woyczynski (1995b).

Other types of random problems for Burgers' equation have also been considered recently in the mathematical literature. Sinai (1992), Albeverio, Molchanov and Surgailis (1994), Molchanov, Surgailis and Woyczynski (1995), Fan (1995), Avellaneda and E (1995),Avellaneda (1995) Hu and Woyczynski (1995), Wehr and Xin (1996), Ryan (1998 a,b) and Bertoin (1998 a,b) considered the statistics of chocks in the zero viscosity limit and relation problems for hyperbolic limiting behavior of solutions of Burgers' equation with random data. This type of scaling is of importance in physical applications. Surgailis (1996) considered intermediate asymptotics between the smooth parabolic limits and shock type of hyperbolic asymptotics.

Recent result on the nonhomogeneous Burgers' equation which account for extermal forcing can be found in Sinai (1991), Holden, Øksendall, Ubøe, Zhang (1996), Saichev and Woyczynski (1996a, 1996b, 1997), Bertini, Carcrini and Jona-Lasinio (1994), Handa (1995), Truman and Zhao (1996), Molchanov, Surgailis and Woyczynski (1997), Kifer (1997).

Some interesting results for fractal Burgers' equation can be found in Biler, Funaki and Woyczynski (1998). These papers are contained some variantes Feynman–Kac formula.

Bossy and Talay (1996, 1997) have studied the rates of convergence for the interacting particle system approximation relation to the propagation of chaos result for the Burgers' equation.

Leonenko, Orsingher and Parkhomenko (1996), Leonenko and Woyczynski (1998a) provides the rate of convergence (in the uniform Kolmogorov's distance) of probability distributions of the parabolically rescaled solutions of the Burgers' equation with random singular Gaussian data.

The classical statistical inference problems (parameter estimation and hypothesis testing) for random fields arising in Burgers' turbulence can be found in Leonenko and Woyczynski(1998b, c, d). Statistical inference for the parameters of equations of phisical oceanography considered by Piterbarg and Rozovskii (1996).

4.2 Hopf–Cole Solution

A crucial property of the Burgers' equation is that it admits a linearization by the so called Hopf–Cole substitution.

We consider the n-dimensional Burgers' equation

$$\frac{\partial u}{\partial t} + (u, \nabla)u = \mu\Delta u, \quad \mu > 0, \tag{4.2.1}$$

subject to the condition

$$u(0, x) = u_0(x) = \nabla\eta(x) \tag{4.2.2}$$

of the gradient form. The equation describes time evolution of the velocity field

$$u(t,x) = [u_1(t,x), \ldots, u_n(t,x)]', \quad (t,x) \in (0,\infty) \times R^n, \quad n \geq 1.$$

The potential $\eta(x)$, $x \in R^n$, is a scalar field, ∇ denotes the gradient operator in R^n, and Δ stands for the n-dimensional Laplacian (see section 1.1).

Equation (4.2.1) is a parabolic equation with quadratic, inertial nonlinearity, which can be viewed as a simplified version of the Navier–Stokes equation with the pressure term ∇p omitted, and with the viscosity coefficient μ corresponding to the inverse of the Reynolds number, (see Rosenblatt (1985), p. 152)). With random initial the problem (4.2.1–4.2.2) is also known as the Burgers turbulence problem.

The Burgers' equation (4.2.1) can be linearized by the Hopf–Cole transformation

$$u(t,x) = -2\mu\nabla \log q(t,x),$$

which reduces (4.2.1) to the linear diffusion equation

$$\frac{\partial q}{\partial t} = \mu \Delta q, \tag{4.2.3.}$$

subject to the initial condition

$$q(0,x) = \exp\left\{-\frac{\eta(x)}{2\mu}\right\}. \tag{4.2.4}$$

Using the well-known formulas for the solution of the initial value problem (4.2.3)–(4.2.4) one obtain the following explicit solution of the Burgers initial-value problem (4.2.1)–(4.2.2):

$$\begin{aligned} u(t,x) &= -2\mu\nabla \log J(t,x) \\ &= \frac{I(t,x)}{J(t,x)}, \end{aligned} \tag{4.2.5}$$

where

$$\begin{aligned} I(t,x) &= \int_{R^n} \frac{x-y}{t}\, g(t,x-y) \exp\left\{-\frac{\eta(x)}{2\mu}\right\} dy, \qquad (4.2.6) \\ J(t,x) &= \int_{R^n} g(t,x-y) \exp\left\{-\frac{\eta(x)}{2\mu}\right\} dy, \end{aligned}$$

and where

$$g(t,x-y)=(4\pi\ \mu\ t)^{-n/2}\exp\left\{-\frac{\|x-y\|^2}{4\mu t}\right\},\qquad x,y\in R^n,\quad t>0,\quad (4.2.7)$$

is the Gaussian (heat) kernel.

We shall call the functional (4.2.5) the Hopf–Cole solution of the initial-value problem (4.2.1)–(4.2.2) (see, for example, Withem (1974), Gurbatov, Malachov and Saichev (1991), Bulinski and Molchanov (1991)).

Note that for $n=1$ the initial value problem (4.2.1)–(4.2.2) becomes

$$\frac{\partial u}{\partial t}+u\frac{\partial u}{\partial x} = \mu\frac{\partial^2 u}{\partial x^2},\qquad \mu>0, \qquad (4.2.8)$$

$$u(0,x) = \frac{\partial}{\partial x}\eta(x). \qquad (4.2.9)$$

Remark 4.2.1 . Zeldovich and his co-authors (see Zeldovich (1970), Zeldovich, Molchanov and Ruzmaikin (1988), Zeldovich and Novikov (1975), Vergassola, Dubrulle, Frish and Noullez (1994)) developed the hydrodynamic approach to evolution of the Universe. According to it, with $n=3$ the evolution of the velocity field $u=u(t,x)$, $(t,x)\in(0,\infty)\times R^n$, of matter is described by the Burgers' equation

$$\begin{cases}\frac{\partial u}{\partial t}+(u,\nabla)u=\mu\Delta u+\nabla\Phi\\ \operatorname{curl} u=0,\\ u(0,x)=\nabla\eta(x),\end{cases}\qquad (4.2.10)$$

while the density of matter $\rho=\rho(t,x)$, $(t,x)\in(0,\infty)\times R^n$, satisfies the continuity equation :

$$\begin{cases}\frac{\partial\rho}{\partial t}+\operatorname{div}(\rho u)=0\\ \rho(0,x)=\rho_0(x),\end{cases}\qquad (4.2.11)$$

where

$$\operatorname{div} u = <\nabla, u>$$

and the potentiality condition

$$\operatorname{curl} u=\nabla\times u=0$$

reflects self-gravitation of the medium, and the viscosity parameter $\mu>0$ its hydrodynamic friction, i.e., the effect of numerous collisions of particles.

The initial fields $\eta(x)$, $x \in R^n$, and $\rho_0(x)$, $x \in R^n$, and the external potential $\Phi(x)$, $x \in R^n$, are assumed to be smooth homogeneous random fields. Unfortunately there is no clarity concerning the important question about the type of distribution of these random fields: one can only hope that asymptotic behavior of $u(t,x)$ and $\rho(t,x)$ as $t \to \infty$ does not depend much on concrete details of the stochastic model. Equation (4.2.10) admit an solution via the Hopf–Cole substitution

$$u(t,x) = -2\mu\nabla \log q(t,x),$$

where the real function $q(t,x)$ satisfies the linear parabolic equation of the Schrödinger type :

$$\begin{cases} \frac{\partial q}{\partial t} = \mu\Delta q + \frac{1}{2\mu}\Phi q, \\ u(0,x) = \exp\{-\frac{\eta(x)}{2\mu}\} \end{cases} .$$

Its solution is given by the Feynmann–Kac formula

$$q(t,x) = E^w\left[\exp\left(\frac{1}{2\mu}\int_0^t \Phi\left\{x+\sqrt{2\mu}w_s\right\}ds - \frac{1}{2\mu}\eta\left\{x+\sqrt{2\mu}w_t\right\}\right)\right], \tag{4.2.12}$$

where $E^w[\cdots]$ is the expectation over trajectories of the standard Wiener process w_s, $s \geq 0$, $w_0 = 0$ in R^n, (see Carmona and Lacroix (1990)). Then we obtain the Feynmann–Kac formula for the solution of Burgers' equation (4.2.10) (see Molchanov, Surgailis and Woyczynski (1997)):

$$\begin{aligned} u(t,x) &= -\left(E^w\left[\exp\left(\frac{1}{2\mu}\int_0^t \Phi(x+\sqrt{2\mu}w_s)ds - \frac{1}{2\mu}\eta(x+\sqrt{2\mu}w_t)\right)\right]\right)^{-1} \\ &\quad \times E^w\left[\int_0^t \left(\nabla\Phi(x+\sqrt{2\mu}w_s)ds - \nabla\eta(x+\sqrt{2\mu}w_t)\right)\right. \\ &\quad \left.\times \exp\left(\frac{1}{2\mu}\int_0^t \Phi(x+\sqrt{2\mu}w_s)ds - \frac{1}{2\mu}\eta(x+\sqrt{2\mu}w_t\right)\right], \\ (t,x) &\in (0,\infty)\times R^n. \end{aligned} \tag{4.2.13}$$

Holden, Øksendel, Ubøe and Zhang (1996) considered Feynmann–Kac formula for stochastic Burgers' equation in context of Wick products.

Remark 4.2.2. We have not discusses the question of uniqueness of the solution (4.2.5). This is not a simple problem. In the homogeneous case (4.2.1) uniqueness was obtained by Hopf (1950), by reducing the problem to the uniqueness of the positive solutions of the heat equation (4.2.3). Dix (1996) shows that in a certain well-defined sense the initial value problem (4.2.1)–(4.2.2) is locally well posed in the Sobolev space H^s for $s > -1/2$ (local existence, uniqueness, and continuous dependence on initial conditions) and that it fails to be well posed (in the same sense) when $s < -1/2$. In particular, consider the well-known 'N-wave' solution of Burgers' equation (4.2.8) with $\mu = 1$ (see Witham (1974))

$$u(t,x) = \frac{x}{t}\frac{\sqrt{a/t}e^{-x^2/4t}}{1+\sqrt{a/t}e^{-x^2/4t}} = -2\frac{\partial}{\partial x}\log[1+\sqrt{a/t}e^{-x^2/4t}],$$

where $a > 0$, $(t,x) \in (0,\infty) \times R^1$. Let $S(R^1)$ be the space of Schwartz tempered test functions ; $S'(R^1)$ is its dual topological vector space, $C(I, X)$, where $I \subset R^1$ and X is a Banach space, denotes the set of all continuous mapping $I \to X$. Dix (1996) makes the following assertions about this solution:

1) for every $\psi \in S(R^1)$, we have

$$\lim_{t\to 0+} \int_{R^1} u(t,x)\psi(x)dx = 0,$$

and thus $u(t,\cdot) \to 0$ in $S'(R^1)$ as $t \to 0+$;

2) for every $1/4 < s < 3/4$, we have

$$\sup_{t>0} t^s \left\|u(t,\cdot)\right\|_{L_2} < \infty;$$

3) $\lim_{t\to 0+} t^{1/4} \left\|u(t,\cdot)\right\|_{L_2} < \infty$;

4) if $s < -1/2$, then

$$\left\|u(t,\cdot)\right\|_{H^s} \to 0 \qquad \text{as } t \to 0+ .$$

Since $a > 0$ can be chosen arbitrary, we see that there are infinitely many solution in $C([0,T], H^s)$, $s < -1/2$, to (4.2.8)–(4.2.9) with $u(0,x) = 0$. This phenomenon can be understood intuitively as follows. If $q(t,x)$ solves the heat equation (4.3.2) with $n = 1$, $\mu = 1$, then

$$u(t,x) = -2\frac{\partial}{\partial x}\log q(t,x)$$

satisfies Burgers' equation. The 'N-wave' solution arises from

$$q(t,x) = 1 + \sqrt{\frac{a}{t}} e^{-x^2/4t},$$

which is a solution of the heat equation (4.2.3) with $n = 1, \mu = 1$ and initial data $q(0,x) = 1 + c\delta$, where $c > 0$ dependents on a and δ is the Dirac delta distribution. When we apply the function log, however, the part tending to $c\delta$ makes no contribution in a distributional sense. The number 1 is not special. We could consider the solution of the heat equation with initial data

$$q(0,x) = \exp\left\{-\frac{1}{2}\int_{-\infty}^{x} u_0(y)dy + c\delta(x)\right\},$$

since

$$u = -2\frac{\partial}{\partial x}\log q$$

would then satisfy Burgers' equation with initial data $u_0(x) \in \mathbf{L}_1$. The initial data would be assumed in the sense of H^s, $s < -1/2$, as one can show using the same method. Thus there are infinitely many nonequivalent solutions emerging from every initial data in $\mathbf{L}_1$. Solutions u of a non-homogeneous Burgers' equation

$$\frac{\partial u}{\partial t} + u\frac{\partial u}{\partial x} = \mu\frac{\partial^2 u}{\partial x^2} + f(t,x)$$

with $f \in \mathbf{L}_1([0,T] \times R)$ can be also expressed in terms of solution q of the heat equation with a potential via the same transformation

$$u(t,x) = -2\frac{\partial}{\partial x}\log q(t,x).$$

Solutions of this variant of the heat equation can also be written down explicitly using Feynmann–Kac formula (4.2.11). If we use the same initial data q_0 as displayed above in this formula then we see that there is nonuniqueness of solutions of the non-homogeneous Burgers' equation of the same type.

The problems of uniqueness and nonuniqueness of stochastic Burgers' equation have been discussed by Bertini, Cancrini and Jona–Lasinio (1994).

Remark 4.2.3. Liu and Liu (1992) proposed the homogeneous Korteweg-deVries-Burgers' equation

$$\frac{\partial u}{\partial t} + u\frac{\partial u}{\partial x} = \mu\frac{\partial^2 u}{\partial x^2} + \nu\frac{\partial^3 u}{\partial x^3}$$

as a normal equation for turbulence, where $\mu > 0$, $\nu > 0$. In particular, they obtained travelling wave solutions $u(t,x) = u(x-ct)$ of the following form:

1) for positive viscosity μ, negative dispersion ν, and $\mu^2 < -4\nu\sqrt{c^2+2A}$,

$$u(t,x) = \begin{cases} u_1 + \frac{u_1-u_2}{2} e^{-\mu(x-ct)/2\nu} \cos\sqrt{\frac{u_1-u_2}{2\nu} - (\frac{\mu}{2\nu})^2}(x-ct), & (x-ct) < 0, \\ u_2 + 3\frac{u_1-u_2}{2} \operatorname{sech}^2 \sqrt{\frac{u_1-u_2}{8\nu}}(x-ct), & (x-ct) \geq 0, \end{cases}$$

and

2) for positive viscosity μ, positive dispersion ν, and $\mu^2 < 4\nu\sqrt{c^2+2A}$,

$$u(t,x) = \begin{cases} u_1 + 3\frac{u_1-u_2}{2} \operatorname{sech}^2 \sqrt{\frac{u_1-u_2}{8\nu}}(x-ct), & (x-ct) < 0, \\ u_2 + \frac{u_1-u_2}{2} e^{-\mu(x-ct)/2\nu} \cos\sqrt{\frac{u_1-u_2}{2\nu} - (\frac{\mu}{2\nu})^2}(x-ct), & (x-ct) \geq 0, \end{cases}$$

where

$$u_1 = c + \sqrt{c^2+2A}, \quad u_2 = c - \sqrt{c^2+2A}$$

correspond to the two steady states in the phase and A is an constant such that $c^2 + 2A > 0$. In case 1), the upper part of the trajectory in the phase (lower part) plane $(u, \dot{u})$ is close to a soliton, whole the constitutes an attenuating oscillatory tail.

Remark 4.2.4 . Saichev and Woyczynski (1996a) have been proposed for the velocity $u(t,x)$ and density $\rho(t,x)$ fields of a polytropic gas the following equations:

$$\begin{aligned} \frac{\partial u}{\partial t} + u\frac{\partial u}{\partial x} + \frac{1}{\rho}\frac{\partial P}{\partial x} &= \mu\frac{\partial^2 u}{\partial x^2} + \nu\frac{\partial^3 u}{\partial x^3} + F((t,x), \\ \frac{\partial \rho}{\partial t} + \frac{\partial}{\partial x}(u\rho) &= \mu\frac{\partial^2 \rho}{\partial x^2} + \nu\frac{\partial^3 \rho}{\partial x^3}, \end{aligned}$$

where $t > 0$, $x \in R^1$, $\mu > 0$, $\nu > 0$, P is the pressure, F is the force, and

$$u = \frac{u_+ + u_-}{2}, \qquad \rho = \frac{\rho_+ + \rho_-}{2},$$

where the auxiliary fields $u_\pm$ satisfy the same nonhomogeneous Korteweg–deVries–Burgers' equation

$$\frac{\partial u_\pm}{\partial t} + u_\pm\frac{\partial u_\pm}{\partial x} = \mu\frac{\partial^2 u_\pm}{\partial x^2} + \nu\frac{\partial^3 u_\pm}{\partial x^3} + F(t,x)$$

with (different) initial conditions : $u_{\pm}(0,x) = u_0(x) \pm c_0(x)$.

Remark 4.2.5. Let (Ω, F, P) be a probability space. Consider the Burgers' equation (4.2.8) with initial condition $u(0,x) = u_0(x)$, $x \in R^1$. In what follows the initial datum $u_0(x) = \xi(w,x)$, $w \in \Omega$, $x \in R^1$, is assumed to be a strictly stationary random process. Rosenblatt (1987) observed the following fact: suppose that

$$\int_0^x u_0(y)dy = o(x^2) \qquad \text{as } x \to \infty.$$

Then, for each $\beta > 0$

$$\begin{aligned}\frac{1}{2\mu}\int_0^\beta u(t,y)dy &= \frac{1}{2\mu}\int_0^\beta u_0(y)dy \\ &\quad -\log\int_{-\infty}^{\infty}\exp\left\{-\frac{1}{2\mu}\int_0^v u_0(\beta+y)\,dy - \frac{v^2}{4\mu t}\right\}dv \\ &\quad +\log\int_{-\infty}^{\infty}\exp\left\{-\frac{1}{2\mu}\int_0^v u_0(y) - \frac{v^2}{4\mu t}\right\}dv.\end{aligned}$$

Furthermore, if $\psi(\beta) \to \infty$ as $\beta \to \infty$, then as $\beta \to \infty$, in probability

$$\frac{1}{\psi(\beta)}\left(\int_0^{\beta x} u(t,y)\,dy - \int_0^{\beta x} u_0(y)\,dy\right) \to 0.$$

This statement shows that the limiting behavior (as $\beta \to \infty$) of the average process

$$\frac{1}{\psi(\beta)}\int_0^{\beta x} u(t,y)dy, \quad x > 0,$$

in the same as that of the average process

$$\frac{1}{\psi(\beta)}\int_0^{\beta x} u_0(y)dy$$

(in the sense of the finite-dimensional distributions). The limit behavior of

$$\frac{1}{\psi(\beta)}\int_0^{\beta x} u_0(x)dx = \frac{1}{\psi(\beta)}\int_0^{\beta x} \xi(x)dx, \quad x > 0$$

itself has been studied in many cases.

Suppose that $\xi(x)$, $x \in R^1$, is a second-order stationary process with

$$\sigma(\beta) = \text{var} \int_0^\beta \xi(y)dy \to \infty \qquad \text{as } \beta \to \infty.$$

Assume also that

$$E\xi(x) = 0, \quad E\,|\xi(x)|^{2+\delta} < \infty,$$

and that $\xi(x)$, $x \in R^n$, is strongly mixing with the mixing coefficient

$$\alpha(d) = O(d^{-(1+\varepsilon)(1+2/\delta)}),$$

where, as usual

$$\alpha(d) = \sup_{\substack{A \in \mathcal{M}^x_{-\infty}, \\ B \in \mathcal{M}^\infty_{x+d}}} |P(A \cap B) - P(A)P(B)|$$

where $\mathcal{M}^b_a = \sigma\{\xi(x),\ a \le x \le b\}$. Then (see Rozanov (1967))

$$\frac{1}{\sigma(\beta)} \int_0^{\beta x} \xi(y)\, dy \xrightarrow{D} w(x), \quad x > 0,$$

where $w(s)$, $s \ge 0$, is a Brownian motion. Therefore (see Rosenblatt (1987)) as $\beta \to \infty$

$$\frac{1}{\sigma(\beta)} \int_0^{\beta y} u(t,y)dy \xrightarrow{D} w(y), \quad y > 0,$$

where t is fixed. For the last development of limit theorems for mixing processes see Bolthausen (1982), Bradley (1985, 1992), Doukhan (1994).

Let $\xi(x) = G(\zeta(x))$, $x \in R^1$, where $\zeta(x)$, $x \in R^1$, is a stationary Gaussian process with $E\zeta(x) = 0$, $E\zeta^2(x) = 1$ and covariance function $B(x) = L(x)/|x|^\alpha$, $0 < \alpha < 1/m$, $L \in \mathcal{L}$, where $m \ge 1$ is such that $C_1 = \cdots = C_{m-1} = 0$, $C_m \ne 0$, where

$$C_m = \int_{-\infty}^{\infty} G(u)H_m(u)\phi(u)du, \quad m = 0, 1, 2 \ldots,$$

and $E\,G^2(\zeta(0)) < \infty$ and $H_m(u)$ are Chebyshev–Hermite polynomials (2.1.5). Then (see Taqqu (1979)) as $\beta \to \infty$

$$\frac{1}{\psi(\beta)} \int_0^{\beta x} \xi(y)dy \xrightarrow{D} Z_m(x),$$

where

$$\psi(\beta) = \frac{C_m}{m!}\left[\frac{2\beta^{2-m\alpha}L^m(\beta)}{(1-m\alpha)(2-m\alpha)}\right]^{1/2},$$

and

$$\begin{aligned} Z_m(x) &= \operatorname{sgn}\{C_m\}\left[\frac{(1-m\alpha)\,(2-m\alpha)}{2m!\,[2\Gamma(\alpha)\cos(\alpha\pi/2)]^m}\right]^{1/2} \\ &\times \int_{R^m}' \frac{e^{i(\lambda_1+\cdots+\lambda_m)x}-1}{i(\lambda_1+\cdots+\lambda_m)} \frac{W(d\lambda_1)\cdots W(d\lambda_m)}{|\lambda_1\cdots\lambda_m|^{(1-\alpha)/2}} \end{aligned}$$

where $\int' \ldots$ denotes the multiple stochastic integral (see section 2.3) on the Gaussian white noise $W(\cdot)$. Therefore (see Rosenblatt (1987))

$$\frac{1}{\psi(\beta)}\int_0^{\beta x} u(t,y)dy \xrightarrow{D} Z_m(x), \quad x > 0,$$

as $\beta \to \infty$ and $t > 0$ is fixed.

Note that Hu and Woyczynski (1995) presented some resent results in this direction.

4.3 Parabolic Asymptotics for Weakly Dependent Random Data : the Gaussian Scenario

We present asymptotic normality of the solutions of the initial value problems for a multi-dimensional Burgers' equation with weakly dependent possible non-Gaussian initial condition.

Let, now, $(\Omega, \mathcal{F}, P)$ be a complete probability space, the initial velocity potential $\eta(x) = \eta(w, x)$, $w \in \Omega$, $x \in R^n$, be a random field, and $u = u(t, x)$, $(t, x) \in (0, \infty) \times R^n$, be the solution of the random Cauchy problem (4.2.1) and (4.2.2). In view of the inner symmetries of Burgers' equation and its connection to the linear diffusion equation via the Hopf–Cole transformation, a study of the limiting behavior of the parabolically rescaled solution random field (see (4.2.5))

$$u = u\left(t\beta, x\sqrt{\beta}\right), \quad \beta > 0, \quad (t, x) \in (0, \infty) \times R^n,$$

as $\beta \to \infty$ is of obvious interest. If $\mu > 0$ is fixed, under some additional conditions on the random field $\xi(x)$, $x \in R^n$, the above rescaled solution obeys asymptotically a 'Gaussian scenario' (see e.g., Bulinski and Molchanov (1991), Surgailis and Woyczynski (1993, 1994), Albeverio, Molchanov and Surgailis (1994), Leonenko and Deriev (1994), Deriev and Leonenko (1997)).

In what follows we need the following assumptions.

A. The random field $\xi(x) = \xi(w, x)$, $w \in \Omega$, $x \in R^n$, is a real homogeneous and isotropic a.s. differentiable Gaussian field with

$$E\xi(x) = 0, \qquad E\xi^2(x) = 1$$

and covariance function

$$B(\|x\|) = \operatorname{cov}(\xi(0), \xi(x)), \qquad x \in R^n.$$

B. There exists a non-random function $F(u), u \in R^1$, such that

$$E[\exp\{-F(\xi(0))/2\mu]^2 < \infty,$$

where $\xi(x), x \in R^n$, is random field satisfying condition A.

Under assumption B, the function

$$f(u) = e^{-F(u)/2\mu} \in \mathbf{L}_2(R^1, \phi(u)du),$$

may be expanded in a series (2.2.9), that is,

$$f(u) = \sum_{k=0}^{\infty} \frac{C_k}{k!} H_k(u), \quad C_k = \int_{-\infty}^{\infty} e^{-F(u)/2\mu} \phi(u) H_k(u) du,$$

where $H_k(u)$ are Chebychev–Hermite polynomials defined by (2.1.5).

C. The function

$$f(u) = \exp\left\{-\frac{F(u)}{2\mu}\right\}, \quad u \in R^1,$$

satisfies assumption B and there exists and integer $m \geq 1$ such that

$$C_1 = \cdots = C_{m-1} = 0, \quad C_m \neq 0.$$

Note that such an $m = H \operatorname{rank} G$ (see section 2.2).

Theorem 4.3.1. Let $u = u(t,x)$, $(t,x) \in (0,\infty)\times R^n$, be a solution of the Cauchy problem (4.2.1)–(4.2.2.) (see (4.2.5)) with random initial condition $\eta(x) = F(\xi(x))$, $x \in R^n$, where the random field $\xi(x)$, $x \in R^n$, and the non-random function $F(\cdot)$ satisfy conditions A, B, C, and

$$\int_{R^n} |B(\|x\|)|^m \, dx < \infty, \quad b = \int_{R^n} \left[\sum_{k=0}^{\infty} \left(\frac{C_k^2}{k!} \right) B^k(\|x\|) \right] dx \neq 0,$$

where

$$m = H \operatorname{rank} G, \quad G(u) = \exp\left\{ -\frac{F(u)}{2\mu} \right\}, u \in R^1.$$

Suppose that random field $\eta(x) = F(\xi(x))$, $x \in R^n$, is a.s. differentiable. Then the finite-dimensional distributions of the random fields

$$U_\beta(t,x) = \beta^{1/2+n/4} u\left(t\,\beta, x\sqrt{\beta}\right), \quad (t,x) \in (0,\infty) \times R^n,$$

converge weakly, as $\beta \to \infty$, to the finite-dimensional distributions of the vector Gaussian random field

$$U(t,x) = [U_1(t,x), \ldots, U_n(t,x)]', \quad (t,x) \in (0,\infty) \times R^n,$$

which is homogeneous in $x \in R^n$, with $EU(t,x) = 0$, and matrix-valued covariance function of the form

$$\begin{aligned} E\,U(t,x)[U(s,y)]' &= R(x-y) \\ &= (R_{kj}(x-y))_{1\le k,j\le n} \\ &= a \exp\left\{ -\frac{\|x-y\|^2}{4\mu(t+s)} \right\} \sum(x-y) \\ &= \int_{R^n} e^{i<\lambda,x-y>} f(\lambda) d\lambda, \end{aligned} \tag{4.3.1}$$

where

$$a = bC_0^{-2} \mu^{1-1/n} \pi^{-n/2} (t+s)^{-(n/2+1)},$$

and the matrix

$$\sum(z) = (\sigma_{kj}^2(z))_{1\le k,j\le n}, \quad z = (z_1, \ldots, z_n)' \in R^n,$$

has elements

$$\begin{aligned}\sigma_{kk}^2(z) &= 1-\frac{z_k^2}{2\mu\ (t+s)}, \quad 1\le k\le n;\\ \sigma_{kj}^2(z) &= -\frac{z_k z_j}{2\mu\ (t+s)}, \quad k\ne j, 1\le k,j\le n,\end{aligned} \tag{4.3.2}$$

and matrix-valued spectral density

$$f(\lambda) = (f_{kj}(\lambda))_{1\le k,j\le n}, \quad \lambda=(\lambda_1 \ldots, \lambda_n)' \in R^n,$$

has elements

$$f_{kj}(\lambda) = (bC_0^{-2}2^{2-n}\pi^{-n}\mu^2)\lambda_k\lambda_j \exp\{-\|\lambda\|^2 \mu (t+s)\}. \tag{4.3.3}$$

Corollary 4.3.1. Let, under the assumptions of Theorem 4.3.1. $F(u) = u$ and $\eta(x) = \xi(x)$, $x \in R^n$, be a homogeneous isotropic Gaussian field with $E\xi(x) = 0$ and covariance function $B(\|x\|)$, $x \in R^n$, such that

$$\int_{R^n} |B(\|x\|)|\, dx < \infty, \qquad \int_{R^n} B(\|x\|)dx \ne 0.$$

Then the statement of Theorem 4.3.1. is true with $m = 1$ and

$$bC_0^{-2} = \int_{R^n} \left(\exp\left\{\frac{B(x)}{4\mu^2}\right\} - 1\right) dx.$$

Corollary 4.3.2 . Under the assumptions of Theorem 4.3.1. let $F(u) = u^2 - 1$, that is, the initial potential $\eta(x) = \xi^2(x) - 1$, where $\xi(x)$, $x \in R^n$, is a homogeneous isotropic Gaussian field with $E\xi(x) = 0$ and covariance function $B(\|x\|)$, $x \in R^n$, such that

$$\int_{R^n} B^2(\|x\|)\, dx < \infty.$$

Then the statement of Theorem 4.3.1. is true with $m = 2$ and

$$a = \mu^{1-1/n}e^{-(1/4\mu^2)} \left(2\sqrt{2\pi}\right)^{-n} \int_{R^n} U(x)dx,$$

$$U(x) = \exp\left\{\frac{1}{\mu}\right\} \mu \left[((1+\mu)^2 - B^2(\|x\|)^{-1/2})) - (1+\mu)^{-1}\right].$$

Corollary 4.3.3 . Let $n = 1$ in the Corollary 4.3.1. Then the covariance function

$$\begin{aligned} R(x-y) &= c\sqrt{\frac{\mu}{\pi}}\frac{1}{(t+s)^{3/2}}\left(1-\frac{(x-y)^2}{2\mu\,(t+s)}\right)\exp\left\{-\frac{(x-y)^2}{4\mu\,(t+s)}\right\} \\ &= \int_{-\infty}^{\infty} e^{i\lambda(x-y)} f(\lambda)\,d\lambda, \end{aligned} \tag{4.3.4}$$

where spectral density

$$f(\lambda) = \frac{2c\mu^2}{\pi}\lambda^2 e^{-\lambda^2\mu(t+s)}, \quad \lambda \in R^1, \tag{4.3.5}$$

and

$$c = \int_{-\infty}^{\infty} (e^{(B(\|x\|)/4\mu^2} - 1)dx.$$

Remark 4.3.1. Note that from (4.3.4) and (4.3.5) we immediately obtain

$$\int_{-\infty}^{\infty} R(z)\,dz = 0, \quad f(0) = 0.$$

Those are singular properties of random field $U(t,x)$ in the case $n = 1$. Using (4.3.1), (4.3.2) and (4.3.3) we obtain for $t = s = 1$:

$$\begin{aligned} \sigma_{ii}^2(0) &= 1, \quad 1 \le i \le n; \\ \sigma_{ii}^2(z) &= 0, \quad \text{if} \quad z_i = 2\sqrt{\mu}, 1 \le i \le n; \quad \det\, R(0) > 0; \end{aligned}$$

and for dimension $n = 1$

$$R(z) = 0, \quad \text{if} \quad |z| = 2\sqrt{\mu};$$

for dimension $n = 2$

$$\det\, R(z) = 0, \quad \text{if} \quad \|z\| = 2\sqrt{\mu};$$

for dimension $n = 3$

$$\det\, R(z) = 0, \quad \text{if} \quad \|z\| = 2\sqrt{\mu} \text{ and } z_1 z_2 z_3 = 0.$$

Next,

$$f_{ij}(0) = 0, \quad 1 \le i,j \le n; \quad \det\, f(\lambda) = 0, \quad \lambda \in R^n \text{ for } n \ge 1.$$

Those are singular properties of random field $U(t,x)$ in the case $n \geq 1$.

For the proof of Theorem 4.3.1., we need a generalized Slutsky's Lemma (see Lemma 1.1.1.), which we recall for the sake of completeness. In the sequel, we shall use the notations '$\xrightarrow{P}$' and '$\xrightarrow{D}$' to denote convergence in probability and the convergence in distributions respectively.

Lemma 4.3.1. Let $\{u_t\}$ and $\{v_t\}$ are families of random vectors from R^n and $\{w_t\}$ is family of random variables and let $u_t \xrightarrow{D} u, \quad v_t \xrightarrow{P} c = (c_1, ..., c_n)$ as $t \to \infty$, where $c_i = \text{const}, i = \overline{1,n}$ and $w_t \xrightarrow{P} d = \text{const}$. Consequently as $t \to \infty$: $u_t + v_t \xrightarrow{D} u + c, \quad w_t u_t \xrightarrow{D} du$ and if $d \neq 0$, then $u_t / w_t \xrightarrow{D} u/d$.

Further the symbols K_i, $i = 1, 2, \ldots$ denote positive constants and unless otherwise specified the $\int \ldots$ means $\int_{R^n} \ldots$.

Now let us introduce some function

$$g^0(x-y,t) = \frac{x-y}{t} g(t, x-y),$$

where $g(t,x-y)$ defined by (4.2.7) and

$$g_i^0(t, x-y) = \frac{x_i - y_i}{t} g(t, x-y), \quad i = 1, \ldots, n.$$

For the simplicity we denote

$$g^0(a,y,t) = g^0\left(a\sqrt{t} - y, t\right).$$

Lemma 4.3.2. For any $i = 1, 2, \ldots, n$ it is true that

1)

$$\left|g\left(a\sqrt{t} - y, t\right)\right| dy = K_1 t^{-1/2};$$

2)

$$\left|g\left(a\sqrt{t} - y, t\right)\right| |N(y)| dy \leq K_2 t^{-(n+1)/2}$$

for any absolutely integrable (scalar-) function $N(y)$.

Proof

We obtain

$$\int \left|g_i^0\left(a\sqrt{t} - y, t\right)\right| dy = \int \left|\frac{a_i\sqrt{t} - y_i}{t}\right| \exp\left\{-\frac{1}{2}\left\|\frac{a\sqrt{t} - y}{\sqrt{2\mu t}}\right\|^2\right\} \frac{dy}{(4\pi\mu t)^{n/2}}.$$

After the transformation

$$z = \frac{a\sqrt{t} - y}{\sqrt{2\mu t}}$$

we have

$$\int \left| g_i^0 \left(a\sqrt{t} - y, t \right) \right| dy = \frac{\sqrt{2\mu}}{\sqrt{t}} \int \frac{1}{(2\pi)^{n/2}} |z_i| \exp\left\{ -\frac{\|z\|^2}{2} \right\} dz = K_i t^{-1/2}.$$

2) is proved in a similar way.
Therefore Lemma 4.3.2. is proved.
For simplicity we shall prove that

$$\begin{aligned} X^t(a) &= U_t(1, a) \\ &= t^{1/2+n/4} u\left(t, a\sqrt{t}\right) \\ &\xrightarrow{D} U(1, a) \end{aligned}$$

as $t \to \infty$ where '$\xrightarrow{D}$' denote the convergence all finite-dimensional distributions. Theorem 4.3.1. in general case $(U_\beta(t, x) \xrightarrow{D} U(t, x)$ as $\beta \to \infty)$ is proved in a similar way.

We shall use Cole–Hopf solution (4.2.5) and Lemma 4.3.1. From (4.2.6) we obtain

$$EI\left(t, a\sqrt{t}\right) = 0 \quad \text{for } a = 0,$$

and

$$EI\left(t, a\sqrt{t}\right) \to 0, \quad \text{as } t \to \infty.$$

Now our task is to investigate the asymptotic behavior as $t \to \infty$ of expansion

$$EI\left(t, a\sqrt{t}\right) \left[I\left(t, b\sqrt{t}\right) \right]', \quad a, b \in R^n.$$

We obtain

$$\begin{aligned} I(x, t) &= \int \frac{x - y}{t} g(t, x - y) f \xi(y)) dy \\ &= \sum_{k=m}^{\infty} \frac{C_k}{k!} \int \frac{x - y}{t} g(t, x - y) H_k(\xi(y)) dy. \end{aligned}$$

From (2.1.8)

$$\begin{aligned} EI\left(t, a\sqrt{t}\right)\left[I\left(t, b\sqrt{t}\right)\right]' &= \sum_{k=m}^{\infty} \frac{C_k}{k!} \int\int g^0(a, y_1, t)\left[g^0(b, y_2, t)\right]' \\ &\quad \times B^k(\|y_1 - y_2\|)dy_1 dy_2 \\ &= \left(\int\int g_i^0(a, y_1, t)\ g_j^0(b, y_2, t) G(\|y_1 - y_2\|)\, dy_1 dy_2\right) \\ &= \left(\sigma_{ij}(a, b, t)\right)_{1\leq i,j\leq n}, \end{aligned}$$

where

$$G(\|x\|\,|) = \sum_{k=m}^{\infty} \frac{C_k^2}{k!} B^k(\|x\|).$$

The integral $\sigma_{ij}(a, b, t)$ can be analyzed better after the transformation

$$w_1 = \frac{y_1 - a\sqrt{t}}{\sqrt{2\mu t}}, \qquad w_2 = \frac{y_2 - b\sqrt{t}}{\sqrt{2\mu t}}.$$

Denote

$$p = \frac{a-b}{\sqrt{2\mu}}.$$

We have

$$\begin{aligned} \sigma_{ij}(a, b, t) &= \frac{2\mu}{t}\int\int w_{1i}w_{2j}\phi(\|w_1\|)\phi(\|w_2\|) \\ &\quad \times G\left(\sqrt{2\mu t}\,\|w_1 - w_2 + p\|\right) dw_1 dw_2, \end{aligned}$$

where $\phi(\|u\|)$ is defined in section 3.4.

Next, making the transformation

$$z_1 = w_1 - w_2, \quad z_2 = w_2,$$

we obtain

$$\begin{aligned} \sigma_{ij}(a, b, t) &= \frac{2\mu}{t(2\pi)^n}\int\int (z_{1i} + z_{2i})z_{2j} \exp\left\{-\frac{1}{2}\left[\|z_1 + z_2\|^2 + \|z_2\|^2\right]\right\} \\ &\quad \times G(\sqrt{2\mu t}\,\|z_1 + p\|)dz_1 dz_2 \\ &= \frac{2\mu}{t}\int G\left(\sqrt{2\mu t}\,\|z_1 + p\|\right) A_{ij}(z_1)\, dz_1, \end{aligned}$$

where

$$A_{ij}(z_1) = (2\pi)^{-n} \int (z_{1i} + z_{2i}) z_{2j} \exp\left\{-\frac{1}{2}\left[\|z_1 + z_2\|^2 + \|z_2\|^2\right]\right\} dz_2.$$

Now calculate the value of $A_{ij}(z_1)$ in the following cases:
a) First, let $i \neq j$:

$$\begin{aligned} A_{ij}(z_1) &= (2\pi)^{-n} \exp\left\{-\|z_1\|^2/2\right\} \\ &\times \int (z_{1i} + z_{2i}) z_{2j} \exp\left\{-\sum_l \left(z_{2l}^2 + z_{2l} z_{1l}\right)\right\} dz_2. \end{aligned}$$

After doing the integration we obtain

$$A_{ij}(z_1) = -\frac{1}{4}(4\pi)^{-n/2} z_{1i}\; z_{1j} \exp\left\{-\frac{1}{4}\|z_1\|^2\right\},$$

and

$$\begin{aligned} \sigma_{ij}(a,b,t) &= -\frac{\mu}{2t}(4\pi)^{-n/2} \\ &\times \int z_{1i} z_{1j} \exp\left\{-\frac{1}{4}\|z_1\|^2\right\} G(\sqrt{2\mu t}\,\|z_1 + p\|) dz_1. \end{aligned}$$

b) Secondly, assume $i = j$. In this case

$$A_{ii}(z_1) = -\frac{1}{2}(4\pi)^{-n/2}(1 - z_{1i}^2/2) \exp\left\{-\frac{1}{4}\|z_1\|^2\right\},$$

and

$$\begin{aligned} \sigma_{ii}(a,b,t) &= \frac{\mu}{t}(4\pi)^{-n/2} \\ &\times \int (1 - z_{1i}^2/2) \exp\left\{-\frac{1}{4}\|z_1\|^2\right\} G(\sqrt{2\mu t}\,\|z_1 + p\|) dz_1. \end{aligned}$$

In both cases after the transformation

$$x = \sqrt{2\mu t}(z_1 + p)$$

we obtain:

$$\begin{aligned}\sigma_{ij}(a,b,t) &= -\frac{\mu}{2t}(4\pi)^{-n/2} \\ &\times \int \left(\frac{x_i}{\sqrt{2\mu t}} - p_i\right)\left(\frac{x_j}{\sqrt{2\mu t}} - p_j\right) \\ &\times \exp\left\{-\frac{1}{4}\left\|\frac{x}{\sqrt{2\mu t}} - p\right\|^2\right\} G(\|x\|)\frac{dx}{(2\mu t)^{n/2}} \\ &= -\frac{\mu}{(8\pi\mu)^{n/2}} t^{-(1+n/2)} \frac{p_i p_j}{2} \exp\{-\|p\|^2/4\} \\ &\times \int G(\|x\|)dx(1+o(1)) \end{aligned} \tag{4.3.6}$$

as $t \to \infty$ when $i \neq j$ and

$$\begin{aligned}\sigma_{ii}(a,b,t) &= \frac{\mu}{t}(4\pi)^{-n/2} \\ &\times \int \left(1 - \frac{1}{2}\left(\frac{x_i}{\sqrt{2\mu t}} - p_i\right)^2\right) \\ &\times \exp\left\{-\frac{1}{4}\left\|\frac{x}{\sqrt{2\mu t}} - p\right\|^2\right\} G(x)\frac{dx}{(2\mu t)^{n/2}} \\ &= \frac{\mu}{(8\pi\mu)^{n/2}} t^{-(1+n/2)} (1 - p_i^2/2) \exp\{-\|p\|^2/4\} \\ &\times \int G(\|x\|)\, dx(1+o(1)) \end{aligned} \tag{4.3.7}$$

as $t \to \infty$.

Expressions (4.3.6)–(4.3.7) may be obtained by some properties of uniform convergence and the Abel–Dirichlet criterion.

So, we obtain a following statement

Lemma 4.3.3. Under the assumptions of the Theorem 4.3.1.

$$\lim_{t\to\infty} EI\left(t, a\sqrt{t}\right)\left[I\left(t, b\sqrt{t}\right)\right]' = C_0 R(a-b),$$

where $R(a-b)$ is defined in (4.3.1) with $t = s = 1$.

Remark 4.3.2. It is clear that limit expressions (4.3.6), (4.3.7) remain true for any integrable function $G(x)$, $x \in R^n$, for example for $G(x) = B^k(\|x\|)$.

Thus, if we denote

$$\begin{aligned}\eta_k(a,t) &= \int\int g^0(a,y_1,t)g^0(a,y_2,t)B^k(\|y_1-y_2\|)dy_1dy_2\\ \sigma_k(a) &= \lim_{t\to\infty} E[t^{1/2+n/4}\eta_k(a,t)]^2,\end{aligned}$$

then

$$C_0R(a)=\sum_{k=m}^{\infty}\frac{C_k^2}{k!}\sigma_k^2(a).$$

We now focus our attention on the asymptotic behavior of $J\left(t,a\sqrt{t}\right)$ as $t\to\infty$.

We have

$$J\left(t,a\sqrt{t}\right)=\int g(a,y,t)f(\xi(y))dy,$$

where

$$g(a,y,t)=g\left(t,a\sqrt{t}-y\right).$$

Next

$$EJ\left(t,a\sqrt{t}\right)=C_0,$$

$$EJ\left(t,a\sqrt{t}\right)=\iint g(a,y_1,t)g(a,y_2,t)Ef(\xi(y_1))f(\xi(y_2))dy_1dy_2,$$

$$Ef(\xi(y_1))f(\xi(y_2))=G(y_1-y_2)+C_0^2.$$

Hence

$$\begin{aligned}\operatorname{var}J\left(t,a\sqrt{t}\right) &= EJ^2\left(t,a\sqrt{t}\right)-\left[EJ\left(t,a\sqrt{t}\right)\right]^2\\ &= U(t)\end{aligned}$$

$$U(t)=\iint g(a,y_1,t)g(a,y_2,t)G(y_1-y_2)dy_1dy_2.$$

After the transformation

$$z_1=\frac{y_1-a\sqrt{t}}{\sqrt{2\mu t}},\qquad z_2=\frac{y_2-b\sqrt{t}}{\sqrt{2\mu t}},$$

we have

$$U(t)=\iint\phi(\|z_1\|)\phi(\|z_2\|)G\left(\sqrt{2\mu t}\,\|z_1-z_2\|\right)dz_1dz_2.$$

Next make the transformation $x_1 = z_1 - z_2$, $x_2 = z_2$ and obtain

$$U(t) = \iint \phi(\|x_1 + x_2\|)\phi(\|x_2\|)G(\sqrt{2\mu t}(\|x_1\|)dx_1 dx_2.$$

It's easy to see that

$$\int \phi(\|x_1 + x_2\|)\phi(\|x_2\|)dx_2 = (4\pi)^{-n/2} \exp\left\{-\frac{1}{4}\|x_1\|^2\right\}.$$

Then

$$\begin{aligned} U(t) &= \int (4\pi)^{-n/2} \exp\left\{-\frac{1}{4}\|x_1\|^2\right\} G\left(\sqrt{2\mu t}\,\|x_1\|\right) dx_1 \\ &= \int \phi\,\|x\|\, G\left(2\sqrt{\mu t}\,\|x\|\right) dx. \end{aligned}$$

Using simple estimation we obtain

$$|U(t)| \leq K \int G\left(2\sqrt{\mu t}\,\|x\|\right) dx \to 0$$

as $t \to \infty$.

Hence $\operatorname{var} J\left(t, a\sqrt{t}\right) \to 0$, as $t \to \infty$.

Using last expression and Chebyshev's inequality we have we obtain next

Lemma 4.3.4. If $t \to \infty$, then

$$J\left(t, a\sqrt{t}\right) \xrightarrow{P} C_0 \tag{4.3.8}$$

Proof of the Teorem 4.3.1.

In view of Slutzky's Lemma 4.3.1. it is sufficient to prove the weak convergence of the finite-dimensional distributions of a vector field

$$\begin{aligned} \tilde{X}^t(a) &= (\tilde{X}_1^t(a), \ldots, \tilde{X}_n^t(a)) \\ &= t^{1/2+n/4} I\left(t, a\sqrt{t}\right), \quad a \in R^n, \end{aligned}$$

as $t \to \infty$ to the finite-dimensional distributions of the Gaussian homogeneous vector field $\tilde{X}(a)$, $a \in R^n$, with $E\tilde{X}(a) = 0$ and the matrix-valued correlation function

$$\begin{aligned} \tilde{T}(a) &= E\,\tilde{X}(a)\tilde{X}(0) \\ &= \tilde{a} \exp\left\{-\frac{\|a\|^2}{8\mu}\right\} \Sigma(a) \end{aligned}$$

where $\Sigma(a)$ is defined in the Theorem 4.3.1. with $t = s = 1$ and

$$\tilde{a} = \frac{\mu^{1-n/2}}{(2\sqrt{2\pi})^n} \int G(\|u\|)du.$$

Let us introduce matrix-valued function

$$\begin{aligned} F(a) &= \exp\left\{-\frac{\|a\|^2}{8\mu}\right\}\Sigma(a) \\ &= (F_{ij}(a))_{1\leq i,j\leq n}. \end{aligned}$$

The proof is similar to the one presented in Ivanov and Leonenko (1989, Theorem 2.3.1.) and based on the Markov Method of Moments and the so called Diagram Formula (see Breuer and Major (1983), Malyshev and Minlos (1985), Ivanov and Leonenko (1989)).

We shall subdivide the proof into five stages.

1. Fix the set of real vectors $\{a_i, i = 1, 2, \ldots, r\} \subset R^n$ and write $M = n \times r$. Next for any integers i, k define the functions:

$$\alpha(k) = i \quad \text{and} \quad \beta(k) = k - (i-1)n \quad \text{if} \quad (i-1)n < k \leq in.$$

Introduce the following notation:

$$\hat{X}_k^t = \tilde{X}_{\beta(k)}^t(a_{\alpha(k)}), \qquad \hat{T}_{ij} = E\hat{X}_i^t\hat{X}_j^t = \hat{T}_{\beta(i)\beta(j)}(a_{\alpha(i)} - a_{\alpha(j)}).$$

For any set of real numbers $\{\lambda_i, i = 1, 2, \ldots, M\}$ consider

$$\zeta_t = \sum_{i=1}^{M} \lambda_i \hat{X}_k^t.$$

By the Markov Moment Method it suffices to show that for any $p \geq 2$

$$\lim_{t\to\infty} E\,\zeta_t^p = \begin{cases} 0, & p = 2\nu + 1, \\ (p-1)!!(\sum_{i,j=1}^{M} \lambda_i\lambda_j\tilde{T}_{ij})^\nu, & p = 2\nu \end{cases} \tag{4.3.9}$$

Let

$$\eta^l(a,t) = t^{1/2+n/4} \int g^0(a,y,t)H_l(\xi(y))dy, \quad \gamma^l(t) = \operatorname{var}\eta^l(a,t).$$

Note that for any $l \geq m$

$$\begin{aligned}\lim_{t\to\infty} \gamma^l(t) &= \left[\frac{\mu^{1-n/2}}{\left(2\sqrt{2\pi}\right)^n} l! \int B^l(\|x\|)dx\right] I_{n\times n} \\ &= k_1(l) I_{n\times n},\end{aligned}$$

where the matrix $I_{n\times n} = [\delta_i^j]_{i,j=1}^n$ and δ_i^j is the usual Kronecker's symbol, and

$$\tilde{a} = \sum_{l=m}^{\infty} \left(\frac{C_l}{l!}\right)^2 k_2(l).$$

Now $\zeta(t)$ can be presented as

$$\zeta_t = \sum_{i=1}^{M} \lambda_i \sum_{l=m}^{\infty} \frac{C_l}{l!} \eta_i^l(t),$$

where

$$\eta_i^l(t) = \eta_{\beta(i)}^l(a_{\alpha(i)}, t).$$

Write

$$\zeta_t = \zeta_t' + \zeta_t'' \quad \text{for any } N \geq r,$$

where

$$\zeta_t' = \sum_{i=1}^{M} \lambda_i \sum_{l=m}^{\infty} \frac{C_l}{l!} \eta_i^l(t), \qquad \zeta_t'' = \sum_{i=1}^{M} \lambda_i \sum_{l>N} \frac{C_l}{l!} \eta_i^l(t). \tag{4.3.10}$$

It is clear, that

$$\lim_{t\to\infty} \operatorname{var} \zeta_t'' \leq K_3 \sum_{i=1}^{M} |\lambda_i|^2 \sum_{l>N} \frac{C_l^2}{l!} \int B(\|x\|)^m dx.$$

It is easy to see that for any $\varepsilon > 0$ there exist N_0, t_0 such that for any $N > N_0$, $t > t_0$ $\operatorname{var} \zeta_t'' < \varepsilon$.

Henceforth we obtain that it is sufficient to show that (4.3.9) holds true provided ζ_t is substituted by ζ_t' and $\tilde{a}$ by a', where

$$a' = \sum_{l=m}^{N} \left(\frac{C_l}{l!}\right)^2 k_2(l).$$

2. A graph Γ with $u_1 + \cdots + u_p$ vertices is called a diagram of order $(u_1, \ldots, u_p)$ if:

a) the set of vertices V of the graph Γ is of the form $V = \cup_{j=1}^{p} W_j$, where $W_j = \{(j, l) : 1 \leq l \leq u_j\}$ is the jth level of the graph Γ, $1 \leq j \leq p$ (if $u_j = 0$ assume $W_j = \emptyset$);

b) each vertex is of degree 1;

c) if $((j_1, l_1),\ (j_2, l_2)) \in \Gamma$ then $j_1 \neq j_2$, that is, the edges of the graph Γ may connect only different levels.

Let $T = T(u_1, \ldots, u_p)$ be a set of diagrams Γ of order $(u_1, \ldots, u_p)$. Denote by $R(V)$ the set of edges of a graph $\Gamma \in T$. For the edge $\omega = ((j_1, l_1),\ (j_2, l_2)) \in R(V)$, $j_1 < j_2$, we set $d_1(\omega) = j_1$, $d_2(\omega) = j_2$. We call a diagram Γ regular if its levels can be split into pairs in such a manner that no edge connects the levels belonging to different pairs. For the set of regular diagrams $T^* \subseteq T(u_1, \ldots, u_p)$ holds. If p is odd, then $T = \emptyset$. The following assertion is called the Diagram Formula.

Let $(\zeta_1, \ldots, \zeta_p)$, $p \geq 2$, be a Gaussian vector with

$$E\zeta_j = 0, \quad E\zeta_j^2 = 1, \quad E\zeta_i\zeta_j = r(i, j), \quad i, j = 1, \ldots, p,$$

and $H_{l_1}(u), \ldots, H_{l_p}(u)$ are the Chebyshev–Hermite polynomials. Then

$$E\left\{\prod_{j=1}^{p} H_{l_j}(\zeta_j)\right\} = \sum_{\Gamma \in T} I_\Gamma,$$

where $T = T(l_1, \ldots, l_p)$ and

$$I_\Gamma = \prod_{\omega \in R(V)} r(d_1(\omega),\ d_2(\omega)).$$

3. The following two parts of proof repeat without any essential differences the ideas of Ivanov and Leonenko (1989).

Let

$$\begin{aligned} D_s = \{(J, N) : \quad & J = (j_1, \ldots, j_s), \quad 1 \leq j_i \leq M, \quad i = 1, \ldots, s; \\ & N = (\nu_1, \ldots, \nu_s), \quad \nu_i = \left(k_1^i, ..., k_1^i\right) \in S_{j_i}, \\ & m \leq j_i \leq N, \quad i = 1, \ldots, s\}. \end{aligned}$$

$$K(J,N) = \prod_{i=1}^{s} \frac{\lambda_{j_i} C_{l_i}}{l_i!}, \qquad \int^{(s)} \dots = \prod_{i=1}^{s} \int \dots .$$

Then according to the Diagram Formula we obtain from (4.3.10) for $s \geq 2$

$$E(\zeta_t')^s = \sum_{D_s} K(J,N) \sum_{\Gamma \in T} F_\Gamma(J,N,t) \tag{4.3.11}$$

where

$$\begin{aligned} F_\Gamma(J,N,t) &= t^{s(1/2+n/4)} \\ &\times \int^{(s)} \prod_{i=1}^{s} g^0_{\beta(j_i)}(a_{\alpha(j_i)}, t, x_i) \\ &\times \prod_{\omega \in R(V)} B(\|x_{d_1(\omega)} - x_{d_2(\omega)}\|)\, dx_1 \dots dx_p. \end{aligned} \tag{4.3.12}$$

Split the sum

$$\sum_{\Gamma \in T} \dots = \sum_{\Gamma \in T^*} \dots + \sum_{\Gamma \in T \setminus T^*} \dots .$$

Then the sum (4.3.11) is subdivided into two parts. The first one corresponding to $\sum_{\Gamma \in T^*}$ is denoted by $\sum_s^*(t)$ and the second one corresponding to $\sum_{\Gamma \in T \setminus T^*}$ is denoted by $\sum_s(t)$. We shall study their behavior as $t \to \infty$ separately.

4. If $p = 2\nu + 1$, $\nu = 1, 2, \dots$, then $T^* = \emptyset$ and $\lim_{t \to \infty} \sum_s^*(t) = 0$. For even $s = 2\nu$ the integral $\int^{(s)}$ in (4.3.12) is subdivided into a product of pairs of integrals such that

$$\begin{aligned} &\lim_{t \to \infty} t^{1+n/2} \int\int g^0_{\beta(i)}(a_{\alpha(i)}, t, u) g^0_{\beta(j)}(a_{\alpha(i)}, t, v) B^{w(i)}(\|u - v\|) du dv \\ = \; & F_{\beta(i)\beta(i)}(a_{\alpha(i)} - a_{\alpha(j)}) k_1(w(i)), \end{aligned}$$

where the ith and the jth levels are joint levels of the diagram of power $w(i)$.

We assume without loss of generality that the diagram joints the $(2r-1)$th level and $2r$-level, $r = 1, \dots, \nu$. We have

$$\lim_{t \to \infty} F_\Gamma(J,N,t) = \prod_{r=1}^{\nu} t^{(1/2+n/4)} \int\int g^0_{\beta j(2r-1)}(a_{\alpha(j_{2r-1})}, t, u)$$

$$\times g^0_{\beta j(2r)}(a_{\alpha(j_{2r})}, t, u) B^{w(2r)}(\|u - v\|) du dv$$

$$= \prod_{r=1}^{\nu} F_{\beta(j_{2r-1})\beta_{(j_{2r})}}) \left(a_{\alpha(j_{2r-1})} - a_{\alpha(j_{2r})}\right) k_2(w(2r)).$$

Then

$$\begin{aligned} I(\Gamma) &= \lim_{t\to\infty} \sum_{D_s} K(J,N) F_\Gamma(J,N,t) \\ &= \left(\sum_{i,j=1}^{M} \lambda_i \lambda_j F_{\beta(i)\beta(j)}(a_{\alpha(i)} - a_{\alpha(j)}) \right)^{\nu} \\ &\quad \times \prod_{r=1}^{\nu} \left[\frac{C^2_{w(2r)}}{w(2r)!^2} k_1(w(2r)) \right] \end{aligned}$$

and

$$\lim_{t\to\infty} \sum\nolimits_p^{*}(t) = \sum_{\Gamma\in T^*} I(\Gamma).$$

The number of regular diagrams with $2m_j$ levels of cardinality $|w(j)|$, $j = 1, \ldots, l$, $\sum_{j=1}^{l} m_j = \nu$ (all the $w(j)$ are different) is equal to

$$\frac{(2\nu - 1)!!\ \nu!}{m_1! \cdots m_l!} \prod_{j=1}^{l} (w(j)!)^{m_j}.$$

For such diagrams (with fixed s and $w(j)$, $j = 1, \ldots, l$) we have

$$I(\Gamma) = \left(\sum_{i,j=1}^{M} \lambda_i \lambda_j F_{\beta(i)\beta(j)}(a_{\alpha(i)} - a_{\alpha(j)}) \right)^{\nu} \prod_{j=1}^{l} \left[\frac{C^2_{w(j)}}{w(j)!^2} k_1(w(j)) \right]^{m_j}.$$

Recalling that $s = 2\nu$ we obtain

$$\begin{aligned} \lim_{t\to\infty} \sum\nolimits_p^{*}(t) &= (p-1)!! \\ &\quad \times \left(\sum_{i,j=1}^{M} \lambda_i \lambda_j F_{\beta(i)\beta(j)}(a_{\alpha(i)} - a_{\alpha(j)}) \right)^{s/2} \\ &\quad \times \left(\sum_{i=m}^{N} \frac{{C_j}^2}{i!^2} k_1(i) \right)^{s/2}. \end{aligned}$$

5. To prove our assertion it remains to show that

$$\lim_{t\to\infty}\sum\nolimits_p(t)=0.$$

This part of proof has essential differences from one described by Ivanov and Leonenko (1989). We shall prove that

$$\begin{aligned} F_\Gamma(J,N,t) &= t^{s(1/2+n/4)} \\ &\quad\times\int^{(s)}\prod_{i=1}^{s} g^0_{\beta(j_i)}(a_{\alpha(j_i)},t,x_i) \\ &\quad\times\prod_{\omega\in R(V)} B(\|x_{d_1(\omega)}-x_{d_2(\omega)}\|)dx_1\dots dx_s \\ &= o(1), \end{aligned}$$

as $t\to\infty$ for a non-regular diagram Γ of any order (odd or even).

It is clear that

$$F_\Gamma(J,L,t)=F_{\pi\Gamma}(\pi(J),\pi(L),t)$$

for any π, permutation of the set $(1,\dots,s)$. Therefore we may ordering the set of levels of diagram in any way. Thus

$$\begin{aligned} F_\Gamma(J,N,t) &= t^{s(1/2+n/4)} \\ &\quad\times\int^{(s)}\prod_{i=1}^{s} g^0_{\beta(j_i)}(a_{\alpha(j_i)},t,x_i) \\ &\quad\times\prod_{i=1}^{s}\prod_{\omega\in R(V)} B(\|x_i-x_{d_2(\omega)}\|)dx_1\dots dx_p. \end{aligned}$$

Denote

$$W(i)=\{\omega\in R(V):d_1(\omega)=i\}.$$

We now proceed to estimate step by step the integrals of the following form:

$$\int g^0_{\beta(j_i)}(a_{\alpha(j_i)},t,x_i)\prod_{W(i)} B\left(\|x_i-x_{d_2(\omega)}\|\right)dx_i$$

and in each step shall obtain the remaining integral of the form

$$\int^{(k)} \prod_{i=1}^{k} g^{0}_{\beta(j_i)}(a_{\alpha(j_i)}, t, x_i) \prod_{i=1}^{k} \prod_{W(i)} B\left(\left\|x_i - x_{d_2(\omega)}\right\|\right) dx_1 \dots dx_k.$$

$(k = p-1, p-2, \dots)$, defined by a diagram that can be obtained from the previous one by excluding all edges with $d_1(w) = i$.

If the diagram is non-regular that there exists an order of integration such that k times will be taken a non-empty subdiagram Γ', where

$$k = \begin{cases} \frac{s}{2}, & s \text{ is even,} \\ \frac{s-1}{2}, & s \text{ is odd} \end{cases}$$

then we estimate $(k+1)$-times integrals of a type

$$\int g^{0}_{\beta(j_i)}(a_{\alpha(j_i)}, t, x_i) \prod_{W(i)} B\left(\left\|x_i - x_{d_2(\omega)}\right\|\right) dx_i.$$

Denote by $q_\Gamma(i)$ the number of edges $\omega \in R(V)$ such that $d_1(\omega) = i$. Then by Lemma 4.3.2. we have

$$\begin{aligned} &\left| \int |g^{0}_{\beta(j_i)}(a_{\alpha(j_i)}, t, x_i)| \prod_{W(i)} B(\left\|x_i - x_{d_2(\omega)}\right\|) dx_i \right| \\ \leq &\int |g^{0}_{\beta(j_i)}(a_{\alpha(j_i)}, t, x_i)| \frac{1}{q_\Gamma(i)} \sum_{W(i)} B\left(\left\|x_i - x_{d_2(\omega)}\right\|\right)^{q_\Gamma(i)} dx_i \\ \leq &\; K_4\, t^{-(n+1)/2}. \end{aligned}$$

The rest of integrals is of the order $O(t^{-1/2})$ as $t \to \infty$ by Lemma 4.3.2. Thus

$$|F_\Gamma(J, L, t)| \leq K_5 t^{s(1/2+n/4)-(k+1)(n+1)/2-(s-k-1)/2}$$

for any non-regular diagram and

$$s\left(\frac{1}{2} + \frac{n}{4}\right) - \frac{(k+1)(n+1)}{2} - \frac{(s-k-1)}{2} = -\frac{(2(k+1)-s)n}{4} < 0$$

because of $k+1 > s/2$. So $|F_\Gamma(J, L, t)| = o(1)$, as $t \to \infty$.

Thus, the Theorem 4.3.1. is proved for the case $U_t(1,a)$. Note that the statement of Theorem 4.3.1. is a generalization of Theorem 3 of a paper Albeverio, Molchanov and Surgailis (1994).

We shall only indicate the main steps of the proof, based on the method of moments and diagram formalism.

Let us consider the case $\eta(x) = \xi(x)$, $x \in R^n$, for simplifying, where $\xi(x)$ satisfies condition A. We obtain from (2.4.6):

$$\begin{aligned}
\operatorname{cov}(J(t_1,x_1)J(t_2,x_2)) &= \int_{R^n}\int_{R^n} g(t_1,x_1-y_1)g(t_2,x_2-y_2) \\
&\quad \times \left[e^{1/4\mu^2}(e^{(B(\|y_1-y_2\|)/4\mu^2} - 1\right] dy_1 dy_2 \\
&= e^{1/4\mu^2}\int_{R^n}\left[\int_{R^n} g(t_2,u)g\left(t_1,u+(x_1-x_2-z)\right)du\right] \\
&\quad \times \left(e^{B(z)/4\mu^2}-1\right)dz \\
&= e^{1/4\mu^2}\int_{R^n} [2\pi\mu(t_1+t_2)]^{-n/2} \\
&\quad \times \exp\left\{-\frac{\|x_1-x_2-z\|^2}{2\mu(t_1+t_2)}\right\}\left(e^{B(z)/4\mu^2}-1\right)dz
\end{aligned}$$

and

$$\begin{aligned}
&\lim_{\beta\to\infty} \beta^{n/2}\operatorname{cov}(J(t_1,x_1)J(t_2,x_2)) \\
&= k_2(8\mu t)^{-n/2}\exp\left\{-\frac{\|x_1-x_2\|^2}{8\mu}\right\}, \quad x_1,x_2\in R^n, \qquad (4.3.13)
\end{aligned}$$

where

$$k_2 = e^{1/4\mu^2}\left[\int_{R^n}(e^{B(z)/4\mu^2}-1)dz\right].$$

Then, by method of moments with diagram formalism (see the proof in the case $U_t(1,a)$ above) it is easy to prove that the finite-dimensional distributions of the

$$\zeta(\beta,x) = \beta^{n/4}\left(J\left(\beta,x\sqrt{\beta}\right) - e^{1/4\mu^2}\right)$$

converge weakly to the finite-dimensional distributions of centrad homogeneous isotropic Gaussian field with covariance function (4.3.13). By functional centrad limit theorem (see Albeverio, Molchanov and Surgailis (1994),

Theorem 2) we obtain that random field

$$\eta(t\beta, x) \xrightarrow{D} \sqrt{k_2} \int_{R^n} g(t, x-y) W(dy) = \sqrt{k_2} w(t, x) \tag{4.3.14}$$

as $\beta \to \infty$ in the sense of finite-dimensional distributions, where $W(\cdot)$ is a Gaussian white noise in R^n. The limiting random field in (4.3.14) has covariance function

$$k_2 \operatorname{cov}(w(t_1, x_1), w(t_2, x_2)) = g(t_1 + t_2, x_1 - x_2).$$

Then asymptotically distributions of the velocity field (4.2.5) can be obtained from the formula (4.2.5), where the denomenator tends in probability to a constant (see Lemma 4.3.4.) while in the numerator the gradient commutes with the passage to the limit.

So that, by Lemma 4.3.1. the random fields

$$U_\beta(t, x) \xrightarrow{D} U(t, x) = -2\mu\sqrt{k_3} \int_{R^n} \nabla\, g(t, x-y) W(dy), \tag{4.3.15}$$

as $\beta \to \infty$, where

$$k_3 = k_2 e^{-1/4\mu^2}.$$

The limiting field $U(t, x)$ has moving average representation (4.3.15). Thus, $U(t, x)$ is homogeneous in x Gaussian random field with covariance function

$$\begin{aligned} R(x-y) &= \int_{R^n} \nabla g(t, x-y) \nabla g(s, y-u) du \\ &= \int_{R^n} e^{i<\lambda, x-y>} f(\lambda) d\lambda, \end{aligned} \tag{4.3.16}$$

where the spectral density

$$f(\lambda) = (2\pi)^{-n} \gamma_t(\lambda) \overline{\gamma_s(\lambda)}, \quad \lambda \in R^n,$$

and

$$\gamma_t(\lambda) = \int_{R^n} e^{-i<\lambda, v>} \nabla g(t, v) dv.$$

Put $\sigma^2 = 2\pi\mu t$ in the formula

$$\int_{R^n} \lambda_j e^{-i<\lambda, v>} \frac{e^{-(\|v\|^2/2\sigma^2)}}{(2\pi\sigma^2)^{n/2}} = (-1)^n i^{-1} \lambda_i\, \sigma^2 e^{-(\|\lambda\|^2 \sigma^2/2)}, \quad j \in \{1, \ldots, n\}.$$

We obtain

$$\gamma_t(\lambda) = 2\mu\sqrt{k_3}\frac{(-1)^n}{i}\exp\left\{-\frac{\|\lambda\|^2\sigma^2}{2}\right\}\lambda_j,$$

and

$$\begin{aligned} f(\lambda) &= (2\pi)^{-n}(2\mu)^2 k_3 \\ &\quad \times \exp\left\{-\|\lambda\|^2\mu(t+s)\right\}(\lambda_k\,\lambda_i)_{1\le k,j\le n}, \qquad (4.3.17) \\ \lambda &\in R^n. \end{aligned}$$

Using (4.3.16), (4.3.17) and the formula

$$\int_{R^n}\lambda_k\,\lambda_i\exp\left\{-<\lambda,x>-\frac{\|\lambda\|^2}{2\kappa^2}\right\}d\lambda$$

$$= \begin{cases} -(2\pi\,\kappa^2)^{n/2}\kappa^4 x_j\,x_k\,\exp\{-\frac{\|x\|^2\kappa^2}{2}\}, & j\neq k \\ -(2\pi\,\kappa^2)^{n/2}\kappa^2(1-\kappa^2\,x_j^2)\,\exp\{-\frac{\|x\|^2\kappa^2}{2}\}, & j=k, \end{cases}$$

we obtain (4.3.1).

The proof of the Theorem in general case is similar (for a more details see Albeverio, Molchanov and Surgailis (1994) and Deriev and Leonenko (1997)).

4.4 Parabolic Limits for Strongly Dependent Random Initial Conditions: the Gaussian Scenario

We study the scaling limit of random fields which are solutions of a multi-dimensional Burgers' equation under initial condition with long-range dependence (singular spectrum) and provide in explicit formula for the spectral density of the limiting Gaussian field.

We shall need the following condition :

A'. Condition A is satisfied with covariance function

$$\begin{aligned} B(\|x\|) &= \|x\|^{-\alpha}L(\|x\|)\searrow 0, \quad \text{as } \|x\|\to\infty, \\ 0 &< \alpha<n,\; x\in R^n, \end{aligned}$$

where the slowly varying function $L \in \mathcal{L}$.

A''. Condition A' is satisfied and there exists a spectral density $f(\lambda)$, $\lambda \in R^n$, which is decreasing from $\|\lambda\| \geq \lambda_0 > 0$ and continuous for all $\lambda \neq 0$.

Under the condition A''

$$\xi(x) = \int_{R^n} e^{i<\lambda,x>} \sqrt{f(\lambda)} W(d\lambda),$$

where $W(\cdot)$ is a Gaussian white noise, and $f(\lambda)$ satisfies (1.4.12) or (2.4.2), that is,

$$f(\|\lambda\|) \sim \|\lambda\|^{\alpha-n} L\left(\frac{1}{\|\lambda\|}\right) c_5(n,\alpha), \quad 0 < \alpha < n, \quad \|\lambda\| \to 0, \tag{4.4.1}$$

where the Tauberian constant $c_5(n,\alpha)$ is given in (1.4.13). This is singular property of initial condition.

Theorem 4.4.1. Let $u(t,x)$, $t > 0$, $x \in R^n$, be a solution of the initial value problem (4.2.1)–(4.2.2) with random initial condition $\eta(x) = F(\xi(x))$, $x \in R^n$, where the random field $\xi(x)$, $x \in R^n$, and the non-random function $F(\cdot)$ satisfy conditions A', B, C and

$$C_1 = \int_{-\infty}^{\infty} \exp\left\{-F(u)\frac{1}{2\mu}\right\} \phi(u)du \neq 0.$$

Suppose that random field $\eta(x) = F(\xi(x))$, $x \in R^n$, is a.s. differentiable.

Then the finite-dimensional distributions of the fields

$$Y_\beta(t,x) = \frac{\beta^{1/2+\alpha/4}}{L^{1/2}\left(\sqrt{\beta}\right)} u\left(t\beta, x\sqrt{\beta}\right), \quad (t,x) \in (0,\infty) \times R^n,$$

converge weakly as $\beta \to \infty$ to the finite-dimensional distributions of trhe vector Gaussian field $Y(t,x)$, $t > 0$, $x \in R^n$, with $EY(t,x) = 0$ and the covariance function of the form

$$\begin{aligned} EY(t,x)[Y(s,y)]' &= V(x-y) \\ &= (V_{i_j}(x-y))_{1\leq i,j\leq n} \\ &= \left(\frac{C_1}{C_0}\right)^2 \frac{(2\mu)^{1-\alpha/2}}{\sqrt{ts}} \\ &\quad \times \left(\int_{R^n}\int_{R^n} w_i z_j \frac{\phi(\|w\|)\phi(\|z\|)dwdz}{\left\|\frac{x-y}{\sqrt{2\mu}} - (w\sqrt{t} - z\sqrt{s})\right\|^\alpha}\right)_{1\leq i,j\leq n} \end{aligned} \tag{4.4.2}$$

where $0 < \alpha < n$; $w = (w_1, \ldots, w_n)' \in R^n$, $z = (z_1, \ldots, z_n)' \in R^n$, $x, y \in R^n$, $t, s > 0$,

$$C_0 = \int_{-\infty}^{\infty} \exp\left\{-F(u)\frac{1}{2\mu}\right\} \phi(u)du$$

and

$$\phi(\|u\|) = (2\pi)^{-n/2} \exp\left\{-\frac{\|u\|^2}{2}\right\}, \quad u \in R^n.$$

In particar, if A" holds then

$$Y(t,x) = \frac{2\mu[c_5(n,\alpha)]^{1/2}}{i}\frac{C_1}{C_0}\int_{R^n} e^{i<\lambda,x-y>}h(\lambda)W(d\lambda), \tag{4.4.3}$$

where $W(\cdot)$ is a complex Gaussian white noise, and

$$h(\lambda) = [h_1(\lambda), \ldots, h(\lambda)]' = \lambda \|\lambda\|^{(\alpha-n)/2} \exp\{-\|\lambda\|^2 \mu t\}, \; x \in R^n.$$

If $F(u) = u$, i.e., $\eta(x) = \xi(x)$, $x \in R^n$, is a Gaussian field, then

$$C_0 = \exp\left\{\frac{1}{8\mu^2}\right\}, \quad C_1 = -\frac{1}{2\mu}\exp\left\{\frac{1}{8\mu^2}\right\}, \quad C_2 = \frac{1}{4\mu^2}\exp\left\{\frac{1}{8\mu^2}\right\} \tag{4.4.4}$$

and

$$V(x-y) = \int_{R^n} e^{i<\lambda,x-y>}q(\lambda)d\lambda,$$

where

$$\begin{aligned} q(\lambda) &= (q_{r_j}(\lambda))_{1\le r,j\le n}, \\ q_{r_j}(\lambda) &= c_5(n,\alpha)\lambda_r\lambda_j e^{-\|\lambda\|^2\mu(t+s)} \|\lambda\|^{(\alpha-n)}, \\ 0 &< \alpha < n, \end{aligned} \tag{4.4.5}$$

Remark 4.4.1 . Using (4.4.5) we obtain

$$\lim \frac{\operatorname{tr} q(\lambda)}{c_5(n,\alpha)} = \begin{cases} 0, & \text{if } \max(0, n-2) < \alpha < n; \\ 1, & \text{if } \alpha = n-2, \quad n \ge 3; \\ \infty, & \text{if } \quad 0 < \alpha < n-2, \quad n \ge 3. \end{cases} \tag{4.4.6}$$

and $\det q(\lambda) = 0$, $\lambda \in R^n$. This is singular properties of the spectrum of rescaled solution $Y(t,x)$ of the Burgers' equation with strongly-dependent initial condition.

Proof.

The proof of Theorem 4.4.1. is based on the expansion in Hilber space $\mathbf{L}_2(\Omega)$ of a Hopf–Cole solution (4.5.2). We have

$$\exp\left\{-\frac{F(\xi(y))}{2\mu}\right\}=\sum_{k=0}^{\infty}\frac{C_k}{k!}H_k(\xi(y)), \tag{4.4.7}$$

where

$$C_k=\int_{-\infty}^{\infty}\exp\left\{-\frac{F(u)}{2\mu}\right\}\phi(u)H_k(u)du, \quad k=0,1,2,\ldots,$$

and $H_k(u)$ are Chebyshev–Hermite polynomials (2.1.5).

Note that for $F(u)=u$ we obtain (4.4.4). From (4.2.6) and (4.4.7) we have

$$\begin{aligned} I\left(\beta t, x\sqrt{\beta}\right) &= C_0\zeta_{0\beta}(t,x)+C_1\zeta_{1\beta}(t,x)+R_{1\beta}+R_{2\beta}, \\ (t,x) &\in (0,\infty)\times R^n,\ \beta>0, \end{aligned} \tag{4.4.8}$$

where

$$\begin{aligned} R_{1\beta} &= \sum_{j\geq 2}\frac{C_j}{j!}\zeta_{j\beta}(t,x), \\ \zeta_{j\beta}(t,x) &= \int_{v(\beta)}\frac{x\sqrt{\beta}-y}{t\beta}g\left(t\beta, x\sqrt{\beta}-y\right)H_j(\xi(y))dy, \quad j=0,1,2,\ldots, \\ R_{2\beta} &= \int_{R^n\setminus v(\beta)}\frac{x\sqrt{\beta}-y}{t\beta}g\left(t\beta, x\sqrt{\beta}-y\right)\exp\left\{-\frac{F(\xi(y))}{2\mu}\right\}dy, \end{aligned}$$

and

$$v(\beta)=\{x\in R^n : \|x\|<\beta\}.$$

Observe that $\zeta_{1\beta}(t,x)$ is a Gaussian random field with $E\zeta_{1\beta}(t,x)=0$, and

$$E\zeta_{0\beta}(t,0)=0, \quad \lim_{\beta\to\infty}\zeta_{0\beta}(t,x)=0.$$

From (2.1.8) we obtain

$$E\zeta_{m\beta}(t,x)[\zeta_{i\beta}(t,x)]'=\delta_m^i(\psi^2_{m,k,r}(\beta))_{1\leq k,r\leq n},$$

where

$$\begin{aligned}\psi^2_{m,k,r}(\beta) &= m! \int_{v(\beta)}\int_{v(\beta)} \frac{x_k\sqrt{\beta}-y_k}{t\beta}\,\frac{x'_k\sqrt{\beta}-y'_k}{s\beta} g\left(t\beta, x\sqrt{\beta}-y\right)\\ &\quad\times g\left(s\beta, x'\sqrt{\beta}-y'\right) B^m(\|y-y'\|)dydy',\end{aligned}$$

and $y=(y_{11},\ldots,y_{1n})'$, $\quad y'=(y'_{11},\ldots,y'_{1n})\in R^n$.

Changing the variables via the transformation

$$\frac{w_k^2}{2}=\frac{\left(x_k\sqrt{\beta}-y_k\right)^2}{4\mu t\beta},\quad \frac{z_r^2}{2}=\frac{\left(x'_r\sqrt{\beta}-y'_r\right)^2}{4\mu s\beta},\quad k,r=1,\ldots,m,$$

and utilizing the basic properties of the slowly varing functions (see Lemma 1.4.12.), we have for $0<\alpha<n/m$, $m\geq 1$, and $\beta\to\infty$

$$\begin{aligned}\psi^2_{m,k,r}(\beta) &= \frac{2\mu m!}{\beta\sqrt{ts}}\int_{D_\beta}\int_{D_\beta} w_k z_r \phi(\|w\|)\phi(\|z\|)\\ &\quad\times B^m\left(\sqrt{2\mu\beta}\left\|\frac{x-x'}{\sqrt{2\mu}}-(w\sqrt{t}-z\sqrt{s})\right\|\right)dwdz\\ &= \frac{2\mu m! L^m(\sqrt{\beta})}{(2\mu)^{m\alpha/2}\beta^{1+m\alpha/2}\sqrt{ts}}\int_{D_\beta}\int_{D_\beta} w_k z_r\\ &\quad\times\frac{L^m\left(\sqrt{2\mu\beta}\left\|\frac{x-x'}{\sqrt{2\mu}}-(w\sqrt{t}-z\sqrt{s})\right\|\right)}{\left\|\frac{x-x'}{\sqrt{2\mu}}-(w\sqrt{t}-z\sqrt{s})\right\|^{\alpha m}L^m(\sqrt{\beta})}dwdz\\ &= (2\mu)^{1-(m\alpha/2)}m!\\ &\quad\times\psi^2_{m,k,r}(x-x')\,\frac{L^m(\sqrt{\beta})}{\beta^{1+m\alpha/2}}\,(1+o(1))\end{aligned}\tag{4.4.9}$$

as $\beta\to\infty$, where

$$\psi^2_{m,k,r}(x-x')=\frac{1}{\sqrt{ts}}\int_{R^n}\int_{R^n}\frac{w_k z_r\phi(\|w\|)\phi(\|z\|)dwdz}{\left\|\frac{x-x'}{\sqrt{2\mu}}-(w\sqrt{t}-z\sqrt{s})\right\|^{\alpha m}},$$

and

$$D_\beta = \left\{y\in R^n:\left\|y+\frac{x}{\sqrt{2\mu t}}\right\|<\sqrt{\frac{\beta}{2\mu t}}\right\}\uparrow R^n,\quad \beta\to\infty.$$

From (4.2.6), (4.2.7) we have

$$J\left(t\beta, x\sqrt{\beta}\right) = \int_{R^n} g\left(t\beta, x\sqrt{\beta} - y\right) \exp\left\{-\frac{F(\xi(y))}{2\mu}\right\} dy \xrightarrow{P} C_0 \quad (4.4.10)$$

as $\beta \to \infty$, by the law of a large number.

From (4.4.8) and (4.4.9) we obtain

$$\frac{C_1}{C_0}\,\zeta_{1\beta}(t,x)\,\frac{\beta^{1/2+\alpha/4}}{L^{1/2}(\sqrt{\beta})} \xrightarrow{D} V(t,x) \quad (4.4.11)$$

as $\beta \to \infty$ in the sence of weak convergence of the finite-dimensional distributions, where $V(t,x)$ is a Gaussian random field with mean zero and covariance function (4.4.2).

In view of Lemma 1.1.3. (Cramer–Wald's arguments), Lemma 4.3.1. (Slutsky's arguments) and (4.2.5), (4.4.10), (4.4.11) it suffices to verify relation

$$\frac{\beta^{1/2+\alpha/4}}{L\left(\sqrt{\beta}\right)}\left(R'_\beta + R''_\beta\right) \xrightarrow{P} 0 \quad (4.4.12)$$

as $\beta \to \infty$, where $0 = (0, \ldots, 0) \in R^n$.

Let

$$\|A\| = \sum_{1 \le i,j \le n} |a_{ij}|$$

be a norm of a matrix $A = (a_{ij})_{1 \le i,j \le n}$. Then it safficies to prove that

$$\lim_{\beta \to \infty} \frac{\beta^{1/2+\alpha/4}}{L\left(\sqrt{\beta}\right)} \left\|ER_{j\beta}R'_{j\beta}\right\| = 0, \quad j = 1, 2, \ldots. \quad (4.4.13)$$

From (4.4.9) we obtain inequality

$$\begin{aligned}\frac{\psi^2_{j,k,r}(\beta)}{j!} - \frac{\psi^2_{i,k,r}(\beta)}{i!} &= \frac{2\mu}{\beta\sqrt{ts}} \int_{D_\beta}\int_{D_\beta} \phi(\|w\|)\phi(\|z\|) \\ &\times \left[B^j\sqrt{2\mu\beta}\left\|\frac{x-x'}{\sqrt{2\mu}} - \left(w\sqrt{t} - z\sqrt{s}\right)\right\|\right. \\ &\left. - B^i\sqrt{2\mu\beta}\left\|\frac{x-x'}{\sqrt{2\mu}} - \left(w\sqrt{t} - z\sqrt{s}\right)\right\|\right] \\ &\times dw_1 \ldots dw_{k-1} d\left(\frac{w_k^2}{2}\right) dw_{k+1} \ldots dw_n\end{aligned}$$

$$\times dz_1 \dots dz_{r-1} d\left(\frac{z_r^2}{2}\right) dz_{r+1} \dots dz_n$$
$$\geq 0,$$

for $j \leq i$. Therefore for $j \leq i$

$$\frac{\psi^2_{i,k,r}(\beta)}{i!} \leq \frac{\psi^2_{j,k,r}(\beta)}{j!}, \tag{4.4.13a}$$

and

$$\begin{aligned} \|ER_{1\beta}R'_{1\beta}\| &= \sum_{1\leq k,r\leq n} \left|\sum_{i\geq 2}\left(\frac{C_i}{i!}\right)^2 \psi^2_{i,k,r}(\beta)\right| \\ &\leq \frac{1}{2}\sum_{1\leq k,r\leq n} \left|\psi^2_{2,k,r}(\beta)\right| \sum_{i\geq 2}\frac{C_i^2}{i!}. \end{aligned} \tag{4.4.14}$$

Under the assumption A' for any $\varepsilon > 0$ there exists a vector $A = (A_{p_1}, \dots A_{p_n})$, $A > 0$, $p_i \geq 0$, $p_1 + \dots + p_n = 1$, such that $B(\|y - y'\|) < \varepsilon$ for $|y_i - y_i'|^2 \geq p_i A^2$, $i = 1, \dots, n$, $\|y - y'\| \geq A$. Now we analyse $\psi^2_{2,k,r}(\beta)$ which we split up in two parts as follows

$$\begin{aligned} \psi^2_{2,k,r}(\beta) &= \left[\iint_{\Delta_1} + \iint_{\Delta_2}\right]\left(\frac{x_k\sqrt{\beta}-y_k}{\beta t}\right)\left(\frac{x_k'\sqrt{\beta}-y_k'}{\beta s}\right) \\ &\quad\times g\left(t\beta, x\sqrt{\beta}-y\right) g\left(s\beta, x'\sqrt{\beta}-y'\right) B^2(\|y-y'\|)dydy' \\ &= S_1 + S_2, \end{aligned}$$

where

$$\Delta_1 = \{(y_1, y_2) \in v(\beta) \times v(\beta) : \|y - y'\| \geq A, \quad |y_i - y_i'| \geq p_i A^2, i = 1, \dots r\}$$

and $\Delta_2 = v(\beta) \times v(\beta) \backslash \Delta_1$. Using the bound $B^2(\cdot) \leq 1$ on a set Δ_2 and the bound $B^2(\cdot) \leq \varepsilon B(\cdot)$ on a set Δ_1, we obtain

$$\begin{aligned} S_2 &\leq \frac{k_1}{\beta}\iint_{\Delta_2'} w_{1i}w_{2i}\phi(\|w_1\|)\phi(\|w_2\|)dw_1dw_2 \\ &\leq \frac{k_2}{\beta}\int_{-a(\beta,t,x)}^{b(\beta,t,x)} w_{1i}\phi(w_{1i})dw_{1i}\int_{-a(\beta,s,x')}^{b(\beta,s,x')} w_{2i}\phi(w_{2i})dw_{2i} \end{aligned}$$

$$\begin{aligned} &\leq \frac{k_3}{\beta}\phi(b(\beta,t,x))\phi(b(\beta,s,x')) \\ &\leq \frac{k_4}{\beta}e^{-(\beta/k_5)}, \end{aligned} \tag{4.4.15}$$

where $k_i > 0, \quad i = 1,2,3,4, \quad \Delta_2' \subset D_\beta \times D_\beta$,

$$a(\beta,t,x) = -\frac{\sqrt{\beta}}{\sqrt{2\mu t}} - \frac{x}{\sqrt{2\mu t}}, \qquad b(\beta,t,x) = \frac{\sqrt{\beta}}{\sqrt{2\mu t}} - \frac{x}{\sqrt{2\mu t}}.$$

Similar

$$\begin{aligned} S_1 &\leq \varepsilon\frac{k_4}{\beta}\iint_{\Delta_1'} w_{1i}w_{2i}\phi(\|w_1\|)\phi(\|w_2\|) \\ &\quad \times B\left(\sqrt{2\mu\beta}\left\|\frac{x-x'}{\sqrt{2\mu}} - (w_1\sqrt{t} - w_2\sqrt{s})\right\|\right) dw_1dw_2 \\ &\leq \varepsilon\frac{k_5\, L(\sqrt{\beta})}{\beta^{1+\alpha/2}}, \end{aligned} \tag{4.4.16}$$

where $k_4, k_5 > 0$.

Combining (4.4.14)–(4.4.16) we obtain

$$\|ER_{1\beta}R'_{1\beta}\| \leq \frac{k_6}{\beta}e^{-(\beta/k_5)} + \varepsilon\frac{L(\sqrt{\beta})}{\beta^{1+\alpha/2}}. \tag{4.4.17}$$

In view of (4.4.17) and (4.4.9) (in the case $m = 1$)

$$\beta^{1+\alpha/2}\|ER_{1\beta}R'_{1\beta}\| \to 0 \tag{4.4.18}$$

as $\beta \to \infty$, since ε is arbitrarily small.

Similar we obtain

$$\begin{aligned} \|ER_{2\beta}R'_{2\beta}\| &\leq \int\limits_{R^n\setminus v(\beta)}\int\limits_{R^n\setminus v(\beta)} \left|\frac{x\sqrt{\beta}-y}{t\beta}\right|\left|\frac{x'\sqrt{\beta}-y'}{s\beta}\right| \\ &\quad \times g\left(t\beta, x\sqrt{\beta}-y\right) g\left(s\beta, x'\sqrt{\beta}-y'\right) \\ &\quad \times E\exp\left\{-\frac{F(\xi(y_1)) + F(\xi(y_2))}{2\mu}\right\} dydy' \\ &\leq \frac{k_7}{\beta}E\exp\left\{-\frac{F(\xi(y_0))}{2\mu}\right\} \end{aligned}$$

$$\times \iint_{\Delta_3} |w_{1_i}\, w_{2_i}|\, \phi(\|w_1\|)\phi(\|w_2\|\, dw_1 dw_2$$
$$\leq k_8\, e^{-(\beta/k_9)},$$

where

$$\Delta_3 = \{w_i \notin D_\beta, i=1,2\} \cup \{w_1 \in D_\beta, w_2 \notin D_\beta\} \cup \{w_1 \notin D_\beta, w_2 \in D_\beta\}.$$

Thus

$$\lim_{\beta\to\infty} \beta^{1+\alpha/2}\ \left\|ER_{2\beta}R'_{2\beta}\right\| \to 0. \tag{4.4.19}$$

From (4.4.18) and (4.4.19) we obtain (4.4.13). We conclude the proof of the first part of the Theorem 4.4.1.

Let us show that the homogeneous field $Y(t,x)$ has spectral density (4.4.5).

Using the self-similar property of the Gaussian white noise $W(\cdot)$

$$W(d(a\lambda)) \stackrel{D}{=} a^{n/2}W(d\lambda), \quad a>0,$$

we obtain

$$\begin{aligned}
\tilde{Y}_\beta(t,x) &= \frac{\beta^{1+\alpha/2}}{L^{1/2}(\sqrt{\beta})}\,\frac{C_1}{C_0}\,\zeta_{1\beta}(t,x)\\
&\stackrel{D}{=} \frac{\beta^{1+\alpha/2}}{L^{1/2}(\sqrt{\beta})}\,\frac{C_1}{C_0}\\
&\times \int_{R^n}\left[\int_{v(\sqrt{\beta})} \frac{x-y}{t\,(4\pi\mu t)^{n/2}}\ \exp\left\{-\frac{\|x-y\|^2}{4\mu t} + i<\lambda,y>\right\}dy\right]\\
&\times\sqrt{f\left(\frac{\|\lambda\|}{\sqrt{\beta}}\right)}\ \beta^{-n/4-1/2}W(d\lambda).
\end{aligned} \tag{4.4.20}$$

Using the identity

$$\begin{aligned}
\int_{R^n} &\frac{\exp\left\{i<\lambda,z> - \frac{\|z-x\|^2}{4\mu t}\right\}}{(4\pi\mu t)^{n/2}}(x-z)dz\\
&= \frac{4\mu t\lambda}{i}\ e^{i<\lambda,x>-\mu t\|\lambda\|^2}, \qquad \lambda\in R^n
\end{aligned} \tag{4.4.21}$$

we have with help (4.4.3) and (4.4.20) the following

$$E\left\|\tilde{Y}_\beta(t,x)-T(t,x)\right\|^2=\frac{\beta^{1+\alpha/2}}{L(\sqrt{\beta})}\left(\frac{2\mu C_1}{C_0}\right)^2\times$$

$$E\left\|\frac{1}{2\mu}\int_{R^n}\left[\int_{v(\sqrt{\beta})}\frac{x-y}{t\,(4\pi\mu t)^{n/2}}\exp\left\{-\frac{\|x-y\|^2}{4\mu t}-i<\lambda,y>\right\}dy\right]\times\right.$$

$$\left.\sqrt{f(\frac{\|\lambda\|}{\sqrt{\beta}})}\beta^{-n/4-1/2}W(d\lambda)-\frac{c_5(n,\alpha)}{i}\int_{R^n}\frac{\lambda\exp\{i<\lambda,x>-\mu t\,\|\lambda\|^2\}}{\|\lambda\|^{(n-\alpha)/2}}W(d\lambda)\right\|^2\le$$

$$\frac{\beta^{1+\alpha/2}}{L(\sqrt{\beta})}\left(\frac{2\mu C_1}{C_0}\right)^2\times$$

$$E\left\|-\int_{R^n}\frac{\lambda\ \exp\{i<\lambda,x>-\mu t\,\|\lambda\|^2\}}{i\ \beta^{n/4+1/2}}\sqrt{f(\frac{\|\lambda\|}{\sqrt{\beta}})}W(d\lambda)+\right.$$

$$\int_{R^n}\left[\int_{R^n\backslash v(\sqrt{\beta})}\frac{(x-y)\exp\{i<\lambda,x>-\frac{\|x-y\|^2}{4\mu t}\}}{2\mu t\ (4\pi\mu t)^{n/2}\beta^{n/4+1/2}}dy\right]\sqrt{f(\frac{\|\lambda\|}{\sqrt{\beta}})}W(d\lambda)+$$

$$\left.\int_{R^n}\frac{\lambda\exp\{i<\lambda,x>-\mu t\,\|\lambda\|^2\}}{i\ \|\lambda\|^{(n-\alpha)/2}\ \beta^{1/2+\alpha/4}}\sqrt{c_5(n,\alpha W(d\lambda)}\right\|^2\le$$

$$\left(\frac{2\mu C_1}{C_0}\right)^2\left[\int_{R^n}\left\|\frac{\lambda\ \exp\{i<\lambda,\lambda>-\mu t\,\|\lambda\|^2\}}{i\ \beta^{n/4+1/2}L^{1/2}(\sqrt{\beta})}\sqrt{f(\frac{\|\lambda\|}{\sqrt{\beta}})}-\right.\right.$$

$$\left.[c_5(n,\alpha)]^{1/2}\frac{\lambda\exp\{i<\lambda,x>-\mu t\,\|\lambda\|^2\}}{\|\lambda\|^{(n-\alpha)/2}}\right\|^2 d\lambda+$$

$$\left.\frac{\beta^{1+\alpha/2}}{L(\sqrt{\beta})}\int_{R^n}\left\|\int_{R^n\backslash v(\sqrt{\beta})}\frac{(x-y)\exp\{i<\lambda,x>-\frac{\|x-y\|^2}{4\mu t}\}}{2\mu t\ (4\pi\mu t)^{n/2}\beta^{n/4+1/2}}dy\right\|^2 f(\frac{\|\lambda\|}{\sqrt{\beta}})d\lambda\right]=$$

$$S_{1\beta}+S_{2\beta}. \tag{4.4.22}$$

Function

$$\psi(\lambda)=\|\lambda\|^2\frac{\exp\left\{i<\lambda,x>-\mu t\,\|\lambda\|^2\right\}}{\|\lambda\|^{(n-\alpha)}},\quad 0<\alpha<n,\quad \lambda\in R^n,$$

is absolutely integrable. Using the Tauberian result (4.4.1) and Lemma 1.4.12., we obtain from (4.4.22)

$$\begin{aligned}\lim_{\beta\to\infty} S_{1\beta} &= \lim_{\beta\to\infty}\int_{R^n}\psi(\lambda)Q_\beta(\lambda)d\lambda \\ &= 0,\end{aligned}$$

where $Q_\beta(\lambda)$ is bounded function such that

$$Q_\beta(\lambda) \sim c_5(n,\alpha)\left|\frac{L^{1/2}\left(\frac{\sqrt{\beta}}{\|\lambda\|}\right)}{L^{1/2}\left(\sqrt{\beta}\right)} - 1\right|^2 \left(\frac{C_1}{C_0}\right)^2$$

as $\beta \to \infty$.

For the $S_{2\beta}$ we have the estimate

$$\begin{aligned} S_{2\beta} &\leq \frac{\left(\frac{2\mu C_1}{C_0}\right)^2}{L\left(\sqrt{\beta}\right)\beta^{(n-\alpha)/2}}\int_{R^n} f\left(\frac{\|\lambda\|}{\sqrt{\beta}}\right)d\lambda \left\|\int_{R^n}\frac{(x-y)\exp\left\{-\frac{\|x-y\|^2}{4\mu t}\right\}}{2\mu t\,(4\pi\mu t)^{n/2}}\right\|^2 \\ &\leq \frac{k_{10}}{L\left(\sqrt{\beta}\right)\beta^{(n-\alpha)/2}}, \qquad 0<\alpha<n, \quad \lambda\in R^n, \end{aligned}$$

so that

$$\lim_{\beta\to\infty}\left\|\tilde{Y}_\beta(t,x) - Y(t,x)\right\|^2 = 0.$$

Applying Lemma 1.1.3. we conclude the statement of Theorem 1.4.1. including remarks.

4.5 Parabolic Limits for Strongly Dependent Random Data : the Non-Gaussian Scenario

We now describe the non-Gaussian limiting distributions of the rescaled solutions of Burgers' equation with singular initial data.

Theorem 4.5.1. Let $u = u(t,x)$, $(t,x) \in (0,\infty)\times R^n$, be a solution of the initial value problem (4.2.1)–(4.2.2.) with random initial condition

$\eta(x) = F(\xi(x))$, $x \in R^n$, where the random fields $\xi(x)$, $x \in R^n$, and the non-random function $F(\cdot)$ satisfy conditions A", B, C with $\alpha \in (0, n/2)$, and

$$\begin{aligned} C_1 &= \int_{-\infty}^{\infty} \exp\left\{-F(u)\frac{1}{2\mu}\right\} \phi(u) u du \\ &= 0, \\ C_2 &= \int_{-\infty}^{\infty} \exp\left\{-F(u)\frac{1}{2\mu}\right\} \phi(u)(u^2-1) du \\ &\neq 0. \end{aligned}$$

Suppose that random field $\eta(x) = F(\xi(x))$, $x \in R^n$, is a.s. differentiable.

Then the finite-dimensional distributions of the random field

$$Z_\beta(t,x) = \frac{\beta^{(1+\alpha)/2}}{L(\sqrt{\beta})} u(\beta t, x\sqrt{\beta}), \quad (t,x) \in (0,\infty) \times R^n, \quad 0 < \alpha < n/2,$$

converge weakly as $\beta \to \infty$ to the finite-dimensional distributions of the vector homogeneous in random field $Z_\beta(t,x)$, $(t,x) \in (0,\infty) \times R^n$, with the following spectral multiple stochastic integral representation:

$$\begin{aligned} Z(t,x) &= \frac{C_2}{2C_0} c_5(n,\alpha) \frac{2\mu}{i} \\ &\times \int_{R^{2n}}' e^{i<\lambda_1+\lambda_2,x>-\mu t\|\lambda_1+\lambda_2\|^2} \\ &\times \frac{(\lambda_1+\lambda_2)}{\|\lambda_1\|^{(n-\alpha)/2} \|\lambda_2\|^{(n-\alpha)/2}} W(d\lambda_1) W(d\lambda_2), \end{aligned} \tag{4.5.1}$$

where the constant $c_5(n,\alpha)$ is defined in (1.4.13), the double stochastic integral $\int' \ldots$ is taken with the respect to the complex Gaussian white noise W in R^n (see section 2.3), and

$$C_0 = \int_{-\infty}^{\infty} \exp\left\{-F(u)\frac{1}{2\mu}\right\} \phi(u) du.$$

<u>Remark 4.5.1.</u> Let $F(u) = u^2 - 1$, then

$$\begin{aligned} C_0 &= e^{1/2\mu}\left(\frac{\mu}{1+\mu}\right)^{1/2} > 0, \\ C_1 &= 0, \\ C_2 &= e^{1/2\mu}\left[\left(\frac{\mu}{1+\mu}\right)^{3/2} - \left(\frac{\mu}{1+\mu}\right)^{1/2}\right] < 0, \end{aligned}$$

and

$$\frac{C_2}{2C_0} = -\frac{\mu}{2(1+\mu)}.$$

Remark 4.5.2. The random field (4.5.1) has a non-Gaussian structure. In principle it is possible to calculate the spectral density of higher order using the diagonal formalism of Dobrushin (1979) and a papers of Terdik and Subba Rao (1989), Terdik and Meaux (1991), Terdik (1992, 1995). Then the statistical estimation of unknown parameters of a discretized version of a non-Gaussian field $Z(t,x)$ can be given thanks to Leonenko, Sikorskii and Terdik (1998).

Proof.

Consider the expansion (4.4.7) in Hilbert space $\mathbf{L}_2(\Omega)$ with $C_1 = 0$, $C_2 \neq 0$. From (4.2.6) and (4.4.7) we have

$$I(\beta t, x\sqrt{\beta}) = C_0\ \zeta_{0\beta}(t,x) + \frac{1}{2}C_2\zeta_{2\beta}(t,x) + R_{2\beta} + R_{3\beta}, \tag{4.5.2}$$

where $(t,x) \in (0,\infty) \times R^n$, $R_{2\beta}$ defined in (4.4.8) and

$$R_{3\beta} = \sum_{j\geq 3} \frac{C_j}{j!}\zeta_{j\beta}(t,x).$$

Using the arguments similar to proof of Theorem 4.4.1., we obtain

$$\lim_{\beta\to\infty} I(\beta t, x\sqrt{\beta}) = 0, \tag{4.5.3}$$

$$\frac{\beta^{(1+\alpha)/2}}{L(\sqrt{\beta})}\,[R_{2\beta} + R_{3\beta}] \xrightarrow{P} 0, \tag{4.5.4}$$

and

$$J(\beta t, x\sqrt{\beta}) \xrightarrow{P} C_0 \quad \text{as } \beta \to \infty. \tag{4.5.5}$$

In view of Lemma 1.1.3., Lemma 4.3.1. and (4.4.9), (4.5.2)–(4.5.5) it suffices to verify the relation

$$\begin{aligned}
&\lim_{\beta\to\infty} E\left\| \frac{C_2}{2C_0}\frac{\beta^{(1+\alpha)/2}}{L(\sqrt{\beta})}\,\zeta_{2\beta}(t,x) - Z(t,x)\right\|^2 \\
&= \left(\frac{C_2}{2C_0}\right)^2 \lim_{\beta\to\infty} E\left\|U_\beta(t,x)\right\|^2 \\
&= 0, \qquad (4.5.6)
\end{aligned}$$

where

$$\begin{aligned}
U_\beta(t,x) = &\frac{\beta^{(1+\alpha)/2}}{L(\sqrt{\beta})}\int_{v(\beta)} \frac{x\sqrt{\beta}-y}{t\beta} g\left(t\beta, x\sqrt{\beta}-y\right) H_2(\xi(y))dy \\
&-\frac{2C_0}{2C_2}Z(t,x) \\
= &\ X_{2\beta} - \frac{2C_0}{2C_2}Z(t,x), \qquad (4.5.7)
\end{aligned}$$

and $Z(t,x)$ is defined in (4.5.1).

Using the Itô-formula (2.3.1) and self-similar property of the Gaussian white noise $W(\cdot)$ we obtain

$$\begin{aligned}
X_{2\beta} &= \frac{\beta^{(1+\alpha)/2}}{L(\sqrt{\beta})}\int_{v(\beta)} \frac{x\sqrt{\beta}-y}{t\beta} g\left(t\beta, x\sqrt{\beta}-y\right) \\
&\times\left[\int'_{R^{2n}} e^{i<\lambda_1+\lambda_2,y>}\left[f(\|\lambda_1\|)f(\|\lambda_2\|)\right]^{1/2} W(d\lambda_1)W(d\lambda_2)\right] dy \\
&= \int'_{R^{2n}}\left[\int_{v(\beta)} e^{i<\lambda_1+\lambda_2,y>}\frac{x\sqrt{\beta}-y}{t\beta} g\left(t\beta, x\sqrt{\beta}-y\right)dy\right] \\
&\times\frac{\beta^{(1+\alpha)/2}}{L(\sqrt{\beta})}\left[f(\|\lambda_1\|)f(\|\lambda_2\|)\right]^{1/2} W(d\lambda_1)W(d\lambda_2) \\
&= \int'_{R^{2n}}\left[\int_{v(\sqrt{\beta})} e^{\sqrt{\beta}i<\lambda_1+\lambda_2,y>}\frac{x-y'}{t\sqrt{\beta}}\,\frac{\exp\left\{-\frac{\|x-y'\|^2}{4\mu t}\right\}}{(4\pi\mu t\beta)^{n/2}}\beta^{n/2}dy'\right] \\
&\times\frac{\beta^{(1+\alpha)/2}}{L(\sqrt{\beta})}\left[f(\|\lambda_1\|)f(\|\lambda_2\|)\right]^{1/2} W(d\lambda_1)W(d\lambda_2)
\end{aligned}$$

$$= \int_{R^{2n}}' \left[\frac{1}{t} \int_{v(\sqrt{\beta})} \frac{\exp\left\{ i < \lambda_1 + \lambda_2, y > - \frac{\|x-y'\|^2}{4\mu t} \right\}}{(4\pi\mu t)^{n/2}} (x-y) dy \right]$$
$$\times \left[f\left(\frac{\|\lambda_1\|}{\sqrt{\beta}} \right) f\left(\frac{\|\lambda_2\|}{\sqrt{\beta}} \right) \right]^{1/2} \frac{\beta^{(\alpha-n)/2}}{L(\sqrt{\beta})} W(d\lambda_1) W(d\lambda_2).$$

Using the identity (4.4.21) we have

$$X_{2\beta} = \int_{R^{2n}}' \left[\frac{2\mu(\lambda_1+\lambda_2)}{i} \exp\left\{ i < \lambda_1+\lambda_2, x > -\mu t \ \|\lambda_1+\lambda_2\|^2 \right\} \right.$$
$$\left. - \int_{R^n \setminus v(\sqrt{\beta})} \frac{\exp\left\{ i < \lambda_1+\lambda_2, y > - \frac{\|x-y'\|^2}{4\mu t} \right\}}{t \ (4\pi\mu t)^{n/2}} (x-y) dy \right]$$
$$\times \left[f\left(\frac{\|\lambda_1\|}{\sqrt{\beta}} \right) f\left(\frac{\|\lambda_2\|}{\sqrt{\beta}} \right) \right]^{1/2} \frac{\beta^{(\alpha-n)/2}}{L(\sqrt{\beta})} W(d\lambda_1) W(d\lambda_2). \quad (4.5.8)$$

From (4.5.1), (4.5.7) and (4.5.8) we obtain

$$E \ \|U_\beta(t,x)\|^2 = [2\mu c_5(n,\alpha)]^2 E \left\| \int_{R^n}' \frac{(\lambda_1+\lambda_2)}{i} e^{i<\lambda_1+\lambda_2,x>-\mu t\|\lambda_1+\lambda_2\|^2} \right.$$
$$\times Q_\beta(\lambda_1,\lambda_2) \ \frac{W(d\lambda_1)W(d\lambda_2).}{(\|\lambda_1\| \cdot \|\lambda_2\|)^{(n-\alpha)/2}}$$
$$- \int_{R^{2n}}' \left[\int_{R^n \setminus v(\sqrt{\beta})} \frac{\exp\left\{ i < \lambda_1+\lambda_2, y > - \frac{\|x-y'\|^2}{4\mu t} \right\} (x-y)}{t(4\pi\mu t)^{n/2}} d \right.$$
$$\left. \times \left[f\left(\frac{\|\lambda_1\|}{\sqrt{\beta}} \right) f\left(\frac{\|\lambda_2\|}{\sqrt{\beta}} \right) \right]^{1/2} \frac{\beta^{(\alpha-n)/2}}{L(\sqrt{\beta})} W(d\lambda_1) W(d\lambda_2) \right\|^2,$$

where

$$Q_\beta(\lambda_1,\lambda_2) = (\|\lambda_1\| \cdot \|\lambda_2\|)^{(n-\alpha)/2} \left[f\left(\frac{\|\lambda_1\|}{\sqrt{\beta}} \right) f\left(\frac{\|\lambda_2\|}{\sqrt{\beta}} \right) \right]^{1/2}$$
$$\times \frac{\beta^{(\alpha-n)/2}}{c_5(n,\alpha) L(\sqrt{\beta})} - 1.$$

Using the Theorem 2.3.1. we have

$$E \, \|U_\beta(t,x)\|^2 \leq [2\mu \ c_5(n,\alpha)]^2$$

$$\times\left\{\int_{R^{2n}}\left\|\frac{(\lambda_1+\lambda_2)}{i}\frac{\exp\left\{i<\lambda_1+\lambda_2,y>-\mu t\ \|\lambda_1+\lambda_2\|^2\right\}}{(\|\lambda_1\|\cdot\|\lambda_2\|)^{(n-\alpha)/2}}\right\|^2\right.$$

$$\times Q_\beta^2(\lambda_1,\lambda_2)d\lambda_1 d\lambda_2$$

$$+\int_{R^{2n}}\left\|\int_{R^n\setminus v(\sqrt{\beta})}\frac{\exp\left\{i<\lambda_1+\lambda_2,y>-\frac{\|x-y'\|^2}{4\mu t}\right\}(x-y)}{t\ (4\pi\mu t)^{n/2}}\right\|^2$$

$$\left.\times f\left(\frac{\|\lambda_1\|}{\sqrt{\beta}}\right)f\left(\frac{\|\lambda_2\|}{\sqrt{\beta}}\right)\frac{\beta^{\alpha-n}}{L^2(\sqrt{\beta})}d\lambda_1 d\lambda_2\right\}$$

$$= S_1(\beta)+S_2(\beta).$$

Note that

$$\begin{aligned} S_2(\beta) &\leq k_1\int_{R^{2n}}\frac{\beta^{\alpha-n}}{L^2(\sqrt{\beta})}f\left(\frac{\|\lambda_1\|}{\sqrt{\beta}}\right)f\left(\frac{\|\lambda_2\|}{\sqrt{\beta}}\right)d\lambda_1 d\lambda_2 \\ &\quad\times\left\|\int_{R^n}\frac{\exp\left\{-\frac{\|x-y'\|^2}{4\mu t}\right\}(x-y)}{t\ (4\pi\mu t)^{n/2}}dy\right\|^2 \\ &\leq \frac{k_2}{\beta^{\alpha-n}L^2(\sqrt{\beta})},\quad 0<\alpha<n/2, \end{aligned}$$

and

$$\lim_{\beta\to\infty}S_2(\beta)=0. \tag{4.5.9}$$

The function

$$\exp\{2i<\lambda_1+\lambda_2,x-2\mu t\ \|\lambda_1+\lambda_2\|^2\}\,\|\lambda_1+\lambda_2\|^2\,(\|\lambda_1\|\cdot\|\lambda_2\|)^{\alpha-n}$$

is absolutely integrable on R^{2n}, since it has the order

$$O\left(\|\lambda_1+\lambda_2\|^2\,(\|\lambda_1\|\cdot\|\lambda_2\|)^{\alpha-n}\right),\quad 0<\alpha<n/2,$$

at zero, and

$$O\left(\exp\{-\|\lambda_1+\lambda_2\|^2\}\,\|\lambda_1+\lambda_2\|^2\,(\|\lambda_1\|\cdot\|\lambda_2\|)^{\alpha-n}\right)$$

at infinity. From the condition A" and (4.4.1) we have that the function $Q_\beta(\lambda_1, \lambda_2)$ is bounded and

$$\lim_{\beta\to\infty} Q_\beta(\lambda_1, \lambda_2) = 0.$$

So that we can pass to the limit under the integral for $0 < \alpha < n/2$:

$$\begin{aligned}\lim S_2(\lambda) &= k_3 \lim_{\beta\to\infty} \int_{R^{2n}} \left\| \frac{1}{i} e^{2i<\lambda_1+\lambda_2, x-2\mu t\|\lambda_1+\lambda_2\|^2} \frac{\lambda_1+\lambda_2}{(\|\lambda_1\|\cdot\|\lambda_2\|)^{(n-\alpha)/2}} \right\|^2 \\ &\quad \times Q_\beta(\lambda_1, \lambda_2) d\lambda_1 d\lambda_2 \\ &= 0. \end{aligned} \tag{4.5.10}$$

From (4.5.7), (4.5.9) and (4.5.10) we obtain (4.5.6). We complete the proof of Theorem 4.5.1.

We now describe the class of non-Gaussian limiting distributions of the Hopf–Cole solution (4.2.5) of Burgers' equation (4.2.1) with singular χ-squared random initial conditions.

D. Let $\xi(x) = [\xi_1(x), \ldots \xi_p(x))'$, $x \in R^n$, be an a.s. differential vector Gaussian field satisfies condition E of section 3.4., and

$$\eta_p(x) = \frac{1}{2} \sum_{i=1}^{p} (\xi_i^2(x) - 1), \quad x \in R^n.$$

Remark 4.5.3. If the constant $\rho_\infty = 0$ in condition E of section 3.4., then $\xi_1(x), \ldots \xi_p(x)$ are independent copies of Gaussian field $\xi(x)$ satisfies condition A'.

Note (see section 3.4. for details) that

$$\eta_p(x) = \frac{1}{2} \sum_{i=1}^{p} \frac{\eta_i^2(x)}{\mu_i^2} - \frac{p}{2},$$

where μ_i, $i = 1, \ldots, p$, are defined in (3.4.2) and $\eta(x) = [\eta_1(x), \ldots \eta_p(x)]'$, $x \in R^n$, is a Gaussian vector field with independent components and covariance functions (3.4.4).

D'. The spectral densities of random fields $\eta_i(x)$, $i = 1, \ldots, p$, exist: $f_i(\|\lambda\|)$, $i = 1, \ldots, p$, and they are supposed to be decreasing for $\|\lambda\| \geq \lambda_0 \geq 0$ and continuous at all $\lambda \neq 0$.

Let $Z_j(t,x)$, $j=1,\dots,p$, $(t,x)\in(0,\infty)\times R^n$, are independent copies of non-Gaussian random field (4.5.1) with

$$\frac{C_2}{2C_0} = -\frac{1}{2}(1+\mu)$$

(see remark 4.5.1.).

Consider the numbers

$$\begin{aligned}
\psi_0 &= C_{0,\dots,0} \\
&= \int_{R^p} F(u)\phi(\|u\|)du \\
&= \prod_{i=1}^{p}\left\{1+\frac{1}{2\mu\mu_j}\right\}^{-1/2}, \\
\psi_1 &= C_{2,0,\dots,0} \\
&= \int_{R^p}(u_1^2-1)F(u)\phi(\|u\|)du, \\
\psi_j &= C_{0,2,0,\dots,0} \\
&= C_{0,\dots,0,2} \\
&= \int_{R^p}(u_2^2-1)F(u)\phi(\|u\|)du,
\end{aligned}$$

$j=1,\dots,p$, where

$$F(u)=\exp\left\{-\frac{1}{4\mu}\left[\left(\frac{u_1}{\mu_1}\right)^2+\dots+\left(\frac{u_p}{\mu_p}\right)^2\right]\right\}, \qquad u\in R^p.$$

Theorem 4.5.2. Let $u(t,x)$, $(t,x)\in(0,\infty)\times R^n$, be a solution of the Cauchy problem (4.2.1)–(4.2.2) with random initial conditions $\eta_p(x)$, $x\in R^n$, satisfying the assumptions D, D' with $\alpha\in(0,n/2)$ (see condition E of section 3.4.). Then the finite-dimensional distributions of the field

$$Z_\beta(t,x)=\frac{\beta^{(1+\alpha)/2}}{L(\sqrt{\beta})}\,u(t\beta,x\sqrt{\beta}), \quad (t,x)\in(0,\infty)\times R^n, \quad \alpha\in(0,n/2),$$

converge weakly as $\beta\to\infty$ to the finite-dimensional distributions of the vector homogeneous in x random field

$$Z(t,x)=\sum_{j=1}^{p}\frac{\psi_j\,Z_j(t,x)}{2\psi_0}, \quad (t,x)\in(0,\infty)\times R^n.$$

The proof of Theorem 4.5.2. is based on the multi-dimensional Chebychev–Hermite expansion (3.4.8) (see Leonenko, Orsingher and Parkhomenko (1995)) or Laguerre expansion (see Leonenko and Li Zhanbing (1994) or Leonenko, Li Zhanbing and Rybasov (1995) for details).

4.6 Exact Parabolic Asymptotics for Singular Burgers' Equation

We study the rate of convergence to the normal law of a solution of the one-dimensional Burgers' equation with singular Gaussian initial data.

We consider the initial value problem (4.2.8)–(4.2.9) which describes the evolution of the velocity field $u(t,x)$, $(t,x) \in (0,\infty) \times R^1$.

The following theorem is proved in section 4.4.

Theorem 4.6.1. Let $u(t,x)$, $(t,x) \in (0,\infty) \times R^1$, be the solution of the Cauchy problem (4.2.8)–(4.2.9) with random initial condition $\eta(x)$, $x \in R^1$, satisfying condition A', that is, $\eta(x)$, $x \in R^1$, is stationary Gaussian process with $E\eta(x) = 0$, $E\eta^2(x) = 1$ and covariance function $B(|x|) = L(|x|)/|x|^\alpha$, $0 < \alpha < 1$, $L \in \mathcal{L}$.

Then finite-dimensional distribution of the process

$$Y_t(a) = \frac{t^{1/2+\alpha/4}}{L^{1/2}(\sqrt{t})} u\left(t, a\sqrt{t}\right), \quad a \in R^1,$$

converge weakly as $t \to \infty$ to the finite-dimensional distribution of stationary Gaussian process $Y(a)$, $a \in R^1$, with $EY(a) = a$ and a covariance function of the form

$$V(\ a-b\ ) = (2\mu)^{-1-\alpha/2} \iint\limits_{R^2} w_1 w_2 \phi(w_1)\phi(w_2) \frac{dw_1 dw_2}{\left|w_1 - w_2 - \frac{a-b}{\sqrt{2\mu}}\right|^\alpha}, \quad 0 < \alpha < 1.$$

Introduce the uniform or Kolmogorov's distance between distributions of r.v.s X and Y via the formula

$$\mathcal{K}(X,Y) = \sup_z |P(X \le z) - P(Y \le z)|.$$

Consider the following function

$$\sigma_t^2 = (2\mu)^{-1-\alpha/2} \int\limits_{A(t,a)} \int\limits_{A(t,a)} w_1 w_2 \phi(w_1)\phi(w_2) \frac{L(|w_1 - w_2|\sqrt{2\mu t})}{L(\sqrt{t})} \frac{dw_1 dw_2}{|w_1 - w_2|^\alpha},$$

where

$$A(t,a) = \left[\frac{a}{\sqrt{2\mu}} - \sqrt{\frac{t}{2\mu}},\quad \frac{a}{\sqrt{2\mu}} + \sqrt{\frac{t}{2\mu}}\right].$$

Note that $\lim_{t\to\infty} \sigma_t^2 = V(0)$.

The main result of this paper describes the rate of convergence (as $t \to \infty$) of the one-point probability distributions of the parabolically rescaled solution of Burgers' equation to the normal law and is contained in the following.

Theorem 4.6.2. Let assumptions of Theorem 2.6.1 hold with $0 < \alpha < 1/2$. Then

$$\overline{\lim_{t\to\infty}} \left(\frac{t^{\alpha/6}}{[L(\sqrt{t})]^{1/3}}\right) \mathcal{K}\left(\frac{1}{\sigma_t} Y_t(a),\, \mathcal{N}\right) \le \frac{3}{2}\nu_1^{2/3}\,\nu_2^{1/3}, \tag{4.6.1}$$

where $\mathcal{N}$ is the normal r.v. with parameters (0,1) and

$$\begin{aligned}
\nu_1 &= 1 + \frac{1}{(2\pi)^{1/2}},\\
\nu_2 &= \theta^2(2\mu)^{2-\alpha/2}\left\{2K\left[e^{1/2\mu^2} - \left(1+\frac{1}{4\mu^2}\right)e^{1/4\mu^2}\right]e^{-1/4\mu^2}\frac{M}{8\mu^4}\right\},
\end{aligned}$$

θ is an arbitrary fixed constant such that $\theta > 1$, and

$$K = \frac{\iint_{R^2} w_1 w_2 \phi(w_1)\phi(w_2)\frac{dw_1 dw_2}{|w_1-w_2|^{2\alpha}}}{\iint_{R^2} w_1 w_2 \phi(w_1)\phi(w_2)\frac{dw_1\, dw_2}{|w_1-w_2|^{\alpha}}},$$

$$\mathrm{M} = \iint_{R^2} \phi(w_1)\phi(w_2)\frac{dw_1 dw_2}{|w_1-w_2|^{\alpha}}.$$

Before proving Theorem 4.6.2. we mention some well-known results.

The following two lemmas due to Petrov (1995) and Michel and Pfanzagl (1971).

Lemma 4.6.1. Let X, Y be two r.v.s such that $\mathcal{K}(X,Y) \le K$. Then for any $\varepsilon > 0$

$$\mathcal{K}(X+Y,\mathcal{N}) \le K + \frac{\varepsilon}{(2\pi)^{1/2}} + P\{|Y| > \varepsilon\}.$$

Lemma 4.6.2. Let X, U be two r.v.s and $U > 0$. Then for any $\varepsilon > 0$

$$\mathcal{K}\left(\frac{X}{U},\mathcal{N}\right) \le \mathcal{K}(X,\mathcal{N}) + P\{|U-1| > \varepsilon\} + \varepsilon.$$

Lemma 4.6.3. Let X, Y, U be r.v.s and $U > 0$. Then for any $\varepsilon > 0$

$$\mathcal{K}\left(\frac{X+Y}{U}, \mathcal{N}\right) \leq \mathcal{K}(X, \mathcal{N}) + P\{|Y| > \varepsilon\} + \frac{\varepsilon}{(2\pi)^{1/2}} + P\{|U-1| > \varepsilon\} + \varepsilon.$$

Lemma 4.6.4. Let W, T be two r.v.s. Then for any $\varepsilon > 0$

$$P\{|W+T| > \varepsilon\} \leq P\{|W| > \varepsilon\delta\} + P\{|T| > \varepsilon(1-\delta)\},\ 0 < \delta < 1.$$

Proof Obvious.

Proof of Theorem 4.6.2.

From (4.4.7) we have the following expansion in the Hilbert space $\mathbf{L}_2(\Omega)$:

$$\exp\left\{-\frac{\eta(y)}{2\mu}\right\} = \sum_{k=0}^{\infty} \frac{C_k}{k!} H_k(\eta(y)),$$

where

$$C_k = \frac{1}{(2\pi)^{1/2}} \int_{R^1} \exp\left\{-\frac{u + u^2\eta(y)}{2\mu}\right\} H_k(u)du, \quad k = 0, 1, 2, \ldots,$$

and

$$\sum_{k=0}^{\infty} \frac{C_k^2}{k!} = \int_{-\infty}^{\infty} e^{-u/\mu}\phi(u)du.$$

Note that C_0 and C_1 are defined in (4.4.4).

We consider the r.v.

$$\zeta_k(t,a) = \int_{-t}^{t} \frac{a\sqrt{t} - y}{t} g\left(t, a\sqrt{t} - y\right) H_k(\eta(y))dy,$$

where $g(t,x)$ is defined by (4.2.7) with $n = 1$.

In order to apply Lemma 4.6.3. to the Hopf–Cole solution (4.2.5), we represent

$$\frac{Y_t(a)}{\sigma_t} = \frac{X_t + Y_t}{U_t}, \tag{4.6.2}$$

where $X_t = e^{-1/8\mu^2} A_t C_1 \zeta_1(t,a)$,

$$\begin{aligned} Y_t &= e^{-1/8\mu^2} A_t \left[\sum_{k\geq 2} \frac{C_k^2}{k!} \zeta_k(t,a) + \int_{|y|>t} \frac{a\sqrt{t} - y}{t}\, g\left(t, a\sqrt{t} - y\right) e^{-\eta(y)/2\mu}\right] dy \\ &= e^{-1/8\mu^2} A_t \left[W_t + T_t\right] \end{aligned}$$

and

$$U_t = J(t, a\sqrt{t})\ e^{-1/8\mu^2}, \qquad A_t = t^{1/2+\alpha/4}[\sigma_t L^{1/2}(\sqrt{t})].$$

We note that the r.v.

$$X_t = \frac{\zeta_1(t,a)}{\sqrt{\operatorname{var}\zeta_1(t,a)}}$$

is a standard normal r.v. for any $t > 0$. So we have

$$\mathcal{K}(X_t, \mathcal{N}) = 0. \tag{4.6.3}$$

From (2.1.8) we have that

$$\operatorname{var}\left[A_t W_t e^{-1/8\mu^2}\right] = A_t^2\ e^{-1/4\mu^2} \sum_{k=2}^{\infty} \left(\frac{C_k}{k!}\right)^2 \operatorname{var}\zeta_k(t,a).$$

Applying inequality (4.4.13a) for $n = 1$, we obtain for $0 < \alpha < 1/2$

$$\begin{aligned} \operatorname{var}[A_t W_t e^{-1/8\mu^2}] &\le A_t^2 \frac{\psi_2^2(t)}{2}\ e^{-1/4\mu^2} \sum_{k=2}^{\infty}\left(\frac{C_k}{k!}\right) \\ &= \frac{L(\sqrt{t})}{t^{\alpha/2}}\ K_t \frac{1}{(2\mu)^{\alpha/2} C_1^2} \sum_{k=2}^{\infty} \frac{C_k^2}{k!}, \end{aligned} \tag{4.6.4}$$

where

$$\begin{aligned} \psi_k^2(t) &= \frac{2\mu k!}{(2\mu)^{k\alpha/2} t^{1+k\alpha/2}} \\ &\quad \times \int\limits_{A(t,a)} \int\limits_{A(t,a)} w_1 w_2 \phi(w_1)\phi(w_2) \frac{L^k(|w_1 - w_2|\sqrt{2\mu t})}{|w_1 - w_2|^{k\alpha}} dw_1 dw_2 \\ &= \psi^2(k,\alpha) \frac{2\mu k!}{(2\mu)^{k\alpha/2}} \frac{L^k(\sqrt{2\mu t})}{t^{1+k\alpha/2}} [1 + o(1)], \\ \psi^2(k,\alpha) &= \iint_{R^2} w_1 w_2 \phi(w_1)\phi(w_2) \frac{dw_1 dw_2}{|w_1 - w_2|^{k\alpha}}, \quad 0 < \alpha < 1/k, \end{aligned}$$

and

$$K_t = \frac{\int\limits_{A(t,a)} \int\limits_{A(t,a)} w_1 w_2 \phi(w_1)\phi(w_2) \frac{L^2(|w_1 - w_2|\sqrt{2\mu t})}{L^2(\sqrt{t})|w_1 - w_2|^{2\alpha}} dw_1 dw_2}{\int\limits_{A(t,a)} \int\limits_{A(t,a)} w_1 w_2 \phi(w_1)\phi(w_2) \frac{L(|w_1 - w_2|\sqrt{2\mu t})}{L(\sqrt{t})|w_1 - w_2|^{\alpha}} dw_1 dw_2}.$$

We note that $\lim_{t\to\infty} K_t = \psi^2(1,\alpha) = K$ and

$$k_1 = C_1^{-2}\sum_{k=2}^{\infty}\frac{C_k^2}{k!} = \left[e^{1/2\mu^2} - \left(1+\frac{1}{4\mu^2}\right)\, e^{1/4\mu^2}\right]\left[\frac{1}{4\mu^2}e^{1/4\mu^2}\right]^{-1} \qquad (4.6.5)$$

Applying Chebychev's inequality, we obtain from (4.6.4), (4.6.5) for any $\varepsilon > 0$, $0 < \delta < 1$, that

$$P\{\left|A_t W_t e^{-1/8\mu^2}\right| > \varepsilon\delta\} \leq \frac{\theta^2}{\varepsilon^2}(2\mu)^{-\alpha/2}k_2\frac{L(\sqrt{t})}{t^{\alpha/2}}K_t, \quad \theta = \frac{1}{\delta} > 1. \qquad (4.6.6)$$

Note that

$$\operatorname{var}[A_t T_t e^{-1/8\mu^2}] \leq \frac{A_t^2}{(2\pi)^{1/2}t}\exp\left\{-\left[\frac{a}{(2\mu)^{1/2}} + (t/2\mu)^{1/2}\right]^2\right\}e^{1/4\mu^2}$$

and for any $\varepsilon > 0$, $0 < \delta < 1$, $(\theta = \frac{1}{\delta})$

$$\begin{aligned} & P\left\{\left|A_t T_t e^{-1/8\mu^2}\right| > \varepsilon(1-\delta)\right\} \\ \leq \quad & \frac{\exp\left\{-\left[\frac{a}{(2\mu)^{1/2}} + (t/2\mu)^{1/2}\right]^2\right\}}{t\varepsilon^2(2\pi)^{1/2}}A_t^2 e^{1/4\mu^2}\left[\frac{\theta}{\theta-1}\right]^2. \end{aligned} \qquad (4.6.7)$$

For any $\varepsilon > 0$, $0 < \delta < 1$ we get from (4.6.2), (4.6.3), (4.6.7), and Lemma 4.6.4. the following estimation:

$$\begin{aligned} \Delta_t &= \mathcal{K}\left(\frac{Y_t(a)}{\sigma_t}, \mathcal{N}\right) \leq \frac{\varepsilon}{(2\pi)^{1/2}} + \varepsilon \\ &+ \frac{1}{\varepsilon^2}\left[\theta^2(2\mu)^{-\alpha/2}k_2\frac{L(\sqrt{t})}{t^{\alpha/2}}K_t + R_t\right] + P\left\{|U_t - 1| > \varepsilon\right\}, \end{aligned} \qquad (4.6.8)$$

where $R_t = o\left(\frac{L(\sqrt{t})}{t^{\alpha/2}}\right)$ as $t \to \infty$.
Similar we obtain

$$P\{|U_t - 1| > \varepsilon\} \leq \frac{1}{\varepsilon^2}\left\{\frac{\theta^2 M_t(\alpha)L(\sqrt{t})}{t^{\alpha/2}(2\mu)^{\alpha/2+2}} + k_3 Q_t\right\}, \qquad (4.6.9)$$

where $k_3 > 0$,

$$\begin{aligned} M_t(\alpha) &= \iint_{R^2} \phi(w_1)\phi(w_2)\frac{dw_1 dw_2}{|w_1 - w_2|^\alpha} \to M, \\ Q_t &= o\left(\frac{L(\sqrt{t})}{t^{\alpha/2}}\right) \qquad \text{as } t \to \infty. \end{aligned}$$

From (4.6.8) and (4.6.9) we have

$$\Delta_t \le \varepsilon\left(1 + \frac{1}{(2\pi)^{1/2}}\right) + \frac{1}{\varepsilon^2}\left[\frac{\theta^2 L(\sqrt{t})}{(2\mu)^{\alpha/2} t^{\alpha/2}}\left(k_2\, K_t + \frac{M_t(\alpha)}{4\mu^2}\right) + R_t + k_3\, Q_t\right].$$

In order to minimize the right-hand side of the last inequality, set

$$\varepsilon = \frac{L^{1/3}(\sqrt{t})}{t^{\alpha/6}}\left(1 + \frac{1}{\sqrt{2\pi}}\right)^{-1/3} (2\mu)^{-\alpha/6}\theta^2 \frac{1}{3}\left(2k_2 K_t + \frac{M_t(\alpha)}{2\mu^2}\right)^{1/3}.$$

Thus we derive the following inequality

$$\Delta_t \le \frac{L^{1/3}(\sqrt{t})}{t^{\alpha/6}}\left[\frac{3}{2}\nu_1^{2/3}\, \nu_2^{1/3}(t) + \frac{t^{\alpha/2}}{L(\sqrt{t})}\,(R_t k_4 + k_5 Q_t)\right], \qquad (4.6.10)$$

where $k_4 > 0$, $k_5 > 0$ and

$$\begin{aligned} \nu_2(t) &= \theta^2(2\mu)^{2-\alpha/2}\left\{2K_t\left[e^{1/2\mu^2} - (1 + \frac{1}{4\mu^2})e^{1/4\mu^2}\right]\, e^{-1/4\mu^2} + \frac{M_t(\alpha)}{8\mu^4}\right\} \\ &\to \nu_2 \qquad \text{as } t \to \infty. \end{aligned} \qquad (4.6.11)$$

From the (4.6.10) and (4.6.11) Theorem 4.6.2. follows.

4.7 Hyperbolic Asymptotics for Rescaled Solutions of Burgers' Equation

We present some results on hyperbolic asymptotics in Burger's turbulence problem. This type of scaling is of importance in physical applications (see Burgers (1974), Gurbatov, Malakhov and Saichev (1991)). The mathematically rigorous results presented by Albeverio, Molchanov and Surgailis (1994) and Molchanov, Surgailis and Woyczynski (1995, 1997). This line of work of shocks in Burgers turbulence or the statistical solutions of the Burgers' equation in the zero viscosity limit $\mu \to 0$ or large Reynolds numbers $R = \text{const}/\mu$ (see Sinai (1992), Avellaneda and E (1995), Fan (1995), Wehr and Xin (1996), Ryan (1998) and Bertoin(1998)).

Consider the one-dimensional Burgers' equation (4.2.8) with smooth Gaussian initial conditions (4.2.9).

E. Let $\eta(x)$, $x \in R^1$, be a zero-mean, stationary, Gaussian, a.s. differentiable process with covariance function $B(x) = E\eta(0)\eta(x)$ satisfying the following two conditions:

$$\begin{aligned} B(x) &= o\left(\frac{1}{\log x}\right), \quad x \to \infty; \\ B(x) &= \lambda_0 - \frac{1}{2}\lambda_2 x^2 + \frac{1}{4!}\lambda_4 x^4 + o(x^4), \quad x \to 0. \end{aligned}$$

Theorem 4.7.1. Let assumptions E hold. If $u(t,x)$, $(t,x) \in (0,\infty) \times R^1$, is the solution of the Burgers equation (4.2.8) with the initial velocity potential

$$-\frac{d}{dx}\eta(x), \qquad x \in R^1,$$

then, as $\beta \to \infty$,

$$\begin{aligned} \beta^2\sqrt{2\ \log\beta} &= \frac{1}{\mu} \\ &\to \infty, \end{aligned}$$

the finite-dimensional distributions of the field

$$V_\beta(t,x) = \beta\sqrt{2\ \log\beta}\,u\left(\beta^2\sqrt{2\ \log\beta}\,t,\ \beta x\right), \quad (t,x) \in (0,\infty) \times R^1,$$

converge weakly to the corresponding distributions of the random field

$$V(t,x) = \frac{x - y_{j^*(t,x)}}{t}, \quad (t,x) \in (0,\infty) \times R^1.$$

Here $y_{j^*(t,x)} = y_{j^*}$ is the abscissa of the point of a Poisson process $(y_j, u_j)_{j\in Z^1}$ on R^2, with the intensity

$$EN\ (dy, du) = e^{-u} dy\ du,$$

where

$$N(A) = \#\{j : (y_j, u_j) \in A\},$$

by maximizing the difference

$$\begin{aligned}\max_j \left(\sqrt{\lambda_0}\, u_j - \frac{(x-y_j)^2}{2t} \right) &= \sqrt{\lambda_0}\, u_{j^*} - \frac{(x-y_{j^*})^2}{2t}, \\ \lambda_0 &= E\eta^2(0) = B(0).\end{aligned}$$

The proof of Theorem 4.7.1. is presented in Molchanov, Surgailis and Woyczynski (1995).

The Poisson process $(y_j, u_j)_{j\in Z}$ is the limit of a suitable rescaled local maximum of the Gaussian process $\eta(x)$. The asymptotics is valid at 'intermediate times' $t \sim \mu^{-1}$ and was termed the 'intermediate asymtotics' in Molchanov, Surgailis and Woyczynski (1995). The limit velocity random field $V(t,x)$ is 'pure shock waves'. Also, notice that the random field $V(t,x)$ does not depend on the viscosity parameter μ in equation (4.2.8), and that its shape is what one usually sees in the study of the Burgers' equation in the zero viscosity limit. The random field $V(t,x)$ has 'saw tooth' trajectories, defined by means of a Poisson process on the plane related to high fluctuations of initial potential (see Burgers (1974)). Note that the random field $V(t,x)$ depends on a single parameter $\lambda_0 = E\eta^2(0)$ of the initial Gaussian process $\eta(x)$, $x \in R^1$. Similar processes appear in Holst, Quine and Robinson (1995).

Molchanov, Surgailis and Woyczynski (1995) obtained the finite-dimensional distributions of the field $V(t,x)$, $(t,x) \in (0,\infty) \times R^1$.

In particular, for $\lambda_0 = 1$ the one-dimensional distributions of the random field $V(t,x)$ are Gaussian

$$V(t,x) \stackrel{D}{=} \mathcal{N}\left(0, \frac{1}{t}\right),$$

but two-dimensional distributions are non-Gaussian.

The covariance function

$$\begin{aligned}\rho(t_1,x_1,t_2,x_2) &= EV(t_1,x_1)V(t_2,x_2) \\ &= \frac{1}{t_1t_2}\int_{R^1}(z-x_1)(z-x_2)A^{-1}(z,x_2,x_2)dz \\ &\quad -\frac{t_1-t_2}{t_1t_2}\int_{R^1}|z|\,(1-e^{2(x_2-x_1)z/(t_2-t_1)})^2A^{-2}(z,x_1,x_2)dz,\end{aligned}$$

where

$$A(z,x_1,x_2) = e^{(z-x_1)^2/2t_1}\int\limits_{|y|<|z|} e^{-(y-x_1)^2/2t_1}dy$$
$$+e^{(z-x_2)^2/2t_2}\int\limits_{|y|>|z|} e^{-(y-x_2)^2/2t_2}dy.$$

The corresponding expression for fixed time ($t_1 = t_2 = t$) was obtained in Gurbatov, Malakhov and Saichev (1991), p. 181, and is somewhat simpler, namely

$$\rho(t,x_1,t,x_2) = \frac{1}{t}\frac{d}{dx}(t\ P_t(x)),$$

where

$$P_t(2x) = \frac{1}{\sqrt{2\pi t}}\int_{R^1}\left[e^{(x+z)^2/2t}\Phi_t(x+z) + e^{(x-z)^2/2t}\Phi_t(x-z)\right]^{-1}dz,$$
$$\Phi_t(x) = \frac{1}{\sqrt{2\pi t}}\int_{-a}^{x} e^{-u^2/2t}du.$$

Surgailis (1996) considered intermediate asymptotics between the smooth parabolic limits (see section 4.3–4.5) and shock type of hyperbolic asymptotics. Surgailis (1996) presents the existence of another asymptotics for $u(t,x)$ at very large times

$$t \sim \frac{\exp\{1/\mu^2\}}{\mu}.$$

E' Let $\eta(x)$, $x \in R^1$, be a zero-mean, stationary, Gaussian, a.s. differentiable process with covariance function $B(x)$ such that

$$B(x) = \lambda_0 - \frac{\lambda_2}{2}x^2 + o(x^2), \quad x \to 0,$$
$$B(x) = o\left(\frac{1}{\log x}\right), \quad x \to \infty.$$

Theorem 4.7.2. Let assumptions E' hold, the constant $\lambda_0 > 1$ and $\beta = e^{1/\mu^2} \to \infty$. Then the finite-dimensional distributions of the field

$$\beta\sqrt{2\log\beta}u\left(\beta^2\sqrt{2\log\beta}t,\ \beta x\right), \quad (t,x) \in (0,\infty)\times R^1,$$

converge weakly to the finite-dimensional distributions of the random field

$$
\begin{aligned}
V_1(t,x) &= \left[\sum_i t^{-1}(x-y_i)\,\exp\left\{\sqrt{\lambda_0}u_i - \frac{(x-y_i)^2}{2t}\right\}\right] \\
&\quad \times \left[\sum_i \exp\left\{\sqrt{\lambda_0}u_i - \frac{(x-y_i)^2}{2t}\right\}\right]^{-1}, \\
(t,x) &\in (0,\infty)\times R^1,
\end{aligned}
$$

where $(y_j, u_j)_{i\in Z^1}$ is the same Poisson process as in Theorem 4.7.1. Theorem 4.7.2. is proved by Surguilis (1996).

Note that $V_1(t,x)$ satisfies Burgers' equation (4.2.8) with the Poisson process (y_j, u_j) playing the role of the 'singular initial potential'. Condition $\lambda_0 > 1$ is necessary in order that the field $V_1(t,x)$ is well-defined; the case $\lambda_0 \leq 1$ reduces to the theorem by a corresponding scaling of t and x.

This asymptotics is 'intermediate' between the saw tooth asymptotics, and the smooth parabolic Gaussian limits. For $t \ll 1$ a typical trajectory of $V_1(t,x)$, $x\in R^1$, resembles a 'smooth saw tooth wave' which is almost linear near high 'amplitudes' u_j.

The results of Albeverio, Molchanov and Surgailis (1994) deals with the initial velocity potential data $\eta(x)$, $x \in R^3$, are of the shot noise type.

E". Let

$$
e^{\eta(x)} = \sum_i e^{\zeta_i}\delta(x-x_i), \qquad x\in R^3,
$$

where $\{\zeta_i\}$ is an i.i.d. sequence independent of the Poisson ensemble $\{x_i\}$.

Consider the Burgers' equation (4.2.1) in case $n=3$. Suppose that initial condition (4.2.2) is of the form $-\nabla\eta(x)$, $x\in R^3$.

Under condition E" formula (4.2.5) for the solution takes the form

$$
u(t,x) = -\frac{2\mu\sum_i \frac{x-x_i}{t}\exp\left\{\xi_i - \frac{(x-x_i)^2}{4\mu t}\right\}}{\sum_i \exp\left\{\xi_i - \frac{(x-x_i)^2}{4\mu t}\right\}}.
$$

Additional technical assumptions are that.

K.

(1) The function

$$
H_\zeta(a) = P\{\exp\{\zeta_i\} > a\}
$$

varies slowly as $a \to \infty$;

(2) The integral

$$\int_{R^3} H_\zeta(e^{x^2})dx < \infty,$$

and

$$\lim_{a\to\infty} \frac{H_\zeta(a\ (\log\ a)^{-1/2})}{H_\zeta(a)} = 1;$$

(3) There exists a strictly increasing and continuous inverse function

$$A(\lambda) = H_\zeta^{-1}\left(\frac{1}{\lambda}\right),$$

and a function $D(\lambda) > 0,\ \lambda \geq 1$, regularly varying at infinity with exponent $\theta \in (0, 2/3)$ such that, for any $u \in R^1$,

$$\lim_{\lambda\to\infty} \lambda\ H_\zeta(A(\lambda) + u\ D(\lambda)) = G_\zeta(u) \in [0, \infty]$$

exists.

Before we formulate the second type of limit result we also need another definition and the concept of ϕ-thinning of a point process $(x_i, \xi_i)_{i\in Z^1}$, where $0 \leq \phi = \phi(x) < \infty,\ x \in R^3$, is a even Borel function.

For simplicity we put $\mu = 1/2$.

For (x_i, ζ_i) from the definition in condition E" of the velocity potential define

$$x_{i^*} = x_{i^*}(x, t) \quad \text{and} \quad \zeta_{i^*} = \zeta_{i^*}(x, t)$$

so that

$$\zeta_{i^*} - \frac{(x - x_{i^*})^2}{2t} = \max_j \left(\zeta_{i^*} - \frac{(x - x_{i^*})^2}{2t}\right), \quad x \in R^3.$$

Also, formally, the ϕ-thinning $(x_i^{(\phi)}, \zeta_i^{(\phi)})_{i\in Z^1}$ of an arbitrary point process $(x_i, \zeta_i)_{i\in Z^1}$ is another point process defined by the following condition: for any continuous function $g(\cdot)$ with compact support

$$\sum_i g(x_i^{(\phi)}, \zeta_i^{(\phi)}) = \sum_i g(x_i, \zeta_i)\chi\left[\zeta_i \geq \zeta_j - \phi(x_i - x_j), \quad \forall j \neq i\right].$$

Notice that for

$$\phi(x) = \phi_t(x) = \frac{x^2}{2t}$$

and x restricted to the set $\{x_i\}$ we also have

$$(x_i^{(\phi_t)}, \zeta_i^{(\phi_t)}) = (x_{i^*}, \zeta_{i^*}).$$

Albeverio, Mochanov and Surgailis (1994) proved the following.

Theorem 4.7.3. Let assumptions E", K hold. If $u(t,x)$ is the solution of the Burgers' equation (2.1) with $n = 3$, $\mu = 1/2$, and the initial data described in condition E". Then, for each $x \in R^3$

$$\frac{u(t,x)}{\frac{x - x_{i^*}}{t}} \to 1, \quad t \to \infty$$

in probability. Moreover, there are constants $b_t = L(t)\ t^{1/(2-3\theta)}$ with $L \in \mathcal{L}$, and $a_t = A(b_t^3)$,

$$\left\{\left(\frac{x_i^{(\phi_t)}}{b_t}, \frac{(\zeta_i^{(\phi_t)} - a_t)\ t}{b_t^2}\right)\right\} \xrightarrow{D} \left\{(x_i^{(\infty)}, \zeta_i^{(\infty)})\right\}, \quad t \to \infty,$$

where $\{(x_i^{(\infty)}, \zeta_i^{(\infty)})\}$ coincides in distribution with the ϕ_1-thinning of a Poisson point process on $R^3 \times (G_\zeta^-, +\infty)$, $G_\zeta^- = \inf\{u \in R^1 : G_\zeta(u) < \infty\}$, with intensity measure $-dxdG_\zeta(u)$, $x \in R^3$, $u \in R^1$.

It is well-known (Leadbetter, Lindgren, Rootzen (1983)) that the limit function $G_\zeta(u)$ can only be of the following two types: either

$$G_\zeta(u) = \begin{cases} \frac{1}{(1+cu)^\gamma} & \text{for} \quad u > -1/c; \\ +\infty & \text{for} \quad u \le -1/c; \end{cases}$$

where $c > 0$, $\gamma > 0$, or $G_\zeta(u) = e^{-cu}$, for all $u \in R^1$, with $c > 0$.

Note that Tatsuma and Kida (1972) have obtained the rate of decay of the shock density of the form $(t_1/t_2)^{-\alpha}$, and Hu and Woyczynski (1995) calculated the rate of change of the expected number of shocks in a fixed finite interval after the saw tooth structure formation.

Remark 4.7.1. According to Theorem 4.7.3., the structure of the velocity field $u(t,x)$ (see remark 4.2.1.) can be pictured as follows: there are points x_{i^*} where the initial field

$$\exp\{\eta(x)\} = \exp\{\zeta_i\}, \qquad x = x_i,$$

takes high values ('local maxima'), from which the velocity $u(t,x)$ is pointed outwards, and which remain stable for some time and serve simultaneously all points x in their neighborhood. In the other words, matter is repulsed from x_{i^*} and should concentrate near the boundaries of 'Voronoi cells' which center on these points (see Møller (1994)). After some time the corresponding 'large scale structure' is destroyed, as the local maxima are replaced by higher local maxima (see the definition of x_{i^*}) resulting in a larger cell structure, and so on. The above interpretation is in part speculative and cannot be rigorously established on the basis of Theorem 4.7.3. The more general picture of the large scale structure of the Universe and quasi-Voronoi (curved boundaries) tessellation of the shock front s in forced Burgers' turbulence was presented by Molchanov, Surgailis and Woyczynski (1987). They used the variational principle. E, Rycov and Sinai (1996) presented an explicit construction of weak solutions of systems of conservation laws arising in adhesion particle dynamics. Wher and Xin (1996) and Bertoin (1998) considered the white noise perturbation of the viscous shock fronts of the one-dimensional Burgers' equation (see also Gurbatov, Simdyankin, Aurell, Frisch and Tóth (1998)).

Hausdorff dimensions of stochastic Burgers' flows have been studied by Sinai (1992), Janicki and Woyczynski (1997), Bertoin (1998).

Chapter 5

STATISTICAL PROBLEMS FOR RANDOM FIELDS WITH SINGULAR SPECTRUM

5.1 Estimation of Mathematical Expectation

1. Historical remarks

The research in parameter estimation for random process with singular spectrum has become very active these last ten years (see Beran (1994) for a review). In particular, results of Adenstedt (1974), Samarov and Taqqu (1988), Yajima (1988,1991), Künsch, Beran and Hampel (1993), Dahlhaus (1995) indicate that the least square estimates (LSE) for the unknown mean and the regression coefficients of the polynomial regression in long-memory noise may still be very good.

We present the asymptotic behavior of the LSE of the regression coefficients of the random processes and fields with singular spectrum. We consider the domain estimates. Frequency domain estimates considered by Robison and Hidalgo (1997). Nonlinear extremum in regression models with long-range dependent errors have been discussed by Koul (1992) and Koul and Muhkerjee (1993).

2. Asymptotic properties of the LSE in a regression model for random processes with singular errors

A. Let us consider the regression model of the form

$$\nu(t) = ag(t) + \eta(t), \quad t \in R^1,$$

where $g(t)$ is the known positive function which is square integrable over each segment $[0, T] \subset R^1$, a is the unknown parameter and $\eta(t)$, $t \in R^1$, is the stationary stochastic process of errors with $E\eta(t) = 0$, $E\eta^2(t) < \infty$.

The problem is to estimate the parameter a using observations of the random process $\nu(t)$ during the time interval $[0, T]$, $T \to \infty$.

We concentrate our attention on the LSE for a and discuss some of its properties.

In the following we restrict our considerations on two different types of errors. The first type is described by means of the assumptions described in the assumption A' and the second type by means of the assumptions described in the assumption A".

A'. The process of errors $\eta(t)$, $t \in R^1$, allows a representation $\eta(t) = G(\xi(t))$, where $\xi(x)$ is mean square continuous, stationary Gaussian process with $E(\xi(t)) = 0$, $E(\xi^2(t)) = 1$, and

$$\begin{aligned} B(|t|) &= \operatorname{cov}(\xi(0), \xi(t)) \\ &= \frac{L(|\, t\,|)}{|\, t\,|^{\alpha}}, \quad 0 < \alpha < 1, \end{aligned}$$

$L \in \mathcal{L}$, and $G(t)$ is a real, possible nonlinear function such that $EG^2(\xi(0)) < \infty$. In spite of all that, we require the function $B(|\, t\,|)$ to have a spectral density $f(|\, \lambda\,|)$ which is a decreasing function for $|\, \lambda\,| \geq \lambda_0$, $\lambda_0 \in R^1_+$, and continuous for all $\lambda \neq 0$.

Using all assumptions described in A' and Tauberian theorem (see Theorem 1.4.4.) we can conclude that the spectral density $f(\lambda)$ has the following asymptotic representation:

$$\begin{aligned} f(|\, \lambda\,|) &\sim \alpha\, |\, \lambda\,|^{\alpha-1} \frac{L\left(\frac{1}{|\lambda|}\right)}{2c_5(1,\alpha)}, \quad \lambda \to 0+, \\ c_5(1,\alpha) &= \Gamma(1+\alpha)\cos\frac{\alpha\pi}{2}. \end{aligned} \tag{5.1.1}$$

It is well-known (see section 2.2) that the function $G(t)$ in our conditions allows the expansion

$$G(t) = \sum_{i=0}^{\infty} \frac{C_i}{i!} H_i(t), \quad C_i = \int_{-\infty}^{\infty} G(t)\, H_i(t)\, \phi(t)\, dt, \; t \in R^1,$$

where $\phi(t)$ is defined in (2.2.3a) and $H_i(t)$ is defined in (2.1.5).

Let $m = H \operatorname{rank} G$ (see condition F in section 2.2).

A". The process of errors $\eta(t)$, $t \in R^1$, allows a representation:

$$\eta(t) = G(\zeta(t)) - EG(\zeta(t)), \qquad \zeta(t) = \frac{1}{2} \sum_{i=1}^{r} \xi_i^2(t),$$

where $\xi_i(t)$, $i = 1, \ldots, r$, are independent copies of a stationary Gaussian process $\xi(t)$ with the same properties as in the assumption A'. The function $G(t)$, $t > 0$, is a real Borel function such that $EG^2(\zeta(0)) < \infty$.

Hence, the process $\zeta(t)$, $t \in R^1$, is a χ-squared process (see example 2.1.3.).

We can express $G(t)$, $t > 0$, in terms of $e_k(t)$, given by (2.1.10):

$$\begin{aligned} G(t) &= \sum_{k=0}^{\infty} D_k\, e_k(u), \\ D_k &= \int_0^{\infty} G(t)\, e_k(t)\, p(t)\, dt, \quad k = 0, 1, 2, \ldots, \\ D_0 &= E(\zeta(0)), \qquad \sum_{k=0}^{\infty} D_k^2 < \infty. \end{aligned}$$

and $p(t)$ is given by (2.1.9) with $\beta = r/2$. Let $m = L \operatorname{rank} G \geq 1$ (see condition F' in the section 2.2).

We can see that our model in both cases (A', A") is the model with stationary errors and the long-term dependence.

In view of the serial expansions for $G(t)$ we have the following expansion for the errors covariance function $E\eta(t)\eta(s)$:

1. In case of A':

$$E\eta(t)\,\eta(s) = \sum_{i=1}^{\infty} \frac{C_i^2}{i!} B^i(|t-s|), \tag{5.1.2}$$

$$K = \sum_{i=1}^{\infty} \frac{C_i^2}{i!} = EG^2(\xi(0)) < \infty;$$

2. In case of A":

$$E\eta(t)\,\eta(s) = \sum_{i=1}^{\infty} D_i^2\, B^{2i}(|t-s|). \tag{5.1.3}$$

The LSE for the parameter a on the continuous time interval $[0, T]$, that is, the estimator which we can get minimizing

$$\int_0^T (\nu(t) - a\,g(t))^2 dt,$$

has the form

$$\hat{a}(T) = Q^{-1}(T) \int_0^T g(t)\nu(t)dt,$$

where

$$Q(T) = \int_0^T g^2(t)dt.$$

So it is an unbiased estimator

$$E\hat{a}(T) = a,$$

and its variance has a form

$$\begin{aligned} \operatorname{var}(\hat{a}(T)) &= \operatorname{var}(\hat{a}(T) - a) \\ &= \operatorname{var}(Q^{-1}(T) \int_0^T g(t)\eta(t)dt \\ &= Q^{-2}(T) \int_0^T \int_0^T E(\eta(t)\eta(s))g(t)g(s)dtds. \end{aligned}$$

As we are interested in asymptotic properties, we shall use the equality (5.1.2) and express the variance as follows.

$$\begin{aligned}
\operatorname{var}(\hat{a}(T)) &= Q^{-2}(T)\int_0^T\int_0^T \sum_{k=1}^{\infty}\frac{C_k^2}{k!}B^k(|\,t-s\,|)g(t)g(s)dtds \\
&= Q^{-2}(T)\frac{C_m^2}{m!}\int_0^T\int_0^T B^m(|\,t-s\,|)g(t)g(s)dtds \\
&\quad + Q^{-2}(T)\sum_{i=m+1}^{\infty}\frac{C_i^2}{i!}\int_0^T\int_0^T B^i(|t-s|)g(t)g(s)dtds \\
&= S_m + S_0, \\
S_m &= Q^{-2}(T)\frac{C_m^2}{m!}\int_0^T\int_0^T B^m(|\,t-s\,|)g(t)g(s)dtds, \\
S_0 &= Q^{-2}(T)\sum_{i=m+1}^{\infty}\frac{C_i^2}{i!}\int_0^T\int_0^T B^i(|\,t-s\,|)g(t)g(s)dtds.
\end{aligned}$$

From A' we obtain

$$S_m = Q^{-2}(T)\frac{C_m^2}{m!}\int_0^T\int_0^T \frac{L^m(|\,t-s\,|)}{|\,t-s\,|^{m\alpha}}g(t)g(s)dtds.$$

Now we need some extra assumptions upon the function $g(t)$.

B. Let us suppose that the following limits exist and are finite:

$$\begin{aligned}
l &= \lim_{T\to\infty}\int_0^1\int_0^1 \frac{g(Tu)g(Tv)}{g^2(T)}\frac{dudv}{|\,u-v\,|^{m\alpha}} \neq 0, \\
L &= \lim_{T\to\infty}\int_0^1 \frac{g^2(uT)}{g^2(T)}du
\end{aligned}$$

It is observable that the assumption B holds if $0<\alpha<1/m$ and if there exists a square integrable over $[0,1]$ function $g'(u)$ such that

$$\left|\frac{g(uT)}{g(T)} - g'(u)\right| \to 0 \qquad \text{as } T\to\infty$$

uniformly for $u \in [0,1]$. In this case

$$l = \int_0^1 \int_0^1 g'(u)g'(v)\frac{dudv}{|u-v|^{m\alpha}},$$
$$L = \int_0^1 [g'(u)]^2 du.$$

For instance, it holds for $g(t) = t^\nu$, $\nu \geq 0$, $0 < \alpha < 1/m$.

If assumptions described in B are fulfilled, we can take following calculations:

$$\begin{aligned} S_m &= Q^{-2}(T)\frac{C_m^2}{m!}\int_0^T \int_0^T \frac{L^m(|\, t-s \,|)}{|\, t-s \,|^{m\alpha}} g(t)g(s)dtds \\ &= Q^{-2}(T)\frac{C_m^2}{m!}T^2 g^2(T)\frac{L^m(T)}{T^{m\alpha}} \\ &\times \int_0^1 \int_0^1 \frac{L^m(T \mid u-v \mid)}{L^m(T)}\frac{g(Tu)g(Tv)}{g^2(T)}\frac{dudv}{|\, u-v \,|^{m\alpha}} \\ &= \frac{C_m^2}{m!}Q^{-2}(T)L^m(T)T^{2-m\alpha}g^2(T)l(1+o(1)). \end{aligned}$$

As $B(|\, t \,|)$ is a positive correlation function, there holds $0 \leq B(|\, t \,|) \leq 1$. The assumption $\lim_{t\to\infty} B(t) = 0$ makes us sure that for all $\varepsilon > 0$ there exists $A_0 \in R^1_+$ such that $0 < B(t) < \varepsilon$ for all $t > A_0$. Dividing the square $[0,T] \times [0,T]$ in two disjoint parts

$$\Delta_1(T) = \{(u,v) \in [0,T] \times [0,T] : |\, u-v \,| \leq A_0\},$$

$$\Delta_2(T) = \{(u,v) \in [0,T] \times [0,T] : |\, u-v \,| > A_0\},$$

we have

$$\begin{aligned} 0 &< B^{m+1}(|\, u-v \,|) \\ &\leq \varepsilon B^m(|u-v|) \qquad \text{for } (u,v) \in \Delta_2(T), \end{aligned}$$

and we can estimate S_0 as follows:

$$
\begin{aligned}
S_0 &\leq Q^{-2}(T) \sum_{i=m+1}^{\infty} \frac{C_i^2}{i!} \int_0^T \int_0^T B^i(| t-s |) g(t) g(s) dtds \\
&\leq Q^{-2}(T) K \int_0^T \int_0^T B^{m+1}(| t-s |) g(t) g(s) dtds \\
&\leq Q^{-2}(T) K \int_{\Delta_1(T)} g(t) g(s) dtds \\
&\quad + Q^{-2}(T) K \varepsilon \int_{\Delta_2(T)} \frac{L^m(| t-s |)}{| t-s |^{m\alpha}} g(t) g(s) dtds \\
&= Q^{-2}(T) K \int_{\Delta_1(T)} g(t) g(s) dtds \\
&\quad + \varepsilon Q^{-2}(T) K T^{2-m\alpha} L^m(T) l g^2(T)(1+o(1)).
\end{aligned}
$$

Now, using existence and finality of L, continuity for integrals and estimation:

$$
\int_{\Delta_1(T)} g(t) g(s) dtds \leq \frac{1}{2} \int_0^{A_0} \int_0^{2T} g\left(\frac{t}{2}+\frac{s}{2}\right) g\left(\frac{s}{2}-\frac{t}{2}\right) dsdt
$$

we can easily confirm the following estimation for S_0:

$$
\begin{aligned}
S_0 &\leq \text{conts}\, Q^{-2}(T) T g^2(T) L(1+o(1)) \\
&\quad + \varepsilon Q^{-2}(T) K T^{2-m\alpha} L^m(T) g^2(T) l(1+o(1)), \\
S_0 &\leq Q^{-2}(T) K g^2(T) l L^m(T) T^{2-m\alpha} \\
&\quad \times \left(\text{conts}\, T^{m\alpha-1} L^{-m}(T)(1+o(1)) + \varepsilon^2(1+o(1))\right).
\end{aligned}
$$

as $T \to \infty$. Therefore we have already proved:

Theorem 5.1.1. Let assumptions A, A' and B hold with $0 < \alpha < 1/m$. Then for the $\text{var}(\hat{a}(T))$ of the LSE of unknown parameter a holds:

$$
\text{var}(\hat{a}(T)) = \frac{C_m^2}{m!} Q^{-2}(T) L^m(T) T^{2-m\alpha} g^2(T) l(1+o(1)) \tag{5.1.4}
$$

as $T \to \infty$.

Remark 5.1.1.

If we concern ourselves with the model from Theorem 5.1.1. but with the function $g(t)$ which is not upon the assumptions B we can sometimes calculate some useful estimations applying the similar methods. For instance, if $g(t)$ is bounded and not upon assumptions B, we can use the fact

$$\begin{aligned}\lim_{T\to\infty}\int_0^1\int_0^1 \frac{L^m(T\,|u-v|)}{L^m(T)}\frac{dudv}{|u-v|^{m\alpha}} &= l' \\ &< \infty\end{aligned}$$

for $0<\alpha<1/m$ and, as before, we can obtain

$$\begin{aligned} S_m &\leq \frac{C_m^2}{m!}Q^{-2}(T)L^m(T)T^{2-m\alpha}M^2l'(1+o(1)), \\ S_0 &\leq KM^2Q^{-2}(T)\left[\int_{\Delta_1(T)} dtds + \varepsilon\int_{\Delta_2(T)}\frac{L^m(|t-s|)}{|t-s|^{m\alpha}}dtds\right], \end{aligned}$$

where, $0<g(t)\leq M$. So, in this case we have for the variance:

$$\operatorname{var}(\hat{a}(T)) \leq \frac{C_m^2}{m!}Q^{-2}(T)L^m(T)T^{2-m\alpha}M^2l'(1+o(1)).$$

Example 5.1.1. If we concern ourselves with a model satisfying all assumptions in Theorem 5.1.1. with $L\neq 0$, then it holds:

$$\begin{aligned} Q^2(T) &= \left[\int_0^T g^2(t)dt\right]^2 \\ &= \left[T\int_0^1 g^2(uT)du\right]^2 \\ &= T^2g^4(T)L(1+o(1)), \end{aligned}$$

$$\operatorname{var}(\hat{a}(T)) = \frac{C_m^2}{m!}L^2(T)T^{-m\alpha}g^{-2}(T)\frac{1}{L}(1+o(1)).$$

Specifically, for $g(t)=t^\nu$, $\nu\geq 0$, we have

$$\operatorname{var}(\hat{a}(T)) = \frac{C_m^2}{m!}L^2(T)T^{-m\alpha-2\nu}\frac{1}{L}(1+o(1)) \to 0,$$

as $T\to\infty$.

Example 5.1.2. Let us take the model with

$$g(t) = \sin t + 1, \qquad 0 < \alpha < 1/m.$$

Using Remark 5.1.1. we can easily obtain $\lim_{T\to\infty} \mathrm{var}(\hat{a}(T)) = 0$.

Now, let us comment upon the model in which assumptions A and A" are fulfilled.

Theorem 5.1.2. For the process $\nu(t) = a\, g(t) + \eta(t)$, $t \in R^1$, assumptions A and A" hold with $0 < \alpha < 1/2m$, and let the following limits exist:

$$l = \lim_{T\to\infty} \int_0^1\int_0^1 \frac{g(Tt)g(Ts)}{g^2(T)} \frac{dtds}{|\, t-s\,|^{m\alpha}} < \infty, \quad l \neq 0,$$

$$L = \lim_{T\to\infty} \int_0^1 \frac{g^2(uT)}{g^2(T)} du < \infty.$$

Then for variance $\mathrm{var}(\hat{a}(T))$ of the LSE of unknown parameter a holds:

$$\mathrm{var}(\hat{a}(T)) = D_m^2 Q^{-2}(T) L^{2m}(T) T^{2-2m\alpha} g^2(T) l (1 + o(1)), \tag{5.1.4a}$$

as $T \to \infty$, where $m = L\,\mathrm{rank}\, G$.

For the proof we can use exactly the same method as the Theorem 5.1.1. but we have to apply (5.1.3).

Remark 5.1.3. If we concern the model from the Theorem 5.1.2., but with the function $g(t)$, for which the limits l (and) or L do not exist but $g(t)$ is bounded, we can use a fact that

$$\lim_{T\to\infty} \int_0^1\int_0^1 \frac{L^{2m}(T\,|u-v|)}{L^{2m}(t)} \frac{dudv}{|u-v|^{2m\alpha}} = l' < \infty,$$

for $0 < \alpha < 1/2m$ and as before, we can obtain

$$S_m \le D_m^2 Q^{-2}(T) L^{2m}(T) T^{2-2m\alpha} M^2 l'(1 + o(1)),$$

$$S_0 \le K M^2 Q^{-2}(T) \left[\int_{\Delta_1(T)} dtds + \varepsilon \int_{\Delta_2(T)} \frac{L^{2m}(|t-s|)}{|t-s|^{2m\alpha}} dtds \right],$$

where $0 < g(t) \le M$. So, in this case we have for variance:

$$\mathrm{var}(\hat{a}(T)) \le D_m^2 Q^{-2}(T) L^{2m}(T) T^{2-2m\alpha} M^2(T) l'(1 + o(1)). \tag{5.1.5}$$

The reduction theorems make it possible to describe an asymptotic distribution for the process

$$X(T) = \frac{1}{\sigma(T)}(\hat{a}(T) - a),$$
$$\sigma^2(T) = \text{var } \hat{a}(T),$$

in our models with that of the process

$$X_m(T) = \frac{C_m}{m!\sigma(T)Q(T)} \int_0^T g(t)H_m(\xi(t))dt,$$
$$Q(T) = \int_0^T g^2(T)\, dt,$$

$m = H\,\text{rank}\ G$, if the model satisfies assumptions A and A', or

$$X'_m(T) = \frac{D_m}{\sigma(T)Q(T)} \int_0^T g(t)e_m(\zeta(t))dt,$$

$m = L\,\text{rank}\ G$, if the model satisfies all the assumptions in the Theorem 5.1.2. In fact, there holds:

Theorem 5.1.3. Let us suppose that the assumptions A, A' and B are fulfilled with $0 < \alpha < 1/m$. Then it holds:

$$\lim_{T\to\infty} E\,|X(T) - X_m(T)|^2 = 0.$$

This means that if one of the asymptotic distributions for $X(T)$ or for $X_m(T)$ exists then the other exists too, and they are the same.

Proof

Using the expansion for $G(u)$ in $L^2(R^1, \phi(u)\, du)$ we can see that

$$X(T) - X_m(T) = \frac{1}{Q(T)\sigma(T)} \int_0^T \eta(t)g(t)dt - \frac{C_m}{m!Q(T)\sigma(T)} \int_0^T g(t)H_m(\xi(t))dt,$$

$$
\begin{aligned}
E\left|X(T)-X_m(T)\right|^2 &= E\left|\sum_{i=m+1}^{\infty}\frac{C_i}{i!Q(T)\sigma(T)}\int_0^T H_i(\zeta(t))g(t)dt\right|^2 \\
&= \sum_{i=m+1}^{\infty}\frac{C_i^2}{i![Q(T)]^2\sigma^2(T)}\int_0^T\int_0^T B^i(\mid t-s\mid)g(t)g(s)dtds.
\end{aligned}
$$

So there holds:

$$E\left|X(T)-X_m(T)\right|^2 = S_0\,\sigma^{-2}(T),$$

and we can use the estimation (5.1.5) to prove that

$$
\begin{aligned}
E\left|X(T)-X_m(T)\right|^2 &= \frac{K\,g^2(T)lL^m(T)T^{2-m\alpha}}{[Q(T)]^2\sigma^2(T)} \\
&\times\left(\text{const}\,T^{m\alpha-1}L^{-m}(T)(1+o(1))+\varepsilon^2(1+o(1))\right).
\end{aligned}
$$

It follows for $\sigma(T)$ from the estimation (5.1.4a)

$$
\begin{aligned}
E\left|X(T)-X_m(T)\right|^2 &\le \frac{Km!}{C_m^2(1+o(1))} \\
&\times\left(\text{const}\,T^{m\alpha-1}L^{-m}(T)(1+o(1))+\varepsilon^2(1+o(1))\right).
\end{aligned}
$$

Now, the property

$$\lim_{T\to\infty} T^{\beta}\,L(T)=\infty \qquad \text{for } \beta>0$$

and Lemma 1.4.12. assure the assertion which we want to prove is true. The Theorem 5.1.3. is proved.

Theorem 5.1.4. Let us suppose that the assumptions A, A' are fulfilled with $0<\alpha<(2m)^{-1}$, and let us suppose that the limits l and L from Theorem 5.1.2. exist, $l\neq 0$, and that they are finite. Then there holds:

$$E\mid X(T)-X'_m(T)\mid^2=0.$$

This means that if one of the asymptotic distributions for $X(T)$ and $X'_m(T)$ exists then the other exists too and they are the same. Here, $m=L\,\text{rank}\,G$.

For the proof we can follow the same method as in the proof of the Theorem 5.1.3. but we must use results from the Theorem.5.1.2.

The distribution theory for linear and non-linear functionals of Gaussian processes and fields was developed in Chapter 2.

We shall prove one representation for the asymptotic distribution of the process

$$X(T) = \frac{\hat{a}(T) - a}{\sigma(T)}$$

in the regression models we are concerned with.

The case of errors in form described upon assumptions A' we consider in the subsection 3 and the case of errors in form of assumptions A" in the subsection 4.

3. Asymptotic distributions if the error is in form of $G(\xi(t))$

Theorem 5.1.5. Let us suppose that the assumptions A, A' and B hold, that there exists the real function $g'(u)$ for which

$$\left|\frac{g(uT)}{g(T)} - g'(u)\right| \to 0 \quad \text{as} \quad T \to \infty \tag{5.1.6}$$

uniformly for $u \in [0,1]$ and

$$\int_{R^m} \left| \int_0^1 g'(u) e^{iu(\lambda_1 + \ldots \lambda_m)} du \right|^2 \frac{d\lambda_1 \ldots d\lambda_m}{|\lambda_1 \ldots \lambda_m|^{1-\alpha}} < \infty, \tag{5.1.7}$$

where $m = H \operatorname{rank} G$. Then the process $X(T)$ converges in distributions as $T \to \infty$ to the random variable

$$\begin{aligned} \kappa_m &= C \int'_{R^m} \left| \int_0^1 g'(u) e^{iu(\lambda_1 + \ldots \lambda_m)} du \right|^2 \frac{W(d\lambda_1) \ldots W(d\lambda_m)}{|\lambda_1 \ldots \lambda_m|^{(1-\alpha)/2}}, \\ C &= \frac{\operatorname{sgn} C_m \sqrt{\alpha^m}}{\sqrt{m! 2^m l \Gamma^m (1+\alpha) \cos^m(\alpha\pi/2)}}, \end{aligned}$$

and $\int'_{R^n} \ldots$ means a multiple stochastic integral (see section 2.3) and $W(\cdot)$ is the complex Gaussian white noise.

Proof

The results from the previous section make it possible to restrict our problem to finding an asymptotic distribution for the process

$$X_m(T) = \frac{C_m}{m! \sigma(T) Q(T)} \int_0^T g(t) H_m(\xi(t)) dt.$$

It is well-known that the process $\xi(t)$ described upon assumptions A" admits a representation

$$\xi(t) = \int_{R^1} e^{i\lambda t}\sqrt{f(\lambda)}W(d\lambda),$$

where $W(d\lambda)$ is one complex Gaussian white noise measure on $(R^1, \mathcal{B}(R^1))$. Using Itô formula (2.3.1) we have:

$$\begin{aligned} X_m(T) &= \frac{C_m}{m!\sigma(T)Q(T)} \\ &\times \int_0^T \int_{R^m}' g(t)e^{it(\lambda_1+\cdots+\lambda_m)}\sqrt{f(\lambda_1)\cdots f(\lambda_m)}W(d\lambda_1)\cdots W(d\lambda_m)dt \\ &= \frac{C_m T}{m!\sigma(T)Q(T)}\int_0^1 \int_{R^m}' g(uT)e^{iuT(\lambda_1+\cdots+\lambda_m)} \\ &\times \sqrt{f(\lambda_1)\cdots f(\lambda_m)}W(d\lambda_1)\cdots W(d\lambda_m)du. \end{aligned}$$

The property of semi-stability with the parameter $\frac{1}{2}$ for a Gaussian white noise and changing the variables λ with $\lambda' = T\lambda$ in our stochastic integral make it possible to obtain the following expressions.

$$\begin{aligned} X_m(T) &= \frac{C_m T^{1-\frac{m}{2}}}{m!\sigma(T)Q(T)}\int_0^1 \int_{R^m}' g(uT)e^{iu(\lambda_1'+\cdots+\lambda_m')} \\ &\times \sqrt{f(\frac{\lambda_1'}{T})\cdots f(\frac{\lambda_m'}{T})}W(d\lambda_1')\cdots W(d\lambda_m')du. \end{aligned}$$

Let us define:

$$\begin{aligned} \kappa_m &= C\int_{R^m}' \left[\int_0^1 g'(u)e^{iu(\lambda_1+\cdots+\lambda_m)}du\right]\frac{W(d\lambda_1)\cdots W(d\lambda_m)}{|\lambda_1\cdots\lambda_m|^{(1-\alpha)/2}}, \\ C &= \frac{\mathrm{sgn} C_m\sqrt{\alpha^m}}{\sqrt{m!2^m l\Gamma^m(1+\alpha)\cos^m\frac{\alpha\pi}{2}}}. \end{aligned}$$

So,

$$E|X_m(T)-\kappa_m|^2 = E\left|\int_{R^m}\left[\frac{C_m T^{(1-m/2)}}{m!\sigma(T)Q(T)}\right.\right.$$

$$\times \int_0^1 g(uT) e^{iu(\lambda_1+\cdots+\lambda_m)} \sqrt{f\left(\frac{\lambda_1}{T}\right)\cdots f\left(\frac{\lambda_m}{T}\right)}\, du$$

$$\left. -C \int_0^1 \frac{g'(u) e^{iu(\lambda_1+\cdots+\lambda_m)}}{|\lambda_1\cdots\lambda_m|^{\frac{1-\alpha}{2}}} du \right] W(d\lambda_1)\cdots W(d\lambda_m)\Bigg|^2$$

$$= C^2 \int_{R^m} \int_0^1 \int_0^1 K(T,u,\lambda_1,\cdots,\lambda_m) K(T,v,\lambda_1,\cdots,\lambda_m)$$

$$\times \frac{e^{i(u-v)(\lambda_1+\cdots+\lambda_m)}}{|\lambda_1\cdots\lambda_m|^{1-\alpha}} du dv d\lambda_1\cdots d\lambda_m, \tag{5.1.8}$$

where $K(T,u,\lambda_1,\cdots,\lambda_m) =$

$$\sqrt{f\left(\frac{\lambda_1}{T}\right)\cdots f\left(\frac{\lambda_m}{T}\right)} \left[\frac{L^m(T)|\lambda_1\cdots\lambda_m|^{\alpha-1}}{2^m\Gamma^m(\alpha)\cos^m(\alpha\pi/2)T^{m(\alpha-1)}}\right]^{-1}$$

$$\times \frac{g(uT)}{g(T)} - g'(u).$$

Let

$$U = \left\{(\lambda_1\cdots\lambda_m|) : |\lambda_j| < T^{1-\alpha}, \quad j=1,\dots m\right\} \quad \text{and } F = R^m\backslash U.$$

Subdivide the integral $\int_{R^m}\cdots$ in (5.1.8) into two parts, the integration in the first (denoted by I_1) is carried out over the set U and over the set F (denote it by I_2) in the second one.

Using continuity and asymptotic behavior of the spectral density (5.1.1) we can make the integral I_1 arbitrary small choosing T large enough. As we assume that the spectral density is a decreasing function after some λ_0 (see condition A') the form of F assures that I_2 tends to zero as $T\to\infty$.

As there holds

$$\int_{-\infty}^{\infty} \frac{e^{i(u-v)\lambda}}{|\lambda|^{1-\alpha}} d\lambda = \frac{2}{|u-v|^\alpha}\int_{-\infty}^{\infty} |t|^{\alpha-1}e^{it}dt$$

$$= \frac{2}{|u-v|^\alpha}\Gamma(\alpha)\cos\left(\frac{\alpha\pi}{2}\right),$$

then

$$\int_0^1\int_0^1\int_{R^m} \frac{e^{i(u-v)(\lambda_1+\cdots+\lambda_m)}}{|\lambda_1\cdots\lambda_m|^{1-\alpha}} d\lambda_1\cdots d\lambda_m du dv < \infty.$$

Now we have

$$\lim_{T\to\infty} E|X_m(T) - \kappa_m|^2 = 0,$$

which we wanted to prove.

Example 5.1.3. Let us take the model with $m = 1$, $\mathrm{sgn} C_1 = 1$, $0 < \alpha < 1/2$ and $g(u) = u^\nu$, where $\nu > 0$ is integer. Then there holds:

$$\begin{aligned} l &= \int_0^1 \int_0^1 \frac{u^\nu v^\nu}{|u-v|^\alpha} du dv, \\ \kappa_1 &= C \int_{R^1} Q(\lambda) \frac{W(d\lambda)}{|\lambda|^{(1-\alpha)/2}}, \\ Q(\lambda) &= \int_0^1 u^\nu e^{i\lambda u} du = \frac{1}{i^\nu} \frac{d^\nu}{d\lambda^\nu} \frac{e^{i\lambda} - 1}{i\lambda}, \\ C &= (2l\Gamma(1+\alpha)\cos(\alpha\pi/2))^{-1/2}, \\ \mathrm{var}\, \kappa_1 &= C^2 \int_{R^1} |Q(\lambda)|^2 \frac{d\lambda}{|\lambda|^{1-\alpha}} \\ &= 2C^2\Gamma(\alpha)\cos(\alpha\pi/2) \int_0^1 \int_0^1 \frac{u^\nu v^\nu}{|u-v|^\alpha} du dv \\ &= 1. \end{aligned}$$

So, κ_1 has normal distribution with parameters $\mu = 0$ and $\sigma^2 = 1$ for all $\nu > 0$.

Example 5.1.4. Let us take the model with $m = 2$, $\mathrm{sgn} C_m = 1$, $0 < \alpha < 1/4$ and $g(u) = u^\nu$, $\nu > 0$. Then it holds:

$$\begin{aligned} l &= \int_0^1 \int_0^1 \frac{u^\nu v^\nu}{|u-v|^{2\alpha}} du dv, \\ \kappa_2 &= C \int_{R^1} Q(\lambda_1 + \lambda_2) W(d\lambda_1) W(d\lambda_2) \left|\lambda_1\lambda_2\right|^{(1-\alpha)/2}, \\ Q(\lambda_1 + \lambda_2) &= \int_0^1 u^\nu e^{i(\lambda_1+\lambda_2)u} du, \\ C &= 2\alpha \left(2l\Gamma^2(1+\alpha)\cos^2(\alpha\pi/2)\right)^{-1/2}, \\ \mathrm{var}\, \kappa_2 &= \frac{C^2}{2} \int_{R^1} |Q(\lambda_1 + \lambda_2)|^2 \frac{d\lambda_1 d\lambda_2}{|\lambda_1\lambda_2|^{1-\alpha}} = \frac{1}{4}. \end{aligned}$$

The distribution of κ_2 is not normal in this case. Namely, using (2.3.2) we can see that κ_2 allows a representation

$$\kappa_2 = C \sum_{k=1}^{\infty} \nu_k (X_k^2 - 1),$$

where X_k, $k \in Z_+^1$, is a sequence of i.i.d. standard normal random variables and ν_k, $k = 1, 2, \ldots$, is a sequence of nonnegative real numbers. This is an immediate consequence of the following theorem:

Theorem 5.1.6. Let us suppose that all the assumptions of the Theorem 5.1.5. are fulfilled with $m = 2$, and $g'(u)$ is a positive function on $[0, 1]$. Then there exists a sequence of nonnegative real numbers ν_k, $k \in Z_+^1$, and a sequence of i.i.d. standard normal random variables X_k, $k \in Z_+^1$, such that

$$\sum_{k=1}^{\infty} \nu_k^2 < \infty, \qquad \kappa_2 = C \sum_{k=1}^{\infty} \nu_k (X_k^2 - 1).$$

Proof

Let us express the random variable κ_2/C as

$$\frac{\kappa_2}{C} = \int_{R^2} H(\lambda_1, \lambda_2) W(d\lambda_1) W(d\lambda_2),$$

$$H(\lambda_1, \lambda_2) = \int_0^1 \frac{g'(u) e^{iu(\lambda_1 + \lambda_2)}}{|\lambda_1|^{(1-\alpha)/2} |\lambda_2|^{(1-\alpha)/2}} du,$$

where $H(\cdot, \cdot)$ is a complex-valued measurable function with the properties

$$H(\lambda_1, \lambda_2) = H(\lambda_2, \lambda_1) = \overline{H(-\lambda_1, \lambda_2)}.$$

As it holds

$$\int_{R^2} \left| \int_0^1 \frac{g'(u) e^{iu(\lambda_1 + \lambda_2)}}{|\lambda_1|^{(1-\alpha)/2} |\lambda_2|^{(1-\alpha)/2}} du \right|^2 d\lambda_1 d\lambda_2 < \infty,$$

then,

$$\int_{R^2} |H(\lambda_1, \lambda_2)|^2 d\lambda_1 d\lambda_2 < \infty.$$

If we consider the real Hilbert space $\mathbf{L}^2(R^1, d\lambda)$ consisting of the complex valued functions $f(x)$, $x \in R^1$, such that

$$f(x) = \overline{f(-x)}, \qquad x \in R^1, \qquad \int_{R^1} | f(x) |^2 \, dx < \infty,$$

with the scalar product

$$(f \mid g) = \int_{R^1} f(x)\overline{g(x)}dx,$$

we can define the integral operator

$$Af(x) = \int_{R^1} H(x, -y) f(y) dy$$

which maps $\mathbf{L}^2(R^1, d\lambda)$ into itself and which is self-adjoint Hilbert-Schmidt operator.

Therefore A has a system of real eigenvalues $\nu_1, \nu_2, \ldots$ such that

$$\sum_{k=1}^{\infty} \nu_k^2 < \infty.$$

Let $\phi_1, \phi_2, \ldots$ be a complete orthonormal system of eigenvectors of the operator A.

We can write by Theorem 1.1.4:

$$H(x, -y) = \sum_{k=1}^{\infty} \nu_k \phi_k(x)\overline{\phi_k(y)},$$

where the convergence is in the $\mathbf{L}^2(R^1, d\lambda) \times \mathbf{L}^2(R^1, d\lambda)$ sense.

Using Itô formula (2.3.1) we have:

$$\begin{aligned} \int_{R^2}' H(\lambda_1, \lambda_2) W(d\lambda_1) W(d\lambda_2) &= \sum_k \int_{R^2}' \nu_k \phi_k(\lambda_1) \phi_k(\lambda_2) W(d\lambda_1) W(\lambda_2) \\ &= \sum_k \nu_k H_2 \left(\int_{R^2}' \phi_k(\lambda) W(d\lambda) \right), \end{aligned}$$

where $H_2(x) = x^2 - 1$ is the second Chebychev–Hermite polynomial with leading coefficient 1.

Let

$$X_k = \int_{R^1} \phi_k(\lambda) W(d\lambda).$$

where $(X_k,\ k \in Z^1_+)$ are jointly Gaussian and independent because of the orthogonality of the ϕ_ks, what proves our statement.

Let us check when the operator A is positive semi-definite. Namely, this positivity assures nonnegativity for all of its eigenvalues $(\nu_k,\ k \in Z^1_+)$.

Using the fact

$$\begin{aligned} H(\lambda_1, -\lambda_2) &= \frac{1}{|\lambda_1|^{(1-\alpha)/2} |\lambda_2|^{(1-\alpha)/2}} H_1(\lambda_1 - \lambda_2), \\ H_1(x) &= \int_0^1 g'(u) e^{iux} du, \end{aligned}$$

we can see that if $g'(u)$ is positive on $[0,1]$, $H(x)$ is positive semi-definite. The Theorem is proved.

4. Asymptotic distributions if the error is in form of $G(\zeta(t))$

Theorem 5.1.7. Let us suppose that the assumptions A, A" are fulfilled with $m = 1$, $m = L \operatorname{rank} G$, $0 < \alpha < 1/2$, limits l and L from the Theorem 5.1.2. exist, $l \neq 0$, and they are finite, and that there exists the real function $g'(u)$ for which

$$\left| \frac{g(uT)}{g(T)} - g'(u) \right| \to 0, \quad \text{as } T \to \infty,$$

uniformly for $u \in [0,1]$ and

$$\int_{R^2} \left| \int_0^1 g'(u) e^{iu(\lambda_1+\lambda_2)} du \right|^2 \frac{d\lambda_1 d\lambda_2}{|\lambda_1 \lambda_2|^{1-\alpha}} < \infty.$$

Then the process $X(T)$ converges in distributions as $T \to \infty$ to the random variable

$$Y = \sum_{i=1}^{r} \kappa'_i,$$

where κ'_i, $i = 1, \dots r$, are independent copies of κ',

$$\kappa' = C \int_{R^2}' \left[\int_0^1 g'(u) e^{iu(\lambda_1+\lambda_2)} du \right] \frac{W(d\lambda_1) W(d\lambda_2)}{|\lambda_1|^{(1-\alpha)/2} |\lambda_2|^{(1-\alpha)/2}},$$

$$C = \frac{-\text{sgn} D_1 \alpha}{2\sqrt{2rl}\Gamma(1+\alpha)\cos(\alpha\pi/2)},$$

and $\int\limits_{R^2}' \cdots$ means a multiple stochastic integral in the sense of section 2.3.

Proof

The results from the section 3 make it possible to restrict our problem to finding asymptotic distributions for the process

$$X_1'(T) = \frac{D_1}{\sigma(T)Q(T)} \int_0^T g(t) e_1(\zeta(t)) dt,$$

where $e_1(u)$ is defined in (2.1.11). So, we have:

$$\begin{aligned} X_1(T) &= \frac{D_1}{\sigma(T)Q(T)} \int_0^T g(t) \left[\frac{r}{2} - \zeta(t)\right] \sqrt{\frac{2}{r}} dt \\ &= \frac{D_1}{\sigma(T)Q(T)\sqrt{2r}} \sum_{i=1}^{r} \int_0^T g(t)(1 - \xi_i^2(t)) dt \\ &= \frac{-D_1}{\sigma(T)Q(T)\sqrt{2r}} \sum_{i=1}^{r} \int_0^T g(t) H_2(\xi_i(t)) dt, \end{aligned}$$

where $H_2(t)$ is the second Chebyshev–Hermite polynomial.

Using the same methods as in the proof of the Theorem 5.1.5. and an asymptotic behavior (5.1.4) for the variance $\sigma^2(T)$ we can obtain:

$$\frac{-D_1}{\sigma(T)Q(T)\sqrt{2r}} \int_0^T g(t) H_2(\xi) dt \to \kappa' \quad \text{as } T \to \infty$$

in $\mathbf{L}_2(\Omega)$ sense.

So, let us form a random variable

$$Y = \sum_{i=1}^{r} \kappa_i',$$

where κ_i' are independent copies of κ'. Then it holds:

$$\sqrt{E\mid X_1'(T)-Y\mid^2} = \sqrt{E\left|\sum_{i=1}^{r}\frac{D_1}{\sigma(T)Q(T)\sqrt{2r}}\int_0^T g(t)(1-\xi_i^2(t))dt - Y\right|^2}$$

$$\leq \sum_{i=1}^{r}\sqrt{E\left|\frac{D_1}{\sigma(T)Q(T)\sqrt{2r}}\int_0^T g(t)(1-\xi_i^2(t))dt - \kappa_i\right|^2},$$

and this tends to zero when $T \to \infty$. The Theorem is proved

5.Estimation of regression coefficients of random fields

We define the long memory regression model for the univariant observed response $\nu(x) \in R^1$ known to be dependent on corresponding inputs $x \in R^n$. Therefore,

$$\nu(x) = \langle a, g(x)\rangle + \eta(x), \quad x \in R^n, \tag{5.1.9}$$

where $g(x) = [g_1(x), \ldots, g_k(x)]'$ is the known vector function, $g : R^n \to R^{nk}$, whose coordinate functions $g_i(x)$, $i = 1, \ldots, k$, form a linearly independent set of real functions, they are positive on U and square integrable over the same set for all bounded $U \subseteq R^n$ containing the origin. $a = (a_1, \ldots, a_k)'$ is the unknown vector of parameters, $\eta(x)$, $x \in R^n$, is a homogeneous and isotropic random field of errors with $E\eta(x) = 0$, $E\eta^2(x) < \infty$ and it is supposed that $\eta(x)$ is a local functional of homogeneous and isotropic Gaussian field (we denote it with $\xi(x)$, $x \in R^n$) with the long range dependence.

Let us introduce the structure of the errors exactly.

C. Let $\xi(x)$, $x \in R^n$, be a real-valued mean square continuous homogeneous and isotropic Gaussian field with the following properties:

It has a non-increasing correlation function $B(x)$ such that

$$\lim_{|x|\to\infty} B(\|x\|) = 0;$$

A correlation function $B(x)$ has the form

$$B(\|x\|) = \frac{L(\|x\|)}{\|x\|^{\alpha}}, \quad \text{where } 0 < \alpha < n,$$

and $L \in \mathcal{L}$.

$$E(\xi(x)) = 0, \quad E(\xi^2(x)) = 1, \;\; x \in R^n;$$

$B(\|x\|)$ has spectral density $f(\|\lambda\|)$ that tends to zero as $\|\lambda\| \to \infty$ and is decreasing on $(\|\lambda_0\|, \infty)$ for some $\lambda_0 \in R^n$ and continuous for $\lambda \neq 0$.

The errors of regression models we are dealing with are constructed from the Gaussian process $\xi(x)$ described in one of the two ways presented in the assumptions C' and C".

C'. The field of errors $\eta(x)$, $x \in R^n$, allows a representation $\eta(x) = G(\xi(x))$, where $\xi(x)$ is random field satisfying assumptions C and $G(t)$ is a real Borel function such that $E(\xi(0)) = 0$ and $E(\xi^2(0)) < \infty$.

C". We assume that $0 < \alpha < n/m$, where $m = H \operatorname{rank} G$.

D. The field of errors, $\eta(x)$, $x \in R^n$, allows a representation:

$$\eta(x) = G(\zeta(x)) - EG(\zeta(x)), \quad \zeta(x) = \frac{1}{2}\sum_{i=1}^{r} \xi_i^2(x),$$

where $\xi_i(x)$, $i = 1, \ldots r$, are independent copies of a field $\xi(x)$ described in the assumption C' and $G(t)$ is a real, possible non-linear function such that $EG^2(\zeta(x)) < \infty$.

D'. We assume that $0 < \alpha < n/2m$, where $m = L \operatorname{rank} G$.

Let a random field be observed on the sets $U(T)$, where $U(T)$ is the image of the set U under the homotetic transformation with the centre at origin and the coefficient $T > 0$. Also, we require that the set U contains the origin, that it is measurable, bounded and convex. In fact, practical situation often claims that U is a ball or parallelepiped but we can permit this weaker condition.

Therefore we can compute the Lebesgue measure of $U(T)$ from the Lebesgue measure of U: $|U(T)| = T^n|U|$.

Also, suppose that $g(x) = (g_1(x), \ldots, g_k(x))'$ is the radial function: $g(x) = g(\|x\|)$,with coordinate functions $g_i(x)$, $x \in R^n$, forming the linearly independent set of positive, continuous, real functions which are square integrable on $U(T)$ for any $T > 0$. $a = (a_1, \ldots, a_k)'$ is the vector of unknown parameters and $\eta(x)$, $x \in R^n$, is the homogeneous isotropic field of errors with the following properties:

E. $E\eta(x) = 0$, $E\eta^2(x) < \infty$, $x \in R^n$; correlation function $E\eta(x)\eta(y)$ satisfies the requirements

$$E\eta(x)\eta(y) = \sum_{i=m}^{\infty} I_i B^{pi}(\|x - y\|),$$

for some p, $m \in Z_+^1$, where I_i are nonnegative constants, $i = 1, 2, \ldots$, $K = \sum_{i=m}^{\infty} I_i < \infty$, $B(x), x \in R^n$, is a real nonnegative radial function, bounded with one and of the form

$$B(\|x\|) = \frac{L(\|x\|)}{\|x\|^{\alpha}}, \quad 0 < \alpha < \frac{n}{pm}, \quad L \in \mathcal{L}.$$

In these conditions the LSE for the unknown vector parameter **a** has an explicit form:

$$\hat{a} = \hat{a}(U(T)) = V(U(T))^{-1} \int\limits_{U(T)} g(x)\nu(x)dx,$$

where

$$V(U(T)) = \int\limits_{U(T)} g(x)g(x)'dx.$$

We can easily check, using linear independence and square integrability of $g_1(x), \ldots, g_k(x)$, that $V(U(T))$ is a real, symmetric, positive definite matrix for all $T > 0$, so the inverse $V(U(T))^{-1}$ exists.

The problem of analyzing an asymptotic behavior of the correlation matrix

$$D\hat{a}(U(T)) = E\hat{a}\hat{a}'$$

of the estimator $\hat{a}(U(T))$ we can treat in the same way as in the section 2. Therefore, we can state the following theorem:

Theorem 5.1.8. Let the regression model (5.1.9) satisfy the assumptions E and let the vector function $g(x)$ be continuous and such that its coordinate functions $g_i(x)$ form a linearly independent set of real functions, they are positive on $U(T)$ and square integrable over the same set for all $T > 0$ and $i = 1, \ldots k$. Also, suppose that the following limits exist and are finite:

$$\begin{aligned} l &= \lim_{T\to\infty} d(T)^{-1} \int_U \int_U g(Tx)g(Ty)' \frac{dxdy}{\|x-y\|^{mp\alpha}} d(T)^{-1}, \\ M &= \lim_{T\to\infty} d(T)^{-1} V_U(T) d(T)^{-1}, \\ V_U(T) &= \int_U \int_U g(Tx)g(Ty)'dxdy, \end{aligned}$$

$d(T) = \operatorname{diag}(g_1(T), \ldots, g_k(T))$ and that l is a regular matrix.

Then for the correlation matrix $D\hat{a}(U(T))$ of the LSE of an unknown vector parameter a there holds:

$$\lim_{T\to\infty} \left|d(T)^{-1}V(U(T))D\hat{a}(U(T))V(U(T))d(T)^{-1}L^{-mp}(T)T^{mp\alpha-2n}I_m^{-1} - l\right| = 0 \qquad (5.1.10)$$

Also, from the fact

$$\int_U\int_U \frac{dx\,dy}{\|x-y\|^{mp\alpha}} < \infty, \quad 0 < \alpha < \frac{n}{mp},$$

for U bounded and containing the origin, we can easily see that the existence of the radial vector function $\bar{g}(x)$ on U to R^n with the properties

$$\left\|d(T)^{-1}g(Tx) - \bar{g}(x)\right\| \to 0 \text{ as } T\to\infty$$

uniformly for $x \in U$ and

$$l = \int_U\int_U \bar{g}(x)[\bar{g}(y)]'\frac{dx\,dy}{\|x-y\|^{mp\alpha}},$$

is a regular matrix, gives a sufficient condition on $g(x)$ for the application of this theorem.

One example for the response function that satisfies these conditions is

$$g(x) = (\|x\|\|^{\mu_1}, \ldots, \|x\|^{\mu_k})' \quad \text{with } 0 < \mu_1 < \ldots < \mu_k.$$

It must be noted that our models described with the assumptions C', C" and D together with the appropriate response function $g(x)$, $x \in R^n$, correspond to the conditions of the theorem (5.1.8) with $p = 1$ or $p = 2$. Therefore, we can use it when searching for the asymptotic distributions of the LSE in these models.

Proof

Let us analyze the asymptotic behavior of the correlation matrix $D\hat{a}(U(T))$ of the estimator $\hat{a}(U(T))$.

$$\begin{aligned} D\hat{a}(U(T)) &= E[\hat{a}(U(T)) - a][\hat{a}(U(T))) - a]' \\ &= V(U(T))^{-1}\int_{U(T)}\int_{U(T)} g(x)g(y)'E(\eta(x)\eta(y))dxdyV(U(T))^{-1}. \end{aligned}$$

We have and express the correlation matrix as follows.

$$\begin{aligned}
D\hat{a}(U(T)) &= V(U(T))^{-1}\left[\sum_{j=m}^{\infty} I_j \int\limits_{U(T)}\int\limits_{U(T)} g(x)g(y)'B^{pj}(\|x-y\|)dxdy)\right] \times \\
V(U(T))^{-1} &= V(U(T))^{-1}(S_m+S_0)V(U(T))^{-1}, \\
S_m &= I_m \int\limits_{U(T)}\int\limits_{U(T)} B^{pm}(\|x-y\|\, g(x)g(y)'dxdy, \\
S_0 &= I_m \int\limits_{U(T)}\int\limits_{U(T)} \frac{L^{pm}(\|x-y\|)}{\|x-y\|^{mp\alpha}} g(x)g(y)'dxdy.
\end{aligned}$$

Using condition C we obtain

$$S_m = I_m T^{2n-mp\alpha} L^{mp}(T) \int\limits_U\int\limits_U \frac{L^{mp}(T\,\|u-v\|)}{L^{mp}(T)} g(Tu)g(Tv)' \frac{du\,dv}{\|u-v\|^{mp\alpha}};$$

$$\begin{aligned}
d(T)^{-1}V(U(T))D(\hat{a}(U(T)))V(U(T))d(T)^{-1} &= I_m L^{mp}(T)T^{2n-mp\alpha} \\
&\times \int\limits_U\int\limits_U \frac{L^{mp}(T\,\|u-v\|)}{L^{mp}(T)} d(T)^{-1} \\
&\times g(Tu)g(Tv)'d(T)^{-1}\frac{dudv}{\|u-v\|^{mp\alpha}} \\
&+ d(T)^{-1}S_0 d(T)^{-1},
\end{aligned}$$

$$\begin{aligned}
&d(T)^{-1}V(U(T))D\hat{a}(U(T))V(U(T))d(T)^{-1}I_m^{-1}L^{-mp}(T)T^{2n-mp\alpha} - l \\
&= o(1) + d(T)^{-1}S_0 d(T)^{-1}I_m^{-1}L^{-mp}(T)T^{mp\alpha-2n},
\end{aligned}$$

where $o(1)$ is the matrix function such that

$$\lim_{T\to\infty} \|o(1)(T)\| = 0.$$

Let us analyze the term

$$d(T)^{-1}S_0 d(T)^{-1}I_m^{-1}L^{-mp}(T)T^{mp\alpha-2n}.$$

As $B(\|x\|)$ is a positive correlation function, it holds $0 \leq B(\|x\|) \leq 1$. The assumption $\lim_{t\to\infty} B(\|x\|) = 0$ makes us sure that for all $\varepsilon > 0$ there exists $A_0 \in R^1_+$ such that $0 < B(\|x\|) < \varepsilon$, for all $|x| > A_0$.Let

$$U_1 = \{(x,y) \in U(T) \times U(T) : \|x-y\| \leq A_0\},$$

$$U_2 = \{(x,y) \in U(T) \times U(T) : \|x-y\| > A_0\}.$$

Now we have

$$0 < B^{m+1}(\|x-y\|) \leq \varepsilon B^m(\|x-y\|) \quad \text{for } (x,y) \in U_2$$

and we can get:

$$\begin{aligned} S_0 &= \sum_{j=m+1}^{\infty} I_j \int\limits_{U(T)} \int\limits_{U(T)} B^{jp}(\|x-y\|) g(x) g(y)' dx dy \\ &\leq K \int\limits_{U(T)} \int\limits_{U(T)} B^{p(m+1)}(\|x-y\|) g(x) g(y)' dx dy; \end{aligned}$$

$$\begin{aligned} d(T)^{-1} S_0 d(T)^{-1} L^{-mp}(T) T^{mp\alpha-2n} &\leq d(T)^{-1} K \left[\int\limits_{U_1(T)} g(x) g(y)' dx dy \right. \\ &\quad \left. + \varepsilon^p \int\limits_{U_2(T)} \frac{L^{mp}(\|x-y\|)}{\|x-y\|^{mp\alpha}} g(x) g(y)' dx dy \right] \\ &\quad \times d(T)^{-1} L^{-mp}(T) T^{mp\alpha-2n} \\ &\leq K d(T)^{-1} \int\limits_{U_1(T)} g(t) g(y)' dx dy d \\ &\quad \times (T)^{-1} L^{-mp}(T) T^{mp\alpha-2n} \\ &\quad + \varepsilon K (l + o(1)). \end{aligned}$$

The sign '$\leq$' means this relationship holds between every term in appropriate matrices, and K is the constant.

Now, using existence and finiteness of M and continuity for the function $g(x)$ we can easily confirm that

$$d(T)^{-1}S_0 d(T)^{-1}L^{-mp}(T)T^{mp\alpha-2n} \leq \text{const}\, L^{-mp}(T)T^{mp\alpha-n}J + K\varepsilon^p(l+o(1)),$$

where J is an $n \times n$ matrix with every term equal 1. So there holds:

$$\lim_{T\to\infty} \left\| d(T)^{-1}S_0 d(T)^{-1}I_m^{-1}L^{-mp}(T)T^{mp\alpha-2n}\right\| = 0,$$

and that proves the theorem.

Corollary 5.1.1. Let the regression model be as in the previous theorem and let the matrix M be regular. Then for the correlation matrix $D(\hat{a}(U(T)))$ of the LSE of an unknown vector parameter a it holds:

$$\lim_{T\to\infty} \left\| \frac{d(T)\, D\hat{a}(U(T))\, d(T) L^{-mp}(T)T^{mp\alpha}m!}{C_m^2} - M^{-1}lM^{-1} \right\| = 0.$$

Example 5.1.5. If we consider a model with

$$g(x) = (\|x\|^{\mu_1}, \ldots, \|x\|^{\mu_k})', \quad \frac{mp\alpha}{2} < \mu_1 < \ldots < \mu_k,$$

we have

$$d(T) = \operatorname{diag} g\,(T^{\mu_1}, \ldots, T^{\mu_k})$$

and M is a regular matrix. So,

$$\|D\hat{a}(U(T))\| \leq \left\|d(T)^{-1}\right\|^2 \|d(T)D\hat{a}(U(T))d(T)\|$$

and the corollary 5.1.3. ensures that

$$\lim_{T\to\infty} \|D\hat{a}(U(T))\ \| = 0.$$

It means that $\hat{a}(U(T))$ is a mean square consistent estimator for the vector parameter a.

Theorem 5.1.9. Let us observe the regression model (5.1.9) on the ball $v(T) = \{x \in R^n : \|x\| \leq T\}$. Here $g(x) = [g_1(x), \ldots, g_k(x)]'$ is the radial function: $g(x) = g(\|x\|)$, with coordinate functions $g_i(x)$, $x \in R^n$, forming a linearly independent set of positive, continuous, real functions which are square integrable on the $v(T)$ for all $T > 0$.

Also, we suppose that the errors $\eta(x)$ satisfy the conditions C', C" and that there exists a function $\bar{g}(x)$ on $v(1)$ to R^k for which

$$\begin{aligned} \left|d(T)^{-1}g(xT) - \bar{g}(x)\right| &\to 0 \quad \text{as } T \to \infty, \\ d(T) &= \operatorname{diag}\left(g_1(T), \dots, g_k(T)\right), \end{aligned}$$

uniformly for $x \in v(1)$ and

$$\int_{R^{nm}} \left\| \int_{v(1)} \bar{g}(x) e^{i\langle x, \lambda_1 + \dots + \lambda_m \rangle} dx \right\|^2 \frac{d\lambda_1 \dots d\lambda_m}{(\|\lambda_1\| \dots \|\lambda_m\|)^{n-\alpha}} < \infty$$

where $m = H \operatorname{rank} G$. Also, we need that

$$l = \int_{v(1)} \int_{v(1)} \bar{g}(x)(\bar{g}(x))' \frac{du dv}{\|u - v\|^{m\alpha}}$$

is a regular matrix.

Then the process

$$\begin{aligned} X(T) &= F(T)^{-1}(\hat{a}(U(T)) - a), \\ F(T) &= \frac{C_m}{\sqrt{m!}} L^{m/2}(T) T^{n - m\alpha/2} V(U(T))^{-1} d(T) l_1, \\ V(U(T)) &= \int_{v(T)} g(x)(g(x))' dx, \end{aligned}$$

converges in distribution to the random vector

$$\begin{aligned} \kappa_m &= C \int_{R^{nm}}' \left[\int_{v(1)} \bar{g}(u) e^{i\langle u, \lambda_1 + \dots + \lambda_m \rangle} du \right] \frac{W(d\lambda_1) \dots W(d\lambda_m)}{(\|\lambda_1\| \dots \|\lambda_m\|)^{(n-\alpha)/2}}, \\ C &= \left[\frac{\Gamma^m((n-\alpha)/2)}{m! 2^{m\alpha} \pi^{mn/2} \Gamma^m(\alpha/2)} \right]^{1/2} l_1^{-1}. \end{aligned}$$

$\int_{R^{nm}}' \cdots$ means a multiple stochastic integral in the sense of section 2.3. The matrix l_1 is a regular matrix such that $l = l_1 \, l_1'$.

We prove this statement with the help of the reduction principle.

Lemma 5.1.1. Let us suppose that the model is upon the assumption C', C" and that the vector response function $g(x)$ is exactly as in the Theorem 5.1.8. with $p = 1$. If one of the asymptotic distributions for

$$X(T) = H(T)^{-1}(\hat{a}(U(T)) - a)$$

or

$$X_m(T) = \frac{C_m}{\sqrt{m!}} F(T)^{-1} V(U(T))^{-1} \int_{v(T)} g(x) H_m(\xi(x)) dx,$$

exists, then the other exists too, and they are the same. The proof is similar to the section 2.

Proof

Using the lemma we can focus our attention on searching for the asymptotic distribution of

$$X_m(T) = \frac{C_m}{m!} F(T)^{-1} V(U(T))^{-1} \int\limits_{v(T)} g(x) H_m(\xi(x)) dx.$$

Using (2.4.1) and Itô formula (2.3.1) we obtain

$$\begin{aligned} X_m(T) &= \frac{C_m}{m!} F(T)^{-1} V(U(T))^{-1} \\ &\times \int\limits_{v(T)} \int\limits_{R^{nm}}' g(x) e^{i\langle x, \lambda_1 + \cdots + \lambda_m \rangle} \sqrt{f(\lambda_1) \cdots f(\lambda_m)} W(d\lambda_1) \cdots W(d\lambda_m) dx \end{aligned}$$

As we have, for $n \geq 2$,

$$\int\limits_{v(T)} f(x) dx = \int\limits_0^T \left(\int\limits_{s(\rho)} f(x) d\sigma(x) \right) d\rho,$$

where $s(T) = \{x \in R^k : \|x\| = T\}$ and $d\sigma(x)$ is an element of the Lebesgue measure on the $s(T)$ (see (1.1.4)), $g(x)$ is the radial function; we can use the fact

$$\int\limits_{s(T)} e^{i\langle x, s \rangle} d\sigma(x) = \left(\frac{2\pi T}{\|s\|} \right)^{n/2} \|s\| \, J_{(n-2)/2} (T \, \|s\|)$$

to obtain the following expression for $X_m(T)$:

$$
\begin{aligned}
X_m(T) &= \frac{1}{\sqrt{m!}} L^{-m/2}(T) T^{(m\alpha/2-n)} l_1^{-1} d(T)^{-1} \\
&\times \int_{R^{nm}}' \int_0^T g(\rho) \left(\frac{2\pi\rho}{\|\lambda_1 + \cdots + \lambda_m\|} \right)^{n/2} J_{(n-2)/2}(\rho \|\lambda_1 + \cdots + \lambda_m\|)\, d\rho \\
&\times \|\lambda_1 + \cdots + \lambda_m\| \sqrt{f(\lambda_1) \cdots f(\lambda_m)} W(d\lambda_1) \ldots W(d\lambda_m),
\end{aligned}
$$

where $J_\nu(z)$ is defined in (1.2.12).

Changing the variable ρ with Tu we have:

$$
\begin{aligned}
X_m(T) &= \frac{1}{\sqrt{m!}} L^{-\frac{m}{2}}(T) T^{(m\alpha/2-n)} l_1^{-1} d(T)^{-1} \\
&\times \int_{R^{nm}}' \int_0^1 g(Tu) \left(\frac{2\pi Tu}{\|\lambda_1 + \cdots + \lambda_m\|} \right)^{n/2} \\
&\times J_{(n-2)/2}(Tu \|\lambda_1 + \cdots + \lambda_m\|) T du \\
&\times \|\lambda_1 + \cdots + \lambda_m\| \sqrt{f(\lambda_1) \cdots f(\lambda_m)} W(d\lambda_1) \ldots W(d\lambda_m).
\end{aligned}
$$

Now, changing the variables λ with $\lambda' = T\lambda$ and using the self-similarity with the parameter $n/2$ for a Gaussian white noise, we can confirm:

$$
\begin{aligned}
X_m(T) &= \frac{1}{\sqrt{m!}} L^{-m/2}(T) T^{\frac{m\alpha}{2}-n} l_1^{-1} d(T)^{-1} \\
&\times \int_{R^{nm}}' \int_0^1 g(Tu)(2\pi u)^{n/2} T^n \|\lambda_1 + \cdots + \lambda_m\|^{1-n/2} \\
&\times J_{(n-2)/2}(u \|\lambda_1 + \cdots + \lambda_m\|) du \\
&\times \sqrt{f(\frac{\lambda_1}{T}) \cdots f(\frac{\lambda_m}{T})} T^{-nm/2} W(d\lambda_1) \ldots W(d\lambda_m) \\
&\overset{D}{=} \frac{1}{\sqrt{m!}} L^{-m/2}(T) (2\pi)^{n/2} T^{m(\alpha-n)/2} l_1^{-1} d(T)^{-1} \\
&\times \int_{R^{nm}}' \int_0^1 g(Tu) u^{n/2} \|\lambda_1 + \cdots + \lambda_m\|^{(1-n/2)} \\
&\times J_{(n-2)/2}(u \|\lambda_1 + \cdots + \lambda_m\|) du
\end{aligned}
$$

$$\times\sqrt{f\left(\frac{\lambda_1}{T}\right)\cdots f\left(\frac{\lambda_m}{T}\right)}W(d\lambda_1)\ldots W(d\lambda_m).$$

Let us define:

$$\kappa_m = C\int\limits_{R^{nm}}\left[\int\limits_{v(1)}\bar{g}(u)e^{i(u,\lambda_1+\cdots+\lambda_m)}du\right]\frac{W(d\lambda_1)\cdots W(d\lambda_m)}{(\|\lambda_1\|\cdots\|\lambda_m\|)^{(n-\alpha)/2}},$$

$$C = \left[\frac{\Gamma^m\left(\frac{n-\alpha}{2}\right)}{m!2^{m\alpha}\pi^{mn/2}\Gamma^m\left(\frac{\alpha}{2}\right)}\right]^{1/2} l_1^{-1},$$

$$K(T,\lambda_1,\ldots,\lambda_m) = \left[\frac{f\left(\frac{\lambda_1}{T}\right)\cdots f\left(\frac{\lambda_m}{T}\right)T^{m(\alpha-n)}2^{\alpha m}\pi^{nm/2}\Gamma^m\left(\frac{\alpha}{2}\right)}{L^m(T)(\|\lambda_1\|\cdots\|\lambda_m\|)^{\alpha-n}\Gamma^m\left(\frac{n-\alpha}{2}\right)}\right]^{1/2}$$

Therefore,

$$\hat{a}(T)-\kappa_m = C(2\pi)^{n/2}\int\limits_{R^{nm}}\int\limits_0^1\left[d(T)^{-1}g(Tu)K(r,\lambda_1,\ldots,\lambda_m)-\bar{g}(u)\right]$$

$$\times u^{n/2}\frac{J_{(n-2)/2}(u\|\lambda_1+\cdots+\lambda_m\|)}{\|\lambda_1\cdots+\lambda_m\|^{(n-2)/2}}\frac{W(d\lambda_1)\ldots W(d\lambda_m)}{(\|\lambda_1\|\cdots\|\lambda_m\|)^{(n-\alpha)/2}}.$$

Now, using the fact that

$$\int\limits_0^1 u^n\int\limits_{R^{nm}}\frac{J_{(n-2)/2}(u\|\lambda_1+\cdots+\lambda_m\|}{\left\|\sum\limits_{j=1}^m\lambda_j\right\|^{n-2}\prod\limits_{j=1}^m\|\lambda_j\|^{n-\alpha}}d\lambda_1\ldots d\lambda_m du<\infty$$

we can easily confirm the statement of the theorem.

As we can see from the integral representation of κ_m it is always normal for $m=1$ but need not to be normal for $m\neq 1$.

Example 5.1.6. If $g(x)=\|x\|^\mu$, $\mu>0$, we can use the fact:

$$\int\limits_{v(T)}\|x\|^\mu e^{i(\lambda,x)}dx=\int\limits_0^T r^\mu\left(\frac{2\pi r}{\|\lambda\|}\right)^{n/2}\|\lambda\|\,J_{(n-2)/2}(r\|\lambda\|)dr$$

to obtain that for $m = 1$, κ_1 has a normal distribution:

$$\kappa_1 = C \int\limits_{R^n} \left[\int\limits_0^1 r^\mu \left(\frac{2\pi r}{\|\lambda\|} \right)^{n/2} \|\lambda\| \, J_{(n-2)/2} \left(r \|\lambda\| \right) dr \right] \frac{W(d\lambda)}{\|\lambda\|^{(n-\alpha)/2}}.$$

Remark 5.1.3. Also, using the same technique, we can find the asymptotic distribution for the process $X'(T) = F'(T)^{-1}(\hat{a}(U(T)) - a)$,

$$F'(t) = D_1 L(T) T^{n-\alpha} V(U(T))^{-1} d(T) l_1,$$

$$V(U(T)) = \int\limits_{v(T)} g(x) g(x)' dx,$$

in the regression model (5.1.9) observed on the ball $v(T) = \{x \in R^n : \|x\| \le T\}$, with the errors as in the assumption D, D' with $m = 1$, and D_1 is defined in subsection 2. Here $g(x) = [g_1(x), \ldots, g_k(x)]'$ is again the radial function with coordinate functions $g_i(x)$, $x \in R^n$, forming the linearly independent set of positive, continuous, real functions which are square integrable on the $v(T)$ for all $T > 0$ and it has to be supposed that:

There exists a function $\bar{g}(x)$ on the $v(1)$ to R^n for which

$$\left\| d(T)^{-1} g(xT) - \bar{g}(x) \right\| \to 0 \quad \text{as } T \to \infty,$$

$d(T) = \operatorname{diag}(g_1(T), \ldots, g_k(T))$, uniformly for $x \in v(1)$;

$$\int\limits_{R^{2n}} \left\| \int\limits_{v(1)} \bar{g}(x) e^{i\langle x, \lambda_1 + \lambda_2 \rangle} dx \right\|^2 \frac{d\lambda_1 d\lambda_2}{(\|\lambda_1\| \cdot \|\lambda_2\|)^{n-\alpha}} < \infty$$

$$l = \int\limits_{v(1)} \int\limits_{v(1)} \bar{g}(u) [\bar{g}(v)]' \frac{du dv}{\|u - v\|^{2\alpha}}$$

is a regular matrix.

Then the process $X'(T)$ converges in distribution to the random vector $Y = \sum_{i=1}^r \kappa_i'$, where κ_i', $i = 1, \ldots r$, are independent copies of κ',

$$\kappa' = C \int_{R^{2n}}' \left[\int\limits_{v(1)} \bar{g}(u) e^{i\langle u, \lambda_1 + \lambda_2 \rangle} du \right] \frac{W(d\lambda_1) W(d\lambda_2)}{(\|\lambda_1\| \cdot \|\lambda_2\|)^{(n-\alpha)/2}},$$

$$C = \left[\frac{\Gamma^2 \left(\frac{n-\alpha}{2} \right)}{2^{2\alpha+1} r \pi^n \Gamma^2 \left(\frac{\alpha}{2} \right)} \right]^{1/2} l_1^{-1}.$$

$\int_{R^{2n}} \cdots$ means a multiple stochastic integral and the matrix l_1 is a regular matrix such that $l = l_1\, l_1'$.

5.2 Estimation of the Covariance Function

The asymptotic distributions of a normilized estimate for the covariance function of Gaussian and χ-squared random fields are studied.

1. The non-Gaussian limit distributions of the correlogram of a Gaussian field with known or unknown mean

K. Let $\xi(x),\ x \in R^n$ be mean square continuous homogeneous isotropic Gaussian random field with $E\xi(x) = m$ and covariance function

$$B(\|x\|) = \operatorname{cov}(\xi(0)\ \xi(x)) = \frac{L(\|x\|)}{\|x\|^{\alpha}}, \quad 0 < \alpha < \frac{n}{2},\ L \in \mathcal{L},$$

such that the spectral density $f(\lambda),\ \lambda \in R^n$, exists and $f(\lambda) = f(\|\lambda\|)$ is supposed to be decreasing for $\|\lambda\| \geq \|\lambda_0\|$, where $\|\lambda_0\| > 0$, and continuous at all $\lambda \neq 0$.

Remark 5.2.1 Under assumptions K the spectral density $f(\lambda),\ \lambda \in R^n$, has singular property (1.4.12) with Tauberian constant $c_5(n, \alpha)$ given in (1.4.13).

Under the assumptions K we have (see formulae (1.2.4), (1.2.5) and (1.2.6)) the following spectral decompositions:

$$\begin{aligned} B(x) &= \int_{R^n} e^{i\langle\lambda,x\rangle} F(d\lambda) \\ &= \sigma^2 \int_{R^n} e^{i\langle\lambda,x\rangle} f(\lambda) d\lambda, \end{aligned} \tag{5.2.1}$$

where $\sigma^2 = B(0)$, and

$$\begin{aligned} \xi(x) - m &= \int_{R^n} e^{i\langle\lambda,x\rangle} Z(d\lambda) \\ &= \sigma \int_{R^n} e^{i\langle\lambda,x\rangle} \sqrt{f(\lambda)} W(d\lambda), \end{aligned} \tag{5.2.2}$$

where

$$EZ(\Delta_1)\overline{Z(\Delta_2)} = F(\Delta_1 \cap \Delta_2), \quad EW(\Delta_1)\overline{W(\Delta_2)} = |\Delta_1 \cap \Delta_2|\,.$$

Let

$$Q_n = \Pi[0,T), \quad \Pi_T = \Pi[0,T+H], \quad U_n = \Pi[0,1),$$

be cubes in R^n where $H > 0$ is fixed and $T \to \infty$ (see section 1.1).

To estimate the correlation function $B(\|x\|)$ by observation of a sample function of the field $\xi(x)$, $x \in \Pi_T$, we use the statistics

$$B_T(h) = T^{-n} \int_{Q_n} (\xi(x) - \hat{m}) (\xi(x+hl) - \hat{m}_+) \, dx \qquad (5.2.3)$$

where $l = 1$ when $n = 1$ and $l \in s(1)$, $s(1)$ is a unit sphere in R^n, $n > 1$, $h \in [0, H]$,

$$\hat{m} = T^{-n} \int_{Q_n} \xi(x) dx, \qquad \hat{m}_+ = T^{-n} \int_{Q_n} \xi(x+hl) dx. \qquad (5.2.4)$$

Also we consider the estimate for the correlation function which is used when the mean is known:

$$\tilde{B}_T(h) = T^{-n} \int_{Q_n} (\xi(x) - m) (\xi(x+hl) - m) \, dx. \qquad (5.2.5)$$

We shall prove that limit distributions of normilized correlograms $B_T(h)$ and $\tilde{B}_T(h)$ are both non-Gaussian, but different (see Theorem 5.2.2. and 5.2.4. below).

It is obvious that

$$E\tilde{B}_T(h) = B_T(h).$$

It turns out that in the case of long range dependence the asymptotics of the variance does not depend on h as $T \to \infty$.

Lemma 5.2.1. Let assumption K be satisfied and $T \to \infty$. Then for $\alpha \in (0, n/2)$ and for a $h \in [0, H]$ and $l \in s(1)$,

$$\operatorname{var} \tilde{B}_T(h) = c_{18}(n,\alpha) L^2(T) T^{-2\alpha} (1 + o(1)),$$

where

$$c_{18}(n,\alpha) = 2^{n+1} \int_{U_n} \prod_{i=1}^{n} (1 - u_i) \, |u|^{-2\alpha} \, du.$$

Proof

It is easy to show that

$$\begin{aligned}\operatorname{var}\tilde{B}_T(h) &= T^{-2n}\int_{Q_n}\int_{Q_n}\left[B^2(\|x-y\|)+B(\|x-y+hl\|)B(\|x-y-hl\|)\right]dx \\ &= 2^nT^{-2n}\int_{Q_n}\prod_{i=1}^{n}(1-z_i)\left[B^2(\|z\|)+B(\|z+hl\|)\,B(\|z-hl\|)\right]dz.\end{aligned}$$

Changing the variables $z_i\,/\,T=u_i$, $i=1,\ldots,n$, we obtain

$$\operatorname{var}\tilde{B}_T(h)=2^nL^2(T)T^{-2\alpha}\int_{U_n}\prod_{i=1}^{n}(1-u_i)R_T(u)\,\|u\|^{-2\alpha}\,du,$$

where

$$R_T(u)=\|u\|^{2\alpha}L^{-2}(T)T^{2\alpha}\left[B^2(\|u\|\,T)+B(\|uT+hl\|)B(\|uT-hl\|)\right].$$

Using the asymptotics of the correlation function as $T\to\infty$ and well-known properties of functions slowly varying at infinity, we get

$$\lim_{T\to\infty}R_T(u)=2.$$

Passing to the limit under the integral (it can be done for $\alpha\in(0,n/2)$) we have

$$\operatorname{var}\tilde{B}_T(h)\sim c_{18}(n,\alpha)L^2(T)T^{-2\alpha}\quad\text{for}\quad T\to\infty.$$

Theorem 5.2.1. Let assumption K be satisfied and

$$|B(\tau+h)-B(\tau)|\le C\,\|h\|^{1+\rho},\ \rho>0,\ C>0.$$

Then

$$P\left\{\lim_{T\to\infty}\sup_{h\in[0,H]}\left|\tilde{B}_T(h)-B(h)\right|=0\right\}=1.$$

The proof of this Theorem is completely analogous to Theorem 4.2.3. (Ivanov and Leonenko (1989)), see Leonenko and Portnova (1993) for details.

In what follows we shall use the following representation

$$\begin{aligned}\tilde{B}_T(h) &= B(h) \\ &\quad+(2T^n)^{-1}\int_{Q_n}[(1+B(h))H_2(\eta_2(x)) \\ &\quad-(1-B(h))H_2(\eta_1(x))]\,dx,\end{aligned}\tag{5.2.6}$$

where $H_2(u) = u^2 - 1$ is the second Chebyshev–Hermite polynomial and

$$\eta_1(x) = \frac{\xi(x + hl) - \xi(x)}{\sqrt{2(1 - B(h))}}, \qquad x \in R^n,$$

$$\eta_2(x) = \frac{\xi(x + hl) - \xi(x)}{\sqrt{2(1 + B(h))}}, \qquad x \in R^n,$$

for fixed h are homogeneous random fields.

Without loss of generality it will be assumed that $\sigma^2 = 1$ and $m = 0$ in (5.2.5).

From (5.2.4) we obtain

$$\begin{aligned}
\eta_1(x) &= 2(1 - B(h))^{-1/2} \int\limits_{R^n} e^{i\langle\lambda,x\rangle} \left(e^{i\langle\lambda,t\rangle} - 1\right) Z(d\lambda) \\
&= \int\limits_{R^n} e^{i\langle\lambda,x\rangle} Z_1(d\lambda), \\
\eta_1(x) &= 2(1 + B(h))^{-1/2} \int\limits_{R^n} e^{i\langle\lambda,x\rangle} \left(e^{i\langle\lambda,t\rangle} + 1\right) Z(d\lambda) \\
&= \int\limits_{R^n} e^{i\langle\lambda,x\rangle} Z_2(d\lambda),
\end{aligned}$$

where $Z_1(\cdot)$ are spectral measures of the homogeneous fields $\eta_1(x)$, $x \in R^n$, and $\eta_2(x)$, $x \in R^n$, respectively.

Using Itô formula (2.3.1), we have

$$H_2(\eta_k(x)) = \int_{R^{2n}}' e^{i\left\langle x,\lambda^{(1)}+\lambda^{(2)}\right\rangle} Z_k(d\lambda^{(1)}) Z_k(d\lambda^{(2)}), \quad k = 1, 2. \tag{5.2.7}$$

Here and in the sequel the symbol $\int_{R^n}' \cdots$ means the multiple stochastic integral in which the hyper-diagonals $\lambda^{(i)} = \pm\lambda^{(j)}$, $i, j = 1, 2$, are excluded from the integration area(see section 2.3). Using (5.2.6) and (5.2.7), we obtain

$$\tilde{B}_T(h) - B(h) = \frac{1 + B(h)}{2T^n} \int\limits_{Q_n} \int_{R^{2n}}' e^{i\left\langle x,\lambda^{(1)}+\lambda^{(2)}\right\rangle} Z_2(d\lambda^{(1)}) Z_2(d\lambda^{(2)}) dx$$

$$-\frac{1-B(h)}{2T^n}\int\limits_{Q_n}\int_{R^{2n}}^{\prime} e^{i\left\langle x,\lambda^{(1)}+\lambda^{(2)}\right\rangle} Z_1(d\lambda^{(1)})Z_1(d\lambda^{(2)})dx$$

$$= \frac{1}{2T^n}\int_{R^{2n}}^{\prime}\left(e^{i\left\langle\lambda^{(1)},l\right\rangle h}+e^{i\left\langle\lambda^{(2)},l\right\rangle h}\right)$$

$$\times\prod_{k=1}^{n}\frac{e^{iT\left(\lambda_k^{(1)}+\lambda_k^{(2)}\right)}-1}{i\left(\lambda_k^{(1)}+\lambda_k^{(2)}\right)}Z(d\lambda^{(1)})Z(d\lambda^{(2)}).$$

By (5.2.2), under assumptions K, we get that

$$\tilde{B}_T(h)-B(h) = \frac{1}{2T^n}\int_{R^{2n}}^{\prime}\left(e^{i\left\langle\lambda^{(1)},l\right\rangle h}+e^{i\left\langle\lambda^{(2)},l\right\rangle h}\right)\prod_{k=1}^{n}\frac{e^{iT\left(\lambda_k^{(1)}+\lambda_k^{(2)}\right)}-1}{i\left(\lambda_k^{(1)}+\lambda_k^{(2)}\right)}$$

$$\times\sqrt{f(\left\|\lambda^{(1)}\right\|)\, f(\left\|\lambda^{(2)}\right\|)}W(d\lambda^{(1)})W(d\lambda^{(2)}), \qquad (5.2.8)$$

where $W(\cdot)$ is a complex Gaussian while noise in R^n defined by $Z(d\lambda) = \sqrt{f(\lambda)}W(d\lambda)$.

Let us consider a second order r.v. R_2 defined by the multiple stochastic integral

$$R_2 = c_5(n,\alpha)\int_{R^{2n}}^{\prime}\prod_{k=1}^{n}\frac{e^{i\left(\lambda_k^{(1)}+\lambda_k^{(2)}\right)}-1}{i\left(\lambda_k^{(1)}+\lambda_k^{(2)}\right)}\frac{W(d\lambda^{(1)})\;W(d\lambda^{(2)})}{\left\|\lambda^{(1)}\right\|^{(n-\alpha)/2}\left\|\lambda^{(2)}\right\|^{(n-\alpha)/2}}. \qquad (5.2.9)$$

The distribution of r.v. R_2 is called Rosenblatt distribution (see Taqqu (1975), Rosenblatt (1979,1985, p. 72)).

The characteristic function of Rosenblatt distribution R_2 is given by (2.3.3) or for $n=1$ by the explicit formula

$$E\exp\{itR_2\} = \exp\left\{\frac{1}{2}\sum_{k=2}^{\infty}\frac{(2it)^k}{k}\psi_k\right\}, \quad t\in R^1,$$

where

$$\psi_k = \int_0^1 \overset{k}{\cdots}\int_0^1 \frac{du_1\ldots du_k}{\left(|u_1-u_2|\,|u_2-u_3|\ldots|u_k-u_1|\right)^{\alpha}}, \quad 0<\alpha<\frac{1}{2}.$$

Theorem 5.2.2. Under assumptions the distributions of the r.v.s

$$\eta_T(h) = \frac{T^\alpha}{L(T)}(\tilde{B}_T(h) - B(h)) \xrightarrow{D} R_2$$

as $T \to \infty$, for every $h \in \Pi[0, H]$.

Proof

Let us prove that $E\left|\eta_T(h) - R_2\right|^2 \to 0$ as $T \to \infty$.

Changing the variables $\lambda^{(i)}T = \mu^{(i)}$, $i = 1, 2$, and using self-similarity of the Gaussian white noise with parameter $n/2$, we have by (5.2.8) and (5.2.9):

$$\begin{aligned}
\frac{T^\alpha}{L(T)}\tilde{B}_T(h) &= \frac{L^{-1}(T)T^{\alpha-n}}{2}\int'_{R^{2n}}\left(e^{i\langle\mu^{(1)},l\rangle h/T} + e^{i\langle\mu^{(2)},l\rangle h/T}\right) \\
&\quad \times \prod_{k=1}^{n} \frac{e^{i\left(\mu_k^{(1)}+\mu_k^{(2)}\right)} - 1}{i\left(\mu_k^{(1)} + \mu_k^{(2)}\right)} \\
&\quad \times \sqrt{f\left(\frac{\|\mu^{(1)}\|}{T}\right) f\left(\frac{\|\mu^{(2)}\|}{T}\right)} W(d\mu^{(1)})W(d\mu^{(2)}) \\
&= c_5(n,\alpha)\int'_{R^{2n}} \prod_{k=1}^{n} \frac{e^{i\left(\mu_k^{(1)}+\mu_k^{(2)}\right)} - 1}{i\left(\mu_k^{(1)} + \mu_k^{(2)}\right)} Q_T\left(\mu^{(1)}, \mu^{(2)}\right) \\
&\quad \times \left\|\mu^{(1)}\right\|^{(\alpha-n)/2} \left\|\mu^{(2)}\right\|^{(\alpha-n)/2} W(d\mu^{(1)})W(d\mu^{(2)}),
\end{aligned}$$

where

$$\begin{aligned}
Q_T(\mu^{(1)}, \mu^{(2)}) &= \frac{e^{i\langle\mu^{(1)},l\rangle h/T} + e^{i\langle\mu^{(2)},l\rangle h/T}}{2L(T)T^{n-\alpha}c_5(n,\alpha)} \\
&\quad \times \sqrt{f\left(\frac{\|\mu^{(1)}\|}{T}\right) f\left(\frac{\|\mu^{(2)}\|}{T}\right)} \left\|\mu^{(1)}\right\|^{(\alpha-n)/2} \left\|\mu^{(2)}\right\|^{(\alpha-n)/2}.
\end{aligned}$$

It follows from Theorem 2.3.1. that

$$\begin{aligned}
E\left|\eta_T(h) - R_2\right|^2 &= c_5^2(n,\alpha)\int'_{R^{2n}} \prod_{k=1}^{n} \frac{\left|e^{i\left(\mu_k^{(1)}+\mu_k^{(2)}\right)} - 1\right|^2}{\left(\mu_k^{(1)} + \mu_k^{(2)}\right)^2} \\
&\quad \times \left|Q_T(\mu^{(1)}, \mu^{(2)}) - 1\right|^2 \left\|\mu^{(1)}\right\|^{(\alpha-n)/2} \left\|\mu^{(2)}\right\|^{(\alpha-n)/2} d\mu^{(1)}d\mu^{(2)}.
\end{aligned}$$

From the asymptotics of the spectral density at zero and properties of slowly varying functions we have

$$\lim_{T\to\infty} Q_T(\mu^{(1)},\mu^{(2)}) = 1.$$

The function

$$g(\mu^{(1)},\mu^{(2)}) = \left\|\mu^{(1)}\right\|^{(\alpha-n)/2}\left\|\mu^{(2)}\right\|^{(\alpha-n)/2}\prod_{k=1}^{n}\frac{\left|e^{i\left(\mu_k^{(1)}+\mu_k^{(2)}\right)}-1\right|^2}{\left(\mu_k^{(1)}+\mu_k^{(2)}\right)^2}$$

is absolutely integrable, since it has the order $r^{-2n+2\alpha-1}$ at infinity and $r^{-1+2\alpha}$ at zero, so we can pass to the limit under the integrable for $0<\alpha<n/2$ (see the proof Theorem 2.4.1. for details).

Thus

$$E\left|\eta_T(h)-R_2\right|^2 \to \infty \quad \text{as } T\to\infty,$$

whence it follows that the distributions of the random processes $\eta_T(h)$ weakly converge to those of the random variable R_2.

Consider now the estimate (5.2.3) for covariance function of the Gaussian field with unknown mean $m = E\xi(x)$.

We shall use the following representation:

$$B_T(h) = \tilde{B}_T(h) - \xi_T(h), \tag{5.2.10}$$

where

$$\xi_T(h) = T^{-2n}\int_{Q_n}\int_{Q_n}(\xi(x)-m)(\xi(y+hl))dxdy. \tag{5.2.11}$$

The estimate (5.2.3) is asymptotically unbiased, namely from (5.2.10) and (5.2.11) we obtain:

$$\begin{aligned} EB_T(h) &= B(h) - T^{-2n}\int_{Q_n}\int_{Q_n} B(x-y-hl)dxdy \\ &= B(h) - \int_{U_n}\prod_{i=1}^{n}(1-u_i)\left\|u\right\|^{\alpha}L(T)T^{-\alpha}R_T(u)du, \end{aligned}$$

where $U_n = [0,1]^n$ is a unit cube in R^n and

$$\begin{aligned} R_T(u) &= L(T)T^{-\alpha}A_h(Tu) \\ &= \|u\|^{\alpha} L(T)T^{-\alpha}\sum_{i=1}^{2^n} B_i(Tu+hl), \end{aligned} \tag{5.2.12}$$

where $B_i(Tu+hl)$ denotes the correlation function calculated at point $\tilde{x}+hl$, $\tilde{x}$ is obtained from x by replacing some i components $x_{k_1},\dots,x_{k_i}$ by $-x_{k_1},\dots,-x_{k_i}$. Since $\lim_{T\to\infty} R_T(u) = 2^n$ the dominated convergence theorem implies that

$$EB_T(h) = B(h) - 2^n k_n L(T)T^{-\alpha}(1+o(1)), \quad T\to\infty, \tag{5.2.13}$$

where

$$k_n = \int\limits_{U_n}\prod_{i=1}^{n}(1-u_i)\,\|u\|^{-\alpha}\,du.$$

The following Lemma gives the asymptotic behavior of variance of $B_T(h)$.

Lemma 5.2.2. Suppose condition K holds, $\alpha \in (o, n/2)$ and $T\to\infty$. Then for any given $h\in[0,H]$, $l\in s(1)$

1)

$$\operatorname{var} B_T(h) = \tilde{c}_n L^2(T)T^{-2\alpha}(1+o(1)),$$

where

$$\begin{aligned} \tilde{c}_n &= 2^{n+1}c_n + 2^{2n+1}k_n^2 + d_n, \\ c_n &= \int_{U_n}\prod_{i=1}^{n}(1-u_i)\ \|u\|^{-2\alpha}\,du, \\ d_n &= 4c_0^2\sigma^4\int\limits_{R^{2n}}\frac{1}{\left\|\mu_k^{(1)}\right\|^{(n-\alpha)}\left\|\mu_k^{(2)}\right\|^{(n-\alpha)}} \\ &\quad\times\prod_{k=1}^{n}\frac{\left(e^{i\left(\mu_k^{(1)}+\mu_k^{(2)}\right)}-1\right)\left(e^{-i\left(\mu_k^{(1)}\right)}-1\right)\left(e^{-i\left(\mu_k^{(2)}\right)}-1\right)}{i\left(\mu_k^{(1)}+\mu_k^{(2)}\right)\mu_k^{(1)}\mu_k^{(2)}}d\mu^{(1)}d\mu^{(2)}; \end{aligned}$$

2)

$$E\left(B_T(h)-B(h)\right)^2 = \tilde{c}_n L^2(T)\,T^{-2\alpha}(1+o(1)),$$

where

$$\tilde{c}_n = 2^{n+1} c_n + 3 \cdot 2^{2n} k_n^2 + d_n.$$

Remark 5.2.2. When $n = 1$

$$c_1 = \frac{1}{2(1-\alpha)(1-2\alpha)}, \qquad k_1 = \frac{1}{(1-\alpha)(2-\alpha)}.$$

Theorem 5.2.3. Let condition K hold and

$$|B(\tau + h) - B(\tau)| \le c_4^2 \, |h|^{1+\rho}, \quad \rho > 0.$$

Whence

$$P\left\{ \lim_{T\to\infty} \sup_{h\in[0,H]} |B_T(h) - B(h)| = 0 \right\} = 1.$$

Let us consider the second order random variables ζ, η which are defined by means of multiple stochastic integrals:

$$\zeta = c_5(n,\alpha)\sigma^2 \int_{R^{2n}}' \prod_{k=1}^{n} \frac{\exp\left(i\left(\mu_k^{(1)} + \mu_k^{(2)}\right)\right) - 1}{i\left(\mu_k^{(1)} + \mu_k^{(2)}\right)} \frac{W(d\mu^{(1)})W(d\mu^{(2)})}{\|\mu^{(1)}\|^{(n-\alpha)/2} \|\mu^{(2)}\|^{(n-\alpha)/2}}$$

$$= \sigma^2 R_2, \eta = c_5(n,\alpha)\sigma^2 \times$$

$$\int_{R^{2n}}' \prod_{k=1}^{n} \frac{\left(\exp\left(i\mu_k^{(1)}\right) - 1\right)\left(\exp\left(i\mu_k^{(2)}\right) - 1\right)}{\mu_k^{(1)} \mu_k^{(2)}} \frac{W(d\mu^{(1)})W(d\mu^{(2)})}{\|\mu^{(1)}\|^{(n-\alpha)/2} \|\mu^{(2)}\|^{(n-\alpha)/2}}$$

Remark 5.2.3. Obviously $E\zeta^2 < \infty$, $E\eta^2 < \infty$, but the distributions of these random variables are not Gaussian.

Our main result is:

Theorem 5.2.4. For every h under the conditions K the distribution of the random processes

1) $T^\alpha L^{-1}(T)\left(B_T(h) - EB_T(h)\right)$ converge weakly as $T \to \infty$ to that of the random variable $\zeta + \eta$;

2) $T^\alpha L^{-1}(T)\left(B_T(h) - B(h)\right)$ converge weakly as $T \to \infty$ to that of the random variable $\zeta + \eta - 2^n k_n$.

Proof of Lemma 5.2.2.

Using representation (5.2.10) we have

$$\operatorname{var} B_T(h) = \operatorname{var} \tilde{B}_T(h) + \operatorname{var} \xi_T(h) - 2\operatorname{cov}(\tilde{B}_T(h),\ \xi_T(h)).$$

From Lemma 5.2.1. we obtain

$$\operatorname{var} \tilde{B}_T(h) = 2^{n+1} c_n L^2(T) T^{-2\alpha}(1 + o(1)), \qquad T \to \infty.$$

By straightforward computation one can obtain:

$$\begin{aligned} E\xi_T^2(h) &= \frac{2^{2n}}{T^{4n}} \left(\int_{Q_n} \prod_{i=1}^{n} (T - x_i) B(x) dx \right)^2 \\ &\quad + \frac{2^2}{T^{4n}} \left(\int_{Q_n} \prod_{i=1}^{n} (T - x_i) A_h(x) dx \right)^2 \\ &= 2^{2n} \frac{L^2(T)}{T^{2\alpha}} \left(\int_{U_n} \prod_{i=1}^{n} (1 - x_i) \|u\|^{-2\alpha} S_T(u) du \right)^2 \\ &\quad + 2^{2n} \frac{L^2(T)}{T^{2\alpha}} \left(\int_{U_n} \prod_{i=1}^{n} (1 - x_i) \|u\|^{-2\alpha} Q_T(u) du \right)^2, \end{aligned}$$

where the function $A_h(x)$ is defined by (5.2.12) and

$$\begin{aligned} S_T(u) &= \frac{T^\alpha \|u\|^\alpha}{L(T)} B(Tu) \to 1, \\ Q_T(u) &= \frac{T^\alpha \|u\|^\alpha}{L(T)} A_h(Tu) \to 2^n, \qquad T \to \infty. \end{aligned}$$

The dominated convergence theorem implies that

$$E\xi_T^2(h) \sim \frac{T^\alpha}{L(T)} k_n^2 \cdot 2^n \cdot 3$$

The last formula and (5.2.13) yield

$$\operatorname{var} \xi_T(h) = \frac{T^\alpha}{L(T)} k_n^2 \cdot 2^{2n+1}(1 + o(1)), \qquad T \to \infty.$$

To calculate $\operatorname{cov}(\tilde{B}_T(h), \xi_T(h))$ we use spectral representation (5.2.1) and diagram formula (see (3.4.6) or Fox and Taqqu (1987)). By all these we have:

$$\operatorname{cov}(\tilde{B}_T(h), \xi_T(h)) = E\left(\tilde{B}_T(h) - B(h)\right) \xi_T(h)$$

$$= -E\frac{\sigma^2}{T^{3n}}\int_{R^{2n}}' \prod_{k=1}^{n} \frac{\left(e^{iT\left(\mu_k^{(1)}+\mu_k^{(2)}\right)}-1\right)}{i\left(\mu_k^{(1)}+\mu_k^{(2)}\right)} e^{i\langle \mu^{(2)},hl\rangle}$$
$$\times\sqrt{f\left(\|\mu^{(1)}\|\right) f\left(\|\mu^{(2)}\|\right)} W(d\mu^{(1)})W(d\mu^{(2)})$$
$$\times \int_{R^{2n}}' \prod_{k=1}^{n} \frac{\left(e^{-iT(\mu_k^{(1)})}-1\right)\left(e^{-iT(\mu_k^{(2)})}-1\right)}{\mu_k^{(1)}\mu_k^{(2)}} e^{i\langle \mu^{(2)},hl\rangle}$$
$$\times\sqrt{f\left(\|\mu^{(1)}\|\right) f\left(\|\mu^{(2)}\|\right)} W(d\mu^{(1)})W(d\mu^{(2)}).$$

Transforming the variables $\mu^{(i)}T = \lambda^{(i)}$, $i = 1,2$, and using self-similarity with the parameter $n/2$ of Wiener random measure we obtain:

$$\operatorname{cov}(\tilde{B}_T(h), \xi_T(h)) = -E\frac{\sigma^2}{T^{3n}}\int_{R^{2n}}' \prod_{k=1}^{n} \frac{\left(e^{i\left(\lambda_k^{(1)}+\lambda_k^{(2)}\right)}-1\right)}{i\left(\lambda_k^{(1)}+\lambda_k^{(2)}\right)} e^{i\langle \lambda^{(2)}/T,hl\rangle}$$
$$\times\sqrt{f\left(\left\|\frac{\lambda^{(1)}}{T}\right\|\right) f\left(\left\|\frac{\lambda^{(2)}}{T}\right\|\right)} W(d\lambda^{(1)})W(d\lambda^{(2)})$$
$$\times \int_{R^{2n}}' \prod_{k=1}^{n} \frac{\left(e^{-i(\lambda_k^{(1)})}-1\right)\left(e^{-i(\lambda_k^{(2)})}-1\right)}{\lambda_k^{(1)}\lambda_k^{(2)}} e^{i\langle \lambda^{(2)}/T,hl\rangle}$$
$$\times\sqrt{f\left(\left\|\frac{\lambda^{(1)}}{T}\right\|\right) f\left(\left\|\frac{\lambda^{(2)}}{T}\right\|\right)} W(d\lambda^{(1)})W(d\lambda^{(2)})$$
$$= -2\sigma^2 c_5(n,\alpha)\frac{L^2(T)}{T^{2\alpha}}$$
$$\times \int_{R^{2n}}' \prod_{k=1}^{n} \frac{P_T(\lambda^{(1)},\lambda^{(2)})}{\left\|\lambda^{(1)}\right\|^{n-\alpha}\left\|\lambda^{(2)}\right\|^{n-\alpha}}$$
$$\times \frac{\left(e^{i\left(\lambda_k^{(1)}+\lambda_k^{(2)}\right)}-1\right)\left(e^{-i(\lambda_k^{(1)})}-1\right)\left(e^{-i(\lambda_k^{(2)})}-1\right)}{i\left(\lambda_k^{(1)}+\lambda_k^{(2)}\right)\lambda_k^{(1)}\lambda_k^{(2)}} d\lambda^{(1)}d\lambda^{(2)},$$

where

$$P_T(\lambda^{(1)}, \lambda^{(2)}) = \frac{T^{2\alpha-2n} \left\|\lambda^{(1)}\right\|^{n-\alpha} \left\|\lambda^{(2)}\right\|^{n-\alpha}}{c_5(n,\alpha)L^2(T)} \sqrt{f\left(\left\|\frac{\lambda^{(1)}}{T}\right\|\right) f\left(\left\|\frac{\lambda^{(2)}}{T}\right\|\right)}.$$

Since

$$\lim_{T\to\infty} P_T(\lambda^{(1)}, \lambda^{(2)}) = 1$$

by the dominated convergence theorem we get the Lemma assertion.

Proof of Theorem 5.2.3.

The proof of the theorem follows essentially the same procedure used Ivanov and Leonenko (1989), Theorem 4.3.2.

From the Theorem 5.2.1.

$$P\left\{\lim_{T\to\infty} \sup_{h\in[0,H]} \left|\tilde{B}_T(h) - B(h)\right| = 0\right\} = 1.$$

To prove the theorem it remains to show that

$$P\left\{\lim_{T\to\infty} \sup_{h\in[0,H]} |\xi_T(h)| = 0\right\} = 1.$$

From (5.2.7) we see that it suffices to prove that

$$P\left\{\lim_{T\to\infty} \sup_{h\in[0,H]} T^{-n} \left|\int_{Q_n} (\xi(x+hl) - m)dx\right| = 0\right\} = 1.$$

The proof of this statement is just the same in Leonenko and Portnova (1993) except for the following estimates.

Denote

$$Y_T(h) = T^{-n} \int_{Q_n} (\xi(x+hl) - m)dx.$$

Choose some $\delta > 0$, $0 < \gamma < \rho/(1+\rho)$, and the sequence $T_k = k^{(1/\alpha\gamma)+\delta}$. Let $S_{nk} = \Pi[0, T_k]$ be a cube in R^n. Using Shwartz inequality we have:

$$E\left|Y_{T_n}(\tau+h) - Y_{T_n}(\tau)\right|^2 \leq \frac{1}{T_k^{2n}} \int_{S_{nk}} \int_{S_{nk}} \left(E\left(\xi(x+(\tau+h)l) - \xi(x+\tau l)\right)^2\right)^{1/2}$$

$$\times \left(E\left(\xi(y+(\tau+h)l)-\xi(y+\tau l)\right)^2\right)^{1/2} dxdy$$
$$\leq 2(1-B(h))$$
$$\leq c_1 |h|^{1+\rho}.$$

From the other hand

$$\begin{aligned} E\left|Y_{T_n}(\tau+h)-Y_{T_n}(\tau)\right|^2 &\leq 2EY_{T_n}^2(\tau+h)+2EY_{T_n}^2(\tau), \\ Y_{T_n}^2(\tau) &= \frac{1}{T_k^{2n}} \int\limits_{S_{nk}} \int\limits_{S_{nk}} B(x-y)dxdy \\ &= \frac{2^n}{T_k^{2n}} \int\limits_{S_{nk}} \prod_{i=1}^n (T_k - x_i) B(x) dx \\ &= \frac{2^n}{T_k^{2n}} \int\limits_{U_n} \prod_{i=1}^n (1-u_i) B(uT_k) du \\ &\leq c_2 T_k^{-\alpha} \end{aligned}$$

and thus

$$E\left|Y_{T_n}(\tau+h)-Y_{T_n}(\tau)\right|^2 \leq k\,\|h\|^{(1+\rho)(1-\gamma)}\,T_k^{-\alpha\gamma}, \quad k>0.$$

This completes the proof of the Theorem 5.2.3.

Proof of Theorem 5.2.4.

From Theorem 5.2.2. we have that

$$\begin{aligned} \zeta_T(h) &= T^\alpha L^{-1}(T)\left(\tilde{B}_T(h)-B(h)\right) \\ &\xrightarrow{D} \zeta, \end{aligned}$$

but from proof of the mentioned theorem we get the mean square convergence. Using representation (5.2.6) we conclude that it suffices to show that

$$\begin{aligned} \eta_T(h) &= -T^\alpha L^{-1}(T)\left(\xi_T(h) - E\,\xi_T(h)\right) \\ &\to \eta \quad \text{in} \quad \mathbf{L}_2(\Omega). \end{aligned}$$

Indeed, From (5.2.7) and spectral representation (5.2.1) by Diagram formula (3.4.6) we obtain:

$$\eta_T(h) = c_0\sigma^2 \int_{R^{2n}}' \prod_{k=1}^n \frac{\left(e^{i(\lambda_k^{(1)})}-1\right)\left(e^{i(\lambda_k^{(2)})}-1\right)}{\lambda_k^{(1)}\lambda_k^{(2)}}$$

$$\times \frac{D_T(\lambda^{(1)}, \lambda^{(2)})}{\left\|\lambda^{(1)}\right\|^{(n-\alpha)/2} \left\|\lambda^{(2)}\right\|^{(n-\alpha)/2}} W(d\lambda^{(1)}) W(d\lambda^{(2)}),$$

where $D_T(\lambda^{(1)}, \lambda^{(2)}) =$

$$\frac{T^{\alpha-n}}{c_0 L(T)} \left\|\lambda^{(1)}\right\|^{(n-\alpha)/2} \left\|\lambda^{(2)}\right\|^{(n-\alpha)/2} e^{i\langle \lambda^{(2)}/T, hl\rangle} \sqrt{f\left(\left\|\frac{\lambda^{(1)}}{T}\right\|\right) f\left(\left\|\frac{\lambda^{(2)}}{T}\right\|\right)}.$$

Hence we have:

$$\begin{aligned} E\left|\eta_T(h) - \eta\right|^2 &= c_0^2 \sigma^4 \int\limits_{R^{2n}} \prod_{k=1}^{n} \frac{\left|e^{i(\lambda_k^{(1)})} - 1\right|^2 \left|e^{i(\lambda_k^{(2)})} - 1\right|^2}{\left\|\lambda_k^{(1)}\right\|^2 \left\|\lambda_k^{(2)}\right\|^2} \\ &\times \frac{\left|D_T(\lambda^{(1)}, \lambda^{(2)}) - 1\right|^2}{\left\|\lambda^{(1)}\right\|^{(n-\alpha)} \left\|\lambda^{(2)}\right\|^{(n-\alpha)}} d\lambda^{(1)} d\lambda^{(2)}. \end{aligned}$$

When $(0 < \alpha < n/2)$ the dominated convergence theorem implies the first assertion of this theorem. To prove the second one we note that

$$\frac{T^\alpha}{L(T)} \left(\tilde{B}_T(h) - B(h)\right) = \zeta_T(h) + \eta_T(h) - \frac{T^\alpha}{L(T)} E\, \xi_T(h).$$

From (5.2.13) we see that

$$\frac{T^\alpha}{L(T)} E\, \xi_T(h) \to 2^n k_n \quad \text{as} \quad T \to \infty.$$

So the required assertion follows.

Remark 5.2.4. Byldigin and Dykhovichny (1996) have been proved that if $\xi(x)$, $x \in R^n$, is a Gaussian homogeneous field with spectral density $f(\lambda) \in \mathbf{L}_2(R^n)$, then

$$T^{n/2} \left(\tilde{B}_T(h) - B(h)\right) \xrightarrow{D} Y(h) \qquad \text{as } T \to \infty,$$

where $Y(h)$, $h \in R^n$, is a Gaussian field with zero mean and covariance function

$$EY(h_1)Y(h_2) = 2(2\pi)^n \int\limits_{R_n} \cos\langle \lambda, h_1\rangle \cos\langle \lambda, h_2\rangle f^2(\lambda) d\lambda.$$

The functional limit theorem for the covariance function estimate considered by Ivanov and Leonenko (1989) in space of continuous functions $C(\Pi[0,H])$, and Buldigin (1996) and Buldigin and Demianenko (1997) in the space $L_p(\Pi[0,H])$, $p>1$, and some other functional spaces.

2. The non-Gaussian limit distributions of the correlogram of a χ-squared random field

We make the following assumptions:

L. Let $\xi(x) = \left[\xi_1(x),\ldots,\xi_{p1}(x)\right]'$, $x \in R^n$, be a mean square continuous homogeneous isotropic vector Gaussian random with $E\xi(x)=0$ and covariance matrix

$$R = E\xi(0)\xi(x)' = (R_{ij}(\|x\|))_{1\le i,j\le p},$$

where

$$R_{ii}(\|x\|) = a(\|x\|) \quad \text{and} \quad R_{ij}(\|x\|) = b(\|x\|), \quad i \neq j,$$

where $a(0)=1$ and $b(0)=\rho_0 \in [0,1)$.

Denote by

$$\zeta_p(x) = \frac{1}{2}\sum_{i=1}^{p}\xi_i^2(x), \quad x \in R^n,$$

a χ-squared random field with possible dependent components. The dependence of vector components is characterized by the following two conditions.

M'. The terms are strongs dependent:

$$a(x) = a(\|x\|) \sim \frac{L(\|x\|)}{\|x\|^{\alpha}}, \quad \|x\| \to \infty,$$

and

$$\begin{aligned} b(x) &= b(\|x\|) \sim \rho_\infty \frac{L(\|x\|)}{\|x\|^{\alpha}}, \\ \rho_\infty &\in [0,1], \ \|x\| \to \infty, \end{aligned}$$

where $0<\alpha<n/2$ and $L(t) \in \mathcal{L}$.

M". The terms are weakly dependent

$$a(x) = a(\|x\|) \sim \frac{L(\|x\|)}{\|x\|^{\alpha}}, \qquad \|x\| \to \infty,$$

$$b(x) = b(\|x\|) = o(a(x)), \qquad \|x\| \to \infty.$$

Let A be an orthogonal matrix ($AA' = I$, where I is the identity matrix) satisfying the condition

$$\begin{aligned} A'R_\xi A &= V \\ &= \operatorname{diag}\{\lambda_1, \dots, \lambda_p\}, \end{aligned}$$

where

$$\begin{aligned} \lambda_1 &= \lambda_1(x) = a(x) + (p-1)b(x), \\ \lambda_2 &= \lambda_2(x) = \dots = \lambda_p = \lambda_p(x) = a(x) - b(x). \end{aligned}$$

Note that $\lambda_1, \dots, \lambda_p$ are roots of the equation $\det(R_\xi - \lambda I) = 0$.

Let $D = \operatorname{diag}\{d_1, \dots, d_p\}$ be a diagonal matrix such that

$$d_1 = \frac{1}{\sqrt{1+(p-1)\rho_0}}, \qquad d_2 = \dots = d_p = \frac{1}{\sqrt{1-\rho_0}}.$$

Consider the vector random field

$$\begin{aligned} \eta(x) &= [\eta_1(x), \dots, \eta_p(x)]' \\ &= T\xi(x) \\ &= DA\xi(x). \end{aligned}$$

The following assumption refers to the spectral representation of the correlation functional and the fields $\eta_k(x)$, $x \in R^n$, $k = 1, \dots, p$.

N. The spectral densities $f_k(\lambda) = f_k(\|\lambda\|)$, $k = 1, \dots, p$, decrease for $\|\lambda\| \geq \|\lambda_0\|$, where $\|\lambda_0\| > 0$ and continuous at all $\lambda \neq 0$.

Lemma 5.2.3. The random field $\eta(x) = T\xi(x)$, $T = DA$, is a homogeneous isotropic Gaussian random field with independent components:

$$\begin{aligned} E\eta(x) &= 0, \quad E\eta(0)\eta(x)' = (\tilde{R}_{ij}(\|x\|), \quad \text{where } \tilde{R}_{ij}(\|x\|) = 0, \quad i \neq j; \\ \tilde{R}_{11}(\|x\|) &= r_1(\|x\|) = r_1(x) = d_1^2\lambda_1(x), \\ \tilde{R}_{ii}(\|x\|) &= r_2(\|x\|) = r_2(x) = d_2^2\lambda_2(x), \quad \text{and } E\eta_i^2(0) = 1, \ i = 1, \dots, p. \end{aligned}$$

Since

$$\xi(x) = T^{-1}\eta(x) = A^{-1}D^{-1}\eta(x),$$

we have

$$\begin{aligned}\zeta(x) &= \frac{1}{2}\xi(x)'\xi(x)\\ &= \frac{1}{2}\eta(x)'(D^{-1})'(D^{-1})\eta(x)\\ &= \frac{1}{2}\sum_{i=1}^{p}\frac{\eta_i^2(x)}{d_i^2}.\end{aligned}$$

Note that

$$p = \sum_{i=1}^{p}\frac{1}{d_i^2}.$$

Let $\{H_k(u)\}_{k=0}^{\infty}$ be Chebychev–Hermite polynomials with the leading coefficient equal to one (see (2.1.4)). Clearly, $E\zeta(x) = p/2$. Using (5.2.1), we obtain

$$\begin{aligned}\gamma(x) &= \gamma(\|x\|)\\ &= \operatorname{cov}(\zeta(0),\zeta(x))\\ &= \frac{1}{4}E\left(\sum_{i=1}^{p}\frac{H_2(\eta_i(0))}{d_i^2}\right)\left(\sum_{k=1}^{p}\frac{H_2(\eta_k(x))}{d_k^2}\right)\\ &= \frac{1}{4}\sum_{i=1}^{p}\frac{1}{d_i^2}EH_2(\eta_i(0))H_2(\eta_i(x))\\ &= \frac{1}{2}\sum_{i=1}^{p}\frac{r_i^2(x)}{d_i^4}.\end{aligned}$$

Put $Q_n = \Pi[0,T]$ and denote by $\Pi_T = [0, T+H]$ be cubes in R^n, where $H > 0$ is a fixed number and $T \to \infty$. To estimate the correlation function $\gamma(\|x\|)$ from observations of a sample function of the field $\xi(x), x \in \Pi_T$, we use the statistic

$$\gamma_T(h) = \frac{1}{T^n}\int\limits_{Q_n}\left(\xi(x) - \frac{p}{2}\right)\left(\xi(x+hl) - \frac{p}{2}\right)dx, \tag{5.2.14}$$

where $l = 1$ for $n = 1$ and $l \in s(1)$ for $n > 1$, where $s(1)$ is the unit sphere in R^n, $n > 1$, $h \in [0, H]$.

It is obvious that

$$E\gamma_T(h) = \gamma(h), \quad h \in [0, H].$$

We use the following Lemma, which is a spectral case of the diagram formula (3.4.6).

Lemma 5.2.4. Let $(\xi_1, \xi_2, \xi_3, \xi_4)$ be a Gaussian vector with

$$\begin{aligned} E\xi_1 &= E\xi_2 = E\xi_3 = E\xi_4 = 0, \\ E\xi_1^2 &= E\xi_2^2 = E\xi_3^2 = E\xi_4^2 = 1, \end{aligned}$$

and

$$r_{ij} = E\xi_i\xi_j, \quad 1 \leq i, \ j \leq 4.$$

Then

$$\begin{aligned} & E\left[H_2(\xi_1)H_2(\xi_2)H_2(\xi_3)H_2(\xi_4)\right] \\ = \;& 4\left(r_{12}^2 r_{34}^2 + r_{13}^2 r_{24}^2 + r_{14}^2 r_{23}^2 + 4r_{12}r_{13}r_{24}r_{34} + 4r_{12}r_{14}r_{23}r_{34} + 4r_{13}r_{14}r_{24}r_{23}\right) \end{aligned}$$

The following lemma describes the asymptotic behavior of the variance of $\gamma_T(h)$.

Lemma 5.2.5.

1) Let the conditions L and M' be fulfilled, let $\alpha \in (o, n/2)$ and $T \to \infty$. Then for every $h \in [0, H]$ and $l \in s(1)$

$$\operatorname{var} \gamma_T(h) = 2^{n+1} L^2(T) c_n(\alpha) T^{-2\alpha} k_1(h)(1 + o(1)), \tag{5.2.15}$$

where

$$c_n(\alpha) = \int\limits_{U_n} \left[\prod_{i=1}^{n} (1 - u_i)\right] \|u\|^{-2\alpha} \, du,$$

and

$$k_1(h) = \lambda_1^2(h)(1 + (p-1)\rho_0)^2 + (p-1)\lambda_2^2(h)(1 - \rho_0)^2.$$

2) Let the conditions L and M" be fulfilled, let $\alpha \in (o, n/2)$ and $T \to \infty$. Then for every $h \in [0, H]$ and $l \in s(1)$

$$\operatorname{var} \gamma_T(h) = 2^{n+1} L^2(T) c_n(\alpha) T^{-2\alpha} k_2(h)(1 + o(1)),$$

where $k_2(h) = \lambda_1^2(h) + (p-1)\lambda_2^2(h)$.

The following theorem establishes the strong consistency of the correlogram.

Theorem 5.2.5. Let the conditions L, M' and M" be fulfilled, let for some $c > 0$

$$|a(\tau + h) - a(\tau)| \leq C \ \|h\|^{1+\rho}, \quad \rho > 0,$$

$$|b(\tau + h) - b(\tau)| \leq C \ \|h\|^{1+\rho}, \quad \rho > 0.$$

Then

$$P\left\{ \lim_{T\to\infty} \sup_{h\in[0,H]} |\gamma_T(h) - \gamma(h)| = 0 \right\} = 1.$$

Consider the random variable R_2 defined by (5.2.9).

Theorem 5.2.6.

1) Under the conditions L, M' and N the finite-dimensional distributions of the random processes

$$\zeta_T(h) = T^{\alpha} L^{-1}(T) \left(\gamma_T(h) - \gamma(h) \right)$$

converge weakly, as $T \to \infty$, to the finite-dimensional of the processes $\alpha(h)$, $h \in [0, H]$, which we represent in the form

$$\alpha(h) = \sum_{i=1}^{p} \frac{r_i(h)}{d_i^4} \zeta_i, \tag{5.2.16}$$

where ζ_i, $i = 1, \ldots, p$, are independent replicas of the random variable R_2.

2) Under the conditions L, M" and N the finite-dimensional distributions of the random processes

$$\zeta_T(h) = T^{\alpha} L^{-1}(T) \left(\gamma_T(h) - \gamma(h) \right)$$

converge weakly, as $T \to \infty$, to the finite-dimensional of the processes $\beta(h)$, $h \in [0, H]$, that we represent as

$$\beta(h) = \sum_{i=1}^{p} \frac{r_i(h)}{d_i^2} \zeta_i, \tag{5.2.17}$$

where ζ_i, $i = 1, \ldots, p$, are the same as before.

Corollary 5.2.1. If the components of the field $\xi(x)$, $x \in R^n$, are independent, then $\rho_0 = 0$ and the limit process $\theta(h)$ is of the form

$$\theta(h) = \sum_{i=1}^{p} a(h)\zeta_i.$$

Proof of Lemma 5.2.5.
It is obvious that

$$\begin{aligned}
E\gamma_T(h) &= \gamma(h) \\
&= \frac{1}{2}\left(\lambda_1^2(h) + (p-1)\lambda_2^2(h)\right), \\
E\gamma_T^2(h) &= \frac{1}{T^{2n}} \int_{Q_n}\int_{Q_n} E\left(\zeta(x) - \frac{p}{2}\right)\left(\zeta(x+hl) - \frac{p}{2}\right) \\
&\quad \times \left(\zeta(y) - \frac{p}{2}\right)\left(\zeta(y+hl) - \frac{p}{2}\right) dxdy \\
&= \frac{1}{16T^{2n}} \int_{Q_n}\int_{Q_n} \sum_{i,j,k,m=1}^{p} \frac{1}{(d_i d_j d_k d_m)^2} \\
&\quad \times EH_2(\eta_i(x))H_2(\eta_k(x+hl)) \\
&\quad \times H_2(\eta_j(x))H_2(\eta_m(x+hl))dxdy.
\end{aligned}$$

Using Lemma 5.2.4. and performing direct calculations, we obtain

$$\begin{aligned}
\operatorname{var}\gamma_T(h) &= \frac{2^{n-2}}{T^{2n}} \int_{Q_n} \left[\prod_{i=1}^{n}(T - z_i)\right] \\
&\quad \times \left[\lambda_1^4(z) + \lambda_1^2(z-hl)\lambda_1^2(z+hl) + 4\lambda_1^2(h)\lambda_1^2(z)\right. \\
&\quad + 4\lambda_1^2(h)\lambda_1(z-hl)\lambda_1(z+hl) + 4\lambda_1^2(z)\lambda_1(z-hl)\lambda_1(z+l) \\
&\quad + (p-1)\left(\lambda_2^4(z) + \lambda_2^2(z-hl)\lambda_2^2(z+hl) + 4\lambda_2^2(h)\lambda_2^2(z)\right. \\
&\quad \left. + 4\lambda_2^2(h)\lambda_2(z-hl)\lambda_2(z+hl) + 4\lambda_2^2(z)\lambda_2(z-hl)\lambda_2(z+hl)\right) \\
&\quad + (p-1)\lambda_1^2(z+hl)\lambda_2^2(z-hl) + (p-1)\lambda_1^2(z-hl)\lambda_2^2(z+hl) \\
&\quad + (p-1)(p-2)\lambda_2^2(z-hl)\lambda_2^2(z+hl) + 2(p-1)\lambda_1^2(z)\lambda_2^2(z) \\
&\quad \left. + (p-1)(p-2)\lambda_2^4(z)\right] dz.
\end{aligned}$$

After the change of variables $z_i/t = u_i$, $i = 1, \ldots n$, we obtain the following expression:

$$\operatorname{var}\gamma_T(h) = \frac{2^{n-2}L^2(T)}{T^{2\alpha}} \int\limits_{Q_n} \frac{\left[\prod_{i=1}^n (1-u_i)\right] R_T(u)}{\|u\|^{2\alpha}} du,$$

where

$$\begin{aligned} R_T(u) &= \frac{\|uT\|^{2\alpha}}{L^2(T)} \left[\lambda_1^4(uT) + \lambda_1^2(uT-hl)\lambda_1^2(uT+hl) + 4\lambda_1^2(h)\lambda_1^2(uT)\right. \\ &\quad +4\lambda_1^2(h)\lambda_1(uT-hl)\lambda_1(uT+hl) + 4\lambda_1^2(uT)\lambda_1(uT-hl)\lambda_1(uT+l) \\ &\quad +(p-1)\left((p-1)\lambda_2^4(uT) + (p-1)\lambda_2^2(uT-hl)\lambda_2^2(uT+hl)\right. \\ &\quad +4\lambda_2^2(h)\lambda_2^2(uT) + 4\lambda_2^2(h)\lambda_2(uT-hl)\lambda_2(uT+hl) \\ &\quad \left.+\, 4\lambda_2^2(uT)\lambda_2(uT-hl)\lambda_2(uT+l)\right) + (p-1)\lambda_1^2(uT+hl)\lambda_2^2(uT-hl) \\ &\quad \left.+\, (p-1)\lambda_1^2(uT-hl)\lambda_2^2(uT+hl) + 2(p-1)\lambda_1^2(uT)\lambda_2^2(uT)\right]. \end{aligned}$$

Under the condition M' the properties of slowly varying functions imply

$$\begin{aligned} \lim_{T\to\infty} R_T(u) &= 8\left(\lambda_1^2(h)(1+(p-1)\rho_0)^2 + (p-1)\lambda_2^2(h)(1-\rho_0)^2\right) \\ &= 8k_1(h). \end{aligned}$$

Under the condition M" we have

$$\lim_{T\to\infty} R_T(u) = 8\left(\lambda_1^2(h)(1+(p-1)\lambda_2^2\right) = 8k_2(h).$$

If $\alpha \in (0, n/2)$, the dominant convergence theorem implies

$$\operatorname{var}\gamma_T(h) = \frac{2^{n+1}L^2(T)}{T^{2\alpha}} k_j(h) \int\limits_{U_n} \left[\prod_{i=1}^n (1-u_i)\right] \|u\|^{-2\alpha} du(1+o(1)),\ T\to\infty,$$

where $j=1$ under the condition M' and $j=2$ under the condition M".

The proof of Lemma 5.2.5. is completed.

Proof of Theorem 5.2.5

Write

$$\begin{aligned} \gamma_T(h) - \gamma(h) &= \frac{1}{T^n}\int\limits_{Q_n} \left[(\zeta(x)-p/2)(\zeta(x+hl)-p/2) - \gamma(h)\right] dx \\ &= \frac{1}{T^n}\int_{Q_n} u(h,x)dx \\ &= Y_T(h), \\ u(h,x) &= \left(\zeta(x)-\frac{p}{2}\right)\left(\zeta(x+hl)-\frac{p}{2}\right) - \gamma(h). \end{aligned}$$

Choose some $\delta > 0$ and $0 < \gamma < \alpha$. For the sequence $T_k = k^{1/(\alpha\gamma)+\delta}$ we shall prove that

$$P\left\{\lim_{k\to\infty} \sup_{\tau\in[0,H]} |Y_{T_k}(\tau)| = 0\right\} = 1.$$

Let $S_{nk}=\Pi[0, T_k]$ be the cube in R^n. Since

$$\begin{aligned} E\left(\zeta(x) - \zeta(y)\right)^4 &= 12(\gamma(x-y) - \gamma(0))^2 + 2(\lambda_1^2(x-y) - \lambda_1^2(0))^2 \\ &\quad +2(p-1)(\lambda_2^2(x-y) - \lambda_2^2(0))^2, \end{aligned}$$

we have

$$\begin{aligned} & E\left(\zeta(x) - \frac{p}{2}\right)\left(\zeta(y) - \frac{p}{2}\right)[\zeta(x+(\tau+h)l) - \zeta(x+\tau l)] \\ & \times [\zeta(y+(\tau+h)l) - \zeta(y+\tau l)] \\ \le\ & c_5 \|h\|^{1+\rho}, \end{aligned} \tag{5.2.18}$$

where $c_5 > 0$. Observe that

$$\begin{aligned} |Y_{T_k}(\tau+h) - Y_{T_k}(\tau)|^2 \le\ & 2\left(\gamma(\tau+h) - \gamma(\tau)\right)^2 \\ & + \frac{2}{T_k^{2n}} \int_{S_{nk}} \int_{S_{nk}} \left(\zeta(x) - \frac{p}{2}\right)\left(\zeta(y) - \frac{p}{2}\right) \\ & \times [\zeta(x+(\tau+h)l) - \zeta(x+\tau l)] \\ & \times [\zeta(y+(\tau+h)l) - \zeta(y+\tau l)]\, dxdy. \end{aligned}$$

On the other hand, from Lemma 5.3.5. it follows that

$$E\left|Y_{T_k}(\tau+h) - Y_{T_k}(\tau)\right|^2 \le c_6 T_k^{-\alpha},$$

and therefore

$$E\left|Y_{T_k}(\tau+h) - Y_{T_k}(\tau)\right|^2 \le c_7 T_k^{-\alpha\gamma} \|h\|^{(1+\rho)(1-\gamma)}.$$

Put

$$\varphi(Y_{T_k}, q) = \sup_{\substack{|h|\le q;\\ h:(\tau+h)\in[0,H]}} \left|Y_{T_k}(\tau+h) - Y_{T_k}(\tau)\right|.$$

Then

$$P\left\{\varphi(Y_{T_k}, q) \ge \varepsilon\right\} \le \varepsilon^{-2} c_7 T_k^{-\alpha\gamma} q^{(1+\rho)(1-\gamma)-1}.$$

Since $\sum_{k=1}^{\infty} T_k^{-\alpha\gamma} < \infty$, we have

$$P\left\{\lim_{k\to\infty} \varphi(Y_{T_k}, q) = 0\right\} = 1.$$

Let $I \in [0, H]$ be a finite set satisfying the following condition: for every $h \in [0, H]$ there exists h_1 such that $|h - h_1| < 1$. Then

$$\sup_{\tau\in[0,H]} |Y_{T_k}(\tau)| \le \varphi(Y_{T_k}, 1) + \max_{\tau_1\in I} |Y_{T_k}(\tau_1)|.$$

Lemma 5.3.5. and the Borel–Cantelli Lemma 1.1.2. imply that

$$P\left\{\lim_{k\to\infty} \varphi(Y_{T_k}, q) = 0\right\} = 1,$$

$$P\left\{\lim_{k\to\infty} \sup_{\tau\in[0,H]} |Y_{T_k}(\tau)| = 0\right\} = 1.$$

It can be proved that

$$\begin{aligned}\sup_{\tau\in[0,H]} |Y_T(\tau)| \le & \sup_{\tau\in[0,H]} \left[\frac{1}{T_k^n} \int_{S_{n(k+1)}\setminus S_{nk}} |u(\tau, x| \, dx\right.\\ & \left. + \left(\frac{1}{T_k^n} - \frac{1}{T_{k+1}^n}\right) \int_{S_{nk}} |u(\tau, x| \, dx\right]\\ & + \sup_{\tau\in[0,H]} |Y_{T_k}(\tau)|.\end{aligned}$$

Put

$$\eta_k(\tau) = \frac{1}{T_k^n} \int_{S_{n(k+1)}\setminus S_{nk}} |u(\tau, x| \, dx.$$

Using (2.5.18), we obtain

$$E\left(\eta_k(\tau + h) - \eta_k(\tau)\right)^2 \le \frac{1}{T_k^{2n}} \left(T_{k+1}^{2n} - T_k^{2n}\right) c_8 \|h\|^{(1+\rho)}.$$

Since

$$\sum_{k=1}^{\infty} \left(\frac{T_{k+1}^{2n}}{T_k^{2n}} - 1\right) < \infty,$$

we can prove in the same way that

$$P\left\{\lim_{k\to\infty}\sup_{\tau\in[0,H]}|\eta_k(\tau)|=0\right\}=1,$$

completing the proof of Theorem 5.2.5.

Proof of Theorem 5.2.6.

For the sake of convenience, we write the correlogram as

$$\begin{aligned}\gamma_T(h) &= \frac{1}{4T^n}\sum_{i=1}^{p}\int_{Q_n}\frac{H_2(\eta_i(x))H_2(\eta_i(x+hl))}{d_i^4}dx \\ &+\frac{1}{4T^n}\sum_{\substack{i,k=1\\ i\neq k}}^{p}\int_{Q_n}\frac{H_2(\eta_i(x))H_2(\eta_k(x+hl))}{d_i^2d_k^2}dx \\ &= D_T(h)+I_T(h).\end{aligned}$$

Let $h\in[0,H]$ be a fixed number. Let us consider the random fields

$$\begin{aligned}\zeta_T(h) &= \frac{\gamma_T(h)-\gamma(h)}{L(T)}T^\alpha \qquad (5.2.19)\\ &= \frac{D_T(h)-\gamma(h)}{L(T)}T^\alpha+\frac{T^\alpha}{L(T)}I_T(h),\end{aligned}$$

and let us analyze the second term in (5.2.19). We shall show that

$$\frac{T^{\alpha-n}}{4L(T)}\int_{Q_n}H_2(\eta_i(x))H_2(\eta_k(x+hl))dx\xrightarrow{P}0 \quad\text{as}\quad T\to\infty,$$

for every $i,k=1,\dots p$, $i\neq k$. To this end it suffices to prove the convergence in $\mathbf{L}_2(\Omega)$. Then

$$\begin{aligned}E\left(\frac{T^{\alpha-n}}{4L(T)}\int_{Q_n}H_2(\eta_i(x))H_2(\eta_k(x+hl))dx\right)^2 &= \frac{T^{2(\alpha-n)}}{16L^2(T)} \\ &\times\int_{Q_n}\int_{Q_n}2r_i^2(x-y)2r_k^2(x-y)dxdy \\ &\times 2^{n-2}\int_{U_n}\left[\prod_{i=1}^{n}(1-u_i)\right]V_T(u)du,\end{aligned}$$

where

$$V_T(u) = \frac{r_i^2(uT)2r_k^2(uT)T^{2\alpha}}{L^2(T)}.$$

Under the conditions M', M"

$$\lim_{T\to\infty} V_T(u) = 0.$$

Since $\prod_{i=1}^n (1-u_i)$ is an absolutely integrable function, we have

$$\frac{T^\alpha}{L(T)} I_T(h) \xrightarrow{P} 0.$$

From the Slutsky Lemma 1.1.1. we infer that the limit distribution $\zeta_T(h)$ is determined by the first term in (5.2.17).

The condition L implies that the correlation functions of the fields $\eta_k(x)$, $x \in R^n$, $r_k(x)$, $k = 1, \ldots, p$, can be written in the form

$$r_k(x) = \int_{R^n} \exp\{i(\lambda, x)\} F_k(d\lambda),$$

$F_k(\cdot)$, $k = 1, \ldots, p$, are spectral measures of the fields $\eta_k(x)$, $k = 1, \ldots, p$.

Then the fields $\eta_k(x)$, $k = 1, \ldots, r$, $x \in R^n$, can be represented in the form

$$\eta_k(x) = \int_{R^n} \exp\{i(\lambda, x)\} Z_k(d\lambda),$$

where $Z_k(\cdot)$, $k = 1, \ldots, r$, are independent Gaussian random measures associated with the measures $F_k(\cdot)$, $k = 1, \ldots, p$:

$$EZ_k(S) = 0, \qquad E\,|Z_k(S)|^2 = F_k(S), \qquad S \in \mathcal{B}(R^n).$$

Using the Itô's formula (2.3.9) we can write

$$\begin{aligned} H_2(\eta_k(x)) &= \int_{R^n}' \exp\{i\left\langle x, \lambda^{(1)} + \lambda^{(2)}\right\rangle\} Z_k(d\lambda^{(1)}) Z_k(d\lambda^{(2)}), \quad (5.2.20) \\ k &= 1, \ldots, p. \end{aligned}$$

By Itô's formula (2.3.1) we can write

$$H_2(\eta_k(x)) H_2(\eta_k(x+hl)) = \int_{R^n}' \exp\{i\left\langle x, \lambda^{(1)} + \lambda^{(2)}\right\rangle\}$$

$$
\begin{aligned}
&\times Z_k(d\lambda^{(1)})Z_k(d\lambda^{(2)})\\
&\times \int_{R^{2n}}' \exp\{i\left\langle x+hl, \lambda^{(1)}+\lambda^{(2)}\right\rangle\}\\
&\times Z_k(d\lambda^{(1)})Z_k(d\lambda^{(2)})\\
= &\int_{R^{4n}}' \exp\{i\left\langle x, \lambda^{(1)}+\lambda^{(2)}+\lambda^{(3)}+\lambda^{(4)}\right\rangle\}\\
&\times \exp\{i\left\langle hl, \lambda^{(3)}+\lambda^{(4)}\right\rangle\}\\
&\times Z_k(d\lambda^{(1)})Z_k(d\lambda^{(2)})Z_k(d\lambda^{(3)})Z_k(d\lambda^{(4)})\\
&+4\int_{R^{2n}}' \exp\{i\left\langle x, \lambda^{(2)}+\lambda^{(4)}\right\rangle\}\exp\{i\left\langle hl, \lambda^{(4)}\right\rangle\}\\
&\times r_k(h)Z_k(d\lambda^{(2)})Z_k(d\lambda^{(4)})+2r_k^2(h). \qquad (5.2.21)
\end{aligned}
$$

Thus we obtain

$$
\begin{aligned}
D_T(h) = &\frac{1}{4T^n}\sum_{k=1}^{p}\int_{Q_n}\left[\frac{1}{d_k^4}\int_{R^{4n}}' \exp\{i\left\langle x, \lambda^{(1)}+\lambda^{(2)}+\lambda^{(3)}+\lambda^{(4)}\right\rangle\}\right.\\
&\times \exp\{i\left\langle hl, \lambda^{(3)}+\lambda^{(4)}\right\rangle\}Z_k(d\lambda^{(2)})Z_k(d\lambda^{(3)})Z_k(d\lambda^{(4)})\\
&+\frac{4r_k(h)}{d_k^4}\int_{R^{2n}}' \exp\{i\left\langle x, \lambda^{(2)}+\lambda^{(4)}\right\rangle\}\\
&\left.\times \exp\{i\left\langle hl, \lambda^{(4)}\right\rangle\}Z_k(d\lambda^{(2)})Z_k(d\lambda^{(4)}\right]dx\\
&+\frac{1}{2}\sum_{k=1}^{p}\frac{r_k^2(h)}{d_k^4}. \qquad (5.2.22)
\end{aligned}
$$

Using (5.2.19)–(5.2.21) and integrating with respect to the variable x appearing in (5.2.22), we obtain

$$
\begin{aligned}
D_T(h)-\gamma(h) = &\frac{1}{4T^n}\sum_{k=1}^{p}\frac{1}{d_k^4}\int_{R^{4n}}'\prod_{j=1}^{n}\frac{\exp\{iT(\lambda_j^{(1)}+\cdots+\lambda_j^{(4)})\}-1}{iT(\lambda_j^{(1)}+\cdots+\lambda_j^{(4)})}\\
&\times \exp\{i\left\langle hl, \lambda^{(3)}+\lambda^{(4)}\right\rangle\}Z_k(d\lambda^{(1)})\cdots Z_k(d\lambda^{(4)})\\
&+\frac{1}{T^n}\sum_{k=1}^{p}\frac{r_k(h)}{d_k^4}\int_{R^{2n}}'\prod_{j=1}^{n}\frac{\exp\{iT(\lambda_j^{(1)}+\lambda_j^{(2)})\}-1}{iT(\lambda_j^{(1)}+\lambda_j^{(2)})}\\
&\times \exp\{i\left\langle hl, \lambda^{(2)}\right\rangle\}Z_k(d\lambda^{(2)}). \qquad (5.2.23)
\end{aligned}
$$

Under the condition N, we can write (5.2.2) in the form

$$
\begin{aligned}
D_T(h)-\gamma(h) &= \frac{1}{4T^n}\sum_{k=1}^{p}\frac{1}{d_k^4}\int_{R^{4n}}' \prod_{j=1}^{n}\frac{\exp\{iT(\lambda_j^{(1)}+\cdots+\lambda_j^{(4)})\}-1}{iT(\lambda_j^{(1)}+\cdots+\lambda_j^{(4)})} \\
&\quad\times \exp\{i\left\langle hl,\lambda^{(3)}+\lambda^{(4)}\right\rangle\} \\
&\quad\sqrt{f_k(\left\|\lambda^{(1)}\right\|)\cdots f_k(\left\|\lambda^{(4)}\right\|)}W_k(d\lambda^{(1)})\cdots W_k(d\lambda^{(4)}) \\
&\quad+\frac{1}{T^n}\sum_{k=1}^{p}\frac{r_k(h)}{d_k^4}\int_{R^{2n}}' \prod_{j=1}^{n}\frac{\exp\{iT(\lambda_j^{(1)}+\lambda_j^{(2)})\}-1}{iT(\lambda_j^{(1)}+\lambda_j^{(2)})} \\
&\quad\times \exp\{i\left\langle hl,\lambda^{(2)}\right\rangle\} \\
&\quad\times\sqrt{f_k(\left\|\lambda^{(1)}\right\|)f_k(\left\|\lambda^{(2)}\right\|)}W_k(d\lambda^{(1)})W_k(d\lambda^{(2)}),
\end{aligned}
$$

Where $W_k(\cdot)$, $k=1,\ldots,p$, are independent replica of the complex Gaussian white noise in R^n.

Under the condition M' the Tauberian theorem 1.1.4. implies the following asymptotic representation:

$$
\begin{aligned}
f_j(\lambda) &= f_j(\|\lambda\|) \\
&\sim \frac{c_5(n,\alpha)\ L(1/\|\lambda\|)}{\|\lambda\|^{n-\alpha}}, \quad \|\lambda\|\to 0, \quad j=1,\ldots,p.
\end{aligned}
$$

If the condition M" is fulfilled we have

$$
\begin{aligned}
f_j(\lambda) &= f_j(\|\lambda\|) \\
&\sim \frac{d_j^2\ c_5(n,\alpha)\ L(1/\|\lambda\|)}{\|\lambda\|^{n-\alpha}}, \quad \|\lambda\|\to 0, \quad j=1,\ldots,p.
\end{aligned}
$$

Making the change of variables $\lambda^{(k)}T=\mu^{(k)}$, $k=1,2,3,4$, and using the self-similarity of the Wiener random measure with parameter $n/2$ (formally, $W(d(a\lambda))\overset{D}{=}a^{n/2}W(d\lambda),\ a\in R^1,\ \lambda\in R^n$), we obtain

$$
D_T(h)-\gamma(h) = \frac{1}{4T^n}\sum_{k=1}^{p}\frac{1}{d_k^4}\int_{R^{4n}}' \prod_{j=1}^{n}\frac{\exp\{i(\mu_j^{(1)}+\cdots+\mu_j^{(4)})\}-1}{i(\mu_j^{(1)}+\cdots+\mu_j^{(4)})}
$$

$$\times \exp\left\{i\left\langle hl, \frac{\mu^{(3)}+\mu^{(4)}}{T}\right\rangle\right\}$$
$$\times\sqrt{f_k\left(\left\|\frac{\mu^{(1)}}{T}\right\|\right)\cdots f_k\left(\left\|\frac{\mu^{(4)}}{T}\right\|\right)}W_k(d\mu^{(1)})\cdots W_k(d\mu^{(4)})$$
$$+\frac{1}{T^n}\sum_{k=1}^{p}\frac{r_k(h)}{d_k^4}\int'_{R^{2n}}\prod_{j=1}^{n}\frac{\exp\{i(\mu_j^{(1)}+\mu_j^{(2)})\}-1}{i(\mu_j^{(1)}+\mu_j^{(2)})}$$
$$\times \exp\left\{i\left\langle hl, \frac{\mu^{(2)}}{T}\right\rangle\right\}$$
$$\times\sqrt{f_k\left(\left\|\frac{\mu^{(1)}}{T}\right\|\right) f_k\left(\left\|\frac{\mu^{(2)}}{T}\right\|\right)}W_k(d\mu^{(1)})W_k(d\mu^{(2)})$$
$$= A_T(h)+B_T(h). \tag{5.2.24}$$

Let us prove that

$$\frac{T^\alpha}{L(T)}A_T(h)\xrightarrow{P}0.$$

It follows from (5.2.24) that

$$E\left(\frac{T^\alpha}{L(T)}A_T(h)\right)^2 \le 2\sum_{k=1}^{p}\frac{1}{d_k^8}\int'_{R^{4n}}\prod_{j=1}^{n}\frac{\left|\exp\{i(\mu_j^{(1)}+\cdots+\mu_j^{(4)})\}-1\right|^2}{\left(\mu_j^{(1)}+\cdots+\mu_j^{(4)}\right)^2}$$
$$\times\frac{Q_{T,k}\left(\mu^{(1)},\mu^{(2)},\mu^{(3)},\mu^{(4)}\right)}{\|\mu^{(1)}\|^{n-\alpha}\cdots\|\mu^{(4)}\|^{n-\alpha}}d\mu^{(1)}\cdots d\mu^{(4)},$$

where

$$Q_{T,k}\left(\mu^{(1)},\mu^{(2)},\mu^{(3)},\mu^{(4)}\right) = \frac{T^{2\alpha-4n}}{16L^2(T)}$$
$$\times f_k\left(\left\|\frac{\mu^{(1)}}{T}\right\|\right)\cdots f_k\left(\left\|\frac{\mu^{(4)}}{T}\right\|\right)\cdot\|\mu^{(1)}\|^{n-\alpha}\cdots\|\mu^{(4)}\|^{n-\alpha}.$$

Since

$$\lim_{T\to\infty}Q_{T,k}\left(\mu^{(1)},\mu^{(2)},\mu^{(3)},\mu^{(4)}\right)=0,\quad k=1,\ldots p,$$

and

$$\prod_{j=1}^{n}\frac{\left|\exp\{i(\mu_j^{(1)}+\cdots+\mu_j^{(4)})\}-1\right|^2}{\left(\mu_j^{(1)}+\cdots+\mu_j^{(4)}\right)^2\|\mu^{(1)}\|^{n-\alpha}\cdots\|\mu^{(4)}\|^{n-\alpha}}$$

is an absolutely integrable function ($0 < \alpha < n/2$), the dominant convergence theorem implies the required assertion. In view of the Slutsky lemma, we may restrict our attention to demonstrating that under the conditions L, M' and M"

$$\frac{T^{\alpha}}{L(T)} B_T(h) \xrightarrow{D} \alpha(h),$$

and under the conditions L, M' and N

$$\frac{T^{\alpha}}{L(T)} B_T(h) \xrightarrow{D} \beta(h).$$

Using the Lemma 1.1.3. we obtain the convergence of all finite-dimensional distributions.

Let us consider the case where the conditions L, M' and N are satisfied. The case of the conditions L, M" and N is analyzed analogously.

It sufficiently to show that

$$E\left|\frac{T^{\alpha}}{L(T)} B_T(h) - \alpha(h)\right|^2 \to 0 \quad \text{as} \quad T \to \infty.$$

We obtain

$$\begin{aligned} E\left|\frac{T^{\alpha}}{L(T)} B_T(h) - \alpha(h)\right|^2 \leq\ & 2c_0^2 \sum_{k=1}^{p} \frac{r_k^2(h)}{d_k^8} \int_{R^{2n}} \\ & \times \prod_{j=1}^{n} \frac{\left|\exp\{i(\mu_j^{(1)} + \mu_j^{(2)})\} - 1\right|^2}{\left(\mu_j^{(1)} + \mu_j^{(2)}\right)^2} \\ & \times \frac{\left|S_T(\mu^{(1)}, \mu^{(2)}) - 1\right|^2}{\|\mu^{(1)}\|^{n-\alpha} \|\mu^{(2)}\|^{n-\alpha}} d\mu^{(1)} d\mu^{(2)}, \end{aligned}$$

where

$$S_T(\mu^{(1)}, \mu^{(2)}) = \frac{\|\mu^{(1)}\|^{n-\alpha} \|\mu^{(2)}\|^{n-\alpha}}{c_5(n,\alpha)\, T^{n-\alpha} L(T)} \sqrt{f_k\left(\left\|\frac{\mu^{(1)}}{T}\right\|\right) f_k\left(\left\|\frac{\mu^{(2)}}{T}\right\|\right)}.$$

The condition N and the properties of slowly varying functions imply

$$\lim_{T\to\infty} S_T(\mu^{(1)}, \mu^{(2)}) = 1.$$

The function

$$\prod_{j=1}^{n} \frac{\left|\exp\{i(\mu_j^{(1)} + \mu_j^{(2)})\} - 1\right|^2}{\left(\mu_j^{(1)} + \mu_j^{(2)}\right)^2 \|\mu^{(1)}\|^{n-\alpha} \|\mu^{(2)}\|^{n-\alpha}}$$

is absolutely integrable for $0 < \alpha < n/2$, and therefore, the application of the dominant convergence theorem completes the proof of the theorem (see the proof for Theorem 2.4.1. for details).

5.3 Efficient Estimation of Regression Coefficients of a Random Fields Observed on the Sphere

We present some surprising results on efficiency of LSE of regression coefficients of homogeneous isotropic random fields observed on the sphere.

Let $x = (\rho, u)$, $\rho \geq 0$, $u \in s(1)$, be a spherical coordinate of a point $x \in R^n$ (see (1.1.3)), $n \geq 2$.

P. Suppose that the random field

$$\xi(\rho, u) = a\ g(\rho, u) + \eta(\rho, u)$$

is observed on the sphere $s(r)$ in R^n, where a is an unknown parameter, $g(\rho, u)$ is a known function, and $\eta(\rho, u)$ is homogeneous isotropic random field with $E\eta(\rho, u) = 0$ and continuous covariance function

$$B(\rho) = E\eta(\rho_1, u_1)\eta(\rho_2, u_2),$$

where

$$\begin{aligned} \rho &= \|x - y\| \\ &= \sqrt{\rho_1^2 + \rho_2^2 - 2\rho_1\rho_2 \cos\gamma}, \\ \cos\gamma &= \frac{\langle x, y\rangle}{\|x\| \cdot \|y\|}, \end{aligned}$$

$$x = (\rho_1, u_1), \quad y = (\rho_2, u_2), \quad \rho_i \geq 0,\ u_i \in s(1),\ i = 1, 2,$$

and spectral measure $G(\cdot)$ defined in (1.2.14).

Later the following functions will be important:

$$b_m(r) = 2^{n-1}\Gamma\left(\frac{n}{2}\right)\pi^{n/2}\int_0^\infty J^2_{m+(n-2)/2}(\lambda r)(\lambda r)^{2-n}G(d\lambda), \quad m = 0, 1, 2, \ldots$$

where $G(\cdot)$ is the spectral measure of a field $\eta(\rho, u)$ (see (1.2.14)).

Remark 5.3.1. The asymptotic behavior of $b_m(r)$ as $r \to \infty$ can be received from the results of section 1.4. If for some $\gamma \in (-1, n-2)$

$$G'(\lambda) = |s(1)|\,\lambda^\gamma h(\lambda),$$

where $h(\lambda)$ is continuous in a neighborhood of zero, $h(\lambda) \neq 0$, $h(\lambda)$ is bounded on $[0, \infty)$, then from Lemma 1.4.5. we obtain as $r \to \infty$:

$$b_m(r) = (2\pi)^n h(0) c_{19}(n, m, \gamma)\,|s(1)|\,r^{-1-\gamma}(1 + o(1)),$$

where $|s(1)|$ is the area of a surface of the unite sphere $s(1)$ and

$$c_{19}(n, m, \gamma) = \frac{\Gamma(n-2-\gamma)\Gamma\left(\frac{2m+\gamma+1}{2}\right)}{2^{n-2-\gamma}\Gamma^2\left(\frac{n-1-\gamma}{2}\right)\Gamma\left(\frac{2m+2n-\gamma-3}{2}\right)}.$$

Let

$$g^l_m(r) = \int_{s(1)} g(r, u) S^l_m(u)\sigma(du),$$

where $\sigma(du)$ is a Lebesgue measure on the sphere and $S^l_m(u)$ is an orthogonal spherical harmonic (see remark 1.2.5). Suppose that

$$\sum_{m=0}^\infty \sum_{l=1}^{h(m,n)} \frac{[g^l_m(r)]^2}{b_m(r)} < \infty.$$

A theorem of Yadrenko (1983, p. 171) shows that best linear unbiased estimate (BLUE) a^*_r of a parameter based on the observations $\xi(\rho, x)$, $x \in s(r)$, is

$$a^*_r = \frac{\int_{s(1)} l(r, u)\xi(r, u)\sigma(du)}{\sum_{m=0}^\infty \sum_{l=1}^{h(m,n)} \frac{[g^l_m(r)]^2}{b_m(r)}},$$

where

$$l(r,u)=\sum_{m=0}^{\infty}\sum_{l=1}^{h(m,n)}\frac{g_m^l(u)S_m^l(u)}{b_m(r)}.$$

The variance of the estimate a_r^* has the form:

$$\operatorname{var} a_r^*=\left[\sum_{m=0}^{\infty}\sum_{l=1}^{h(m,n)}\frac{[g_m^l(r)]^2}{b_m(r)}\right]^{-1}c_1^2(n),$$

where $c_1^2(n)$ is defind in (1.2.19).

Since the BLUE often has a complicated form, statisticians considers the LSE for a, i.e., value of a which minimized the functional:

$$Q(a)=\int\limits_{s(r)}[\xi(\rho,u)-ag(\rho,u)]^2\,\sigma(du),$$

where $\sigma(du)$ is a Lebesgue measure on the sphere $s(r)$.

The LSE has the following form:

$$\hat a_r=\frac{\int\limits_{s(r)}\xi(\rho,u)\;g(\rho,u)\;\sigma(du)}{\int\limits_{s(r)}g^2(\rho,u)\;\sigma(du)}.$$

Using results of the book (Ivanov and Leonenko (1989), p. 124), we receive the expression for a variance of the LSE:

$$\operatorname{var}\hat a_r=\frac{c_1^2(n)\sum\limits_{m=0}^{\infty}\sum\limits_{l=1}^{h(m,n)}A_{m,l}(r)\;b_m(r)}{\sum\limits_{m=0}^{\infty}\sum\limits_{l=1}^{h(m,n)}[g_m^l(r)]^2},$$

where

$$A_{m,l}(r)=\frac{\left[g_m^l(r)\right]^2}{\sum\limits_{m=0}^{\infty}\sum\limits_{l=1}^{h(m,n)}[g_m^l(r)]^2}.$$

Clearly,

$$A_{m,l}(r)\geq 0,\qquad \sum_{m=0}^{\infty}\sum_{l=1}^{h(m,n)}A_{m,l}(r)=1.$$

Thus for every r, a discrete probability distribution $\{A_{m,l}(r)\}$, $(m,l) \in T$, on the set $T = \{0,1,\ldots\} \times \{1,2,\ldots,h(m,n)\}$ is defined. The efficiency of the LSE is defined as

$$\text{eff}(r,B) = \frac{\text{var } a_r^*}{\text{var } \hat{a}_r}.$$

The asymptotic efficiency is

$$\text{eff}(\infty,B) = \lim_{r\to\infty} \text{eff}(r,B).$$

It is obvious that

$$0 \le \text{eff}(r,B) \le 1, \qquad 0 \le \text{eff}(\infty,B) \le 1.$$

Theorem 5.3.1. Let assumptions P hold, the efficiency $\text{eff}(r,B)$ of LSE $\hat{a}_r$ is of the form:

$$\begin{aligned}
\text{eff}(r,B) &= \left[\sum_{m=0}^{\infty}\sum_{l=1}^{h(m,n)} \left(g_m^l(r)\right)^2\right] \\
&\times \left[\left(\sum_{m=0}^{\infty}\sum_{l=1}^{h(m,n)} \frac{[g_m^l(r)]^2}{b_m(r)}\right)\left(\sum_{m=0}^{\infty}\sum_{l=1}^{h(m,n)} \left[g_m^l(r)\right]^2 b_m(r)\right)\right]^{-1} \\
&= \left[\left(\sum_{m=0}^{\infty}\sum_{l=1}^{h(m,n)} A_{m,l}(r) b_m(r)\right)\left(\sum_{m=0}^{\infty}\sum_{l=1}^{h(m,n)} \frac{A_{m,l}(r)}{b_m(r)}\right)\right]^{-1} \qquad (5.3.1)
\end{aligned}$$

From (5.3.1) we receive the following statement.

Theorem 5.3.2. Let assumptions P hold. Suppose that the discrete distribution $\{A_{m,l}(r)\}$, $(m,l) \in T$, is concentrated only at one point $(m_0,l_0) \in T$ for all $r > 0$ (that is $A_{m_0,l_0}(r) = 1$ and the rest $A_{m,l}(r) = 0$). Then

$$\text{eff}(r,B) = 1$$

for every correlation function B and every $r > 0$.

In particular, if

$$g(r,u) = \psi_1(r)\psi_2(u), \qquad r \ge 0, \qquad u \in s(1), \qquad (5.3.2)$$

then

$$\begin{aligned} g^l_m(r) &= \psi_1(r)\tilde{g}^l_m(r), \\ \tilde{g}^l_m(r) &= \int\limits_{s(1)} \psi_2(u)S^l_m(u)\sigma(du) \end{aligned}$$

and

$$\begin{aligned} \text{eff}(r,B) &= \left[\sum_{m=0}^{\infty}\sum_{l=1}^{h(m,n)} \left(\tilde{g}^l_m(r)\right)^2\right] \\ &\times \left[\left(\sum_{m=0}^{\infty}\sum_{l=1}^{h(m,n)} \frac{[\tilde{g}^l_m(r)]^2}{b_m(r)}\right)\left(\sum_{m=0}^{\infty}\sum_{l=1}^{h(m,n)} [\tilde{g}^l_m(r)]^2 b_m(r)\right)\right]^{-1} \\ &= \left[\left(\sum_{m=0}^{\infty}\sum_{l=1}^{h(m,n)} \tilde{A}_{m,l} b_m(r)\right)\left(\sum_{m=0}^{\infty}\sum_{l=1}^{h(m,n)} \frac{\tilde{A}_{m,l}}{b_m(r)}\right)\right]^{-1}, \end{aligned}$$

where

$$\begin{aligned} \tilde{A}_{m,l} &= \frac{\left(\tilde{g}^l_m(r)\right)^2}{\sum\limits_{m=0}^{\infty}\sum\limits_{l=1}^{h(m,n)} \tilde{g}^l_m(r)} \\ &\geq 0, \\ \sum_{m=0}^{\infty}\sum_{l=1}^{h(m,n)} \tilde{A}_{m,l} &= 1. \end{aligned}$$

Thus the discrete probability distribution $\{A_{m,l}(r)\}$, $(m,l) \in T$,does not depend on $r > 0$.

Theorem 5.3.3. Suppose assumptions P hold and the regression function $g(r,u)$ satisfies (5.3.2). Suppose that the discrete probability distribution $\{A_{m,l}(r)\}$, $(m,l) \in T$, is concentrated only at one point $(m_0, l_0) \in T$. Then $\text{eff}(r,B) = 1$ for every correlation function B and every $r > 0$. Suppose that there exists a function $\mu(r) > 0$ such that the limit

$$\lim_{r\to\infty} [b_m(r)\mu(r)] = b \in (0,\infty)$$

exists and b does not depend on m. If there exists

$$\lim_{r\to\infty} A_{m,l}(r) = \tilde{A}_{m,l}(m,l) \in T,$$

then the asymptotic efficiency $\mathrm{eff}(r,B) = 1$.

If regression function $g(r,u)$ satisfies (5.3.2) then the asymptotic efficiency $\mathrm{eff}(\infty,B) = 1$ for every covariance function B.

5.4 Estimation in the Frequency Domain

This section presents some results on statistical estimation of unknown parameters of random fields in the frequency domain.

1. Minimum contrast estimation

We present a general result on consistency and asymptotic normality of the minimum contrast estimations under the Gaussian–Whittle contrast of a parameter of the random field with discrete parameter (see subsection 1 of section 1.2).

These results can be used for estimation of parameters of random fields with continuous parameter via discretization procedure (see subsection 3 of section 1.2).

Note that the quasi-likelihood, or minimum contrast, method of spectral density parameter fitting for discrete-time processes and fields was first proposed by Whittle (1951, 1953), and was later developed by Walker (1964), Ibragimov (1967), Hannan (1973), Rice (1997), Guyon (1982, 1985), Dzaparidze (1986), Fox and Taqqu (1986), Dahlhaus (1984), Giraitis and Surgailis (1990), Heyde and Gay (1989, 1993), Beran (1994), Leonenko and Woyczynski (1997, 1998 b, c, d, f), Leonenko, Sikorskii and Terdik (1998), among others.

Anh and Lunney (1995) considered the problem of approximations of maximum likelihood estimators and maximum entropy estimations by minimum contrast estimations.

We introduce the following general conditions:

Q. Let $\zeta(x)$, $x \in Z^n$, be a homogeneous central random field with covariance function

$$R(x) = \mathrm{cov}(\zeta(0), \zeta(x)), \qquad x \in Z^n,$$

and spectral density

$$f(\lambda,\theta), \quad \theta \in \Theta, \quad \lambda \in \Pi = [-\pi,\pi)^n,$$

where Θ is a compact set and the true value of parameter θ_0 is in the interior of $\Theta \subset R^r$. If $\theta \neq \theta'$, then $f(\lambda, \theta)$ differs from $f(\lambda, \theta')$ on a set of positive Lebesgue measure.

Now consider a parametric statistical model of distributions P_θ, $\theta \in \Theta$, and put $P_0 = P_{\theta_0}$.

We assume that random field $\zeta(x)$ is observed on the set

$$[1, T] \times \cdots \times [1, T] = [1, T]^n \subset Z^n.$$

In order to make edge effects negligible, the unbiased periodogram

$$I_T(\lambda) = (2\pi)^{-n} \sum_{\substack{|x_j| \leq T-1, \\ j=1,\dots n}} R_T(x) \exp\{-i \langle \lambda, x \rangle\}, \ \lambda \in \Pi, \tag{5.4.1}$$

where

$$R_T(x) = \prod_{j=1}^n (T - |x_j|)^{-1} \sum_{y, y+x \in [1,T]^n} \zeta(y)\zeta(y+x), \ x \in Z^n, \ T \in Z^1,$$

should be used in (5.4.1) (see Guyon, p. 139).

A contrast function for θ_0 is a deterministic mapping $K(\theta_0, \cdot) : \Theta \to [0, \infty)$, which has a unique minimum at $\theta = \theta_0$. For given $K(\theta_0, \cdot)$ let us consider contrast field $S_T(\theta)$, $T \in Z^1$, adapted to $\zeta(x)$, $x \in [1, T]^n$, defined for all $\theta \in \Theta$, and such that

$$\lim_{T \to \infty} \inf [S_T(\theta) - S_T(\theta_0)] \geq K(\theta_0, \theta)$$

in probability P_0. The minimum contrast estimator θ_T minimizes the value of S_T, i.e.,

$$\theta_T = \arg \min_{\theta \in \Theta} S_T(\theta).$$

We shall call a Gaussian–Whittle contrast, defined by the contrast process

$$\begin{aligned} S_T(\theta) &= S_T(\theta, I_T(\lambda), \ f(\lambda)) \\ &= \frac{1}{2(2\pi)^n} \int_\Pi \left[\log \ f(\lambda, \theta) + \frac{I_T(\lambda)}{f(\lambda, \theta)} \right] d\lambda, \quad \theta \in \Theta, \end{aligned} \tag{5.4.2}$$

and the contrast function

$$K(\theta_0, \theta) = (2\pi)^{-n} \frac{1}{2} \int_\Pi \left[\frac{f(\lambda, \theta_0)}{f(\lambda, \theta)} - 1 - \log \frac{f(\lambda, \theta_0)}{f(\lambda, \theta)} \right] d\lambda,$$

and the associated minimum contrast estimator

$$\theta_T = \arg\min_{\theta\in\Theta} S_T(\theta), \tag{5.4.3}$$

where $S_T(\theta)$ is defined in (5.4.2).

Remark 5.4.1. In general the system of equations $\nabla_\theta S_T(\theta) = 0$ is nonlinear and it is not easy to solve it directly. Kiively (1992) have considered the numerical procedure based on the fast Fourier transform algorithm.

Theorem 5.4.1. Assume that the assumptions Q holds, function $f^{-1}(\lambda,\theta)$ is continuous on $\Pi \times \Theta$, and the sample covariance $R_T(x) \to R(x)$ in P_0-probability as $T \to \infty$. Then, in P_0-probability,

$$\begin{aligned} S_T(\theta) &\to S(\theta) \\ &= (2\pi)^{-n}\frac{1}{2}\int_\Pi \left[\log f(\lambda,\theta) + \frac{f(\lambda,\theta_0)}{f(\lambda,\theta)}\right] d\lambda \end{aligned}$$

and minimum contrast estimator $\theta_T \to \theta_0$, as $T \to \infty$.

The proof of this theorem is based on the arguments of Ibragimov (1967) (see Leonenko and Woyczynski (1998 f, c) for details).

Assumptions in the above theorem are weaker than those in the comparable result of Guyon (1995, p. 145).

R.

$$\int_\Pi \log f(\lambda,\theta) d\lambda > -\infty.$$

Under conditions Q and R we have the Wald decomposition in terms of the innovations ε_x, $x \in Z^n$:

$$\zeta(x) = \sum_{s\geq 0} a_s \varepsilon_{t-s},$$

$a_0 = 1$, where ε_y, $y \in Z^n$, are i.i.d.r.v. (Gaussian, if the field ζ is Gaussian) with mean zero and

$$\operatorname{var} \varepsilon_x = \exp\left\{ (2\pi)^{-n} \int_\Pi \log\ f(\lambda,\theta) d\lambda \right\},$$

and $y \geq 0$ means lexicographic order on Z^n.

P. The spectral density $f(\lambda, \theta)$, $\lambda \in \Pi$, $\theta \in \Theta$, and the vector-valued function

$$A(\lambda, \theta) = -\nabla_\theta \frac{1}{f(\lambda, \theta)}, \quad \lambda \in \Pi, \quad \theta \in \Theta,$$

satisfy the following conditions P1-P7:

P1. The parameter integral $\int_\Pi \log f(\lambda, \theta) d\lambda (\equiv 0)$ is twice differentiable with respect to parameter θ.

P2. The function $f^{-1}(\lambda, \theta)$, $\lambda \in \Pi$, $\theta \in \Theta$, is twice differentiable with respect to θ and the derivatives $\partial f^{-1}(\lambda, \theta)/\partial \theta_j$ and $\partial^2 f^{-1}(\lambda, \theta)/\partial \theta_k \theta_j$ are continuous for all $\theta \in \Theta$.

P3. The vector-function $A(\lambda, \theta)$ is symmetric about $\lambda = 0$, for $\lambda \in \Pi$, $w \in \Theta_0 \supset \Theta$ where Θ_0 is an open set.

P4. For all $\theta \in \Theta$, the function $A(\lambda, \theta) \in \mathbf{L}_1(\Pi)$.

P5. For all $\theta \in \Theta$, $f^{-1}(\lambda, \theta)\ A(\lambda, \theta) \in \mathbf{L}_1(\Pi) \cap \mathbf{L}_2(\Pi)$.

P6. There exists a $_1, \ldots, _n$, with $|_i| \leq 1$, $i = 1, \ldots, n$, such that $B(\lambda) f^{-1}(\lambda, \theta)$ is bounded, and $A(\lambda, \theta)/B(\lambda) \in \mathbf{L}_2(\Pi)$, where $B(\lambda) = \prod_{j=1}^n |\lambda_j|^{j}$.

P7. The $r \times r-$ matrix function $\nabla_\theta A'(\lambda, \theta) \in \mathbf{L}_1(\Pi)$ for all $\theta \in \Theta \supset \Theta_0$.

Theorem 5.4.2. Let $\zeta(x)$, $x \in Z^n$, be a Gaussian field satisfying assumptions of Theorem 5.4.1. and conditions R and P.

If the matrix

$$\Sigma = (2\pi)^{-n} \int_\Pi \nabla_\theta \log f(\lambda, \theta_0) \left[\nabla_\theta \log f(\lambda, \theta_0)\right]' d\lambda \tag{5.4.4}$$

is nonsingular, then the minimum contrast estimations $\hat{\theta}_T$ are asymptotically normal as $T \to \infty$,

$$T^{n/2}(\theta_T - \theta_0) \xrightarrow{D} \mathcal{N}(0, 2\Sigma^{-1}).$$

Proof of this theorem is based on the results of Heyde and Gay (1993) and Taylor's development (see Leonenko and Woyczynski (1998 f) for details).

Consider the random fields with continuous parameter. We use the notations of subsection 3 of section 1.2.

S. Let $\xi(x)$, $x \in R^n$, be a homogeneous centre random field with spectral density

$$f_c(\lambda) = f_c(\lambda, \theta), \quad \theta \in \Theta, \quad \lambda \in R^n,$$

where Θ is compact set and the true value of parameter θ_0 is in the interior of $\Theta \subset R^n$. Suppose that assumptions of Theorems 5.4.1.–5.4.2. hold for homogeneous random fields

$$\xi_{d1}(x) = \xi(xh), \quad x \in Z^n, \quad h > 0,$$

(see (1.2.7)) and

$$\xi_{d2}(x) = h^{-n} \int_{A_h} \xi(y)dy,$$

$$A_h = \left\{ y \in R^n : \left(x_i - \frac{1}{2}\right) h \leq y_i < \left(x_i + \frac{1}{2}\right) h, \quad i = 1,, \ldots, n \right\}$$

$x \in Z^n$, $h > 0$ (see (1.2.9)) with spectral densities $f_{d1}(\lambda, \theta)$, $\lambda \in \Pi$, and $f_{d2}(\lambda, \theta)$, $\lambda \in \Pi$, given by the relations (1.2.8) and (1.2.11) respectively with $f_c(\lambda) = f_c(\lambda, \theta)$, $\theta \in \Theta$, $\lambda \in R^n$.

From Theorems 5.4.1.–5.4.2. we obtain

Theorem 5.4.3. Suppose that $i = 1$ or $i = 2$. Let assumptions S hold for homogeneous random field $\xi(x)$, $x \in R^n$, with continuous parameter. Then the minimum contrast estimators $\theta_{i_T} \to \theta$ as $T \to \infty$ in P_0-probability, and

$$T^{n/2}(\theta_{i_T} - \theta_0) \xrightarrow{D} \mathcal{N}(0, 2\Sigma_i^{-1}).$$

where θ_{i_T} and Σ_i are defined in (5.4.2)–(5.4.3) and (5.4.4) respectively with $f_{di}(\lambda, \theta)$, $\lambda \in \Pi$, in places of $f(\lambda) = f(\lambda, \theta)$.

The detailed proofs will appear in Leonenko and Woyczynski (1998 f).

Remark 5.4.2. The limiting covariance matrix Σ_i in Theorem 5.4.3. can be consistently estimated by

$$\hat{\Sigma}_i = \frac{1}{(2T)^n} \sum_{\lambda_s \in [0,T]^n} \nabla_\theta \log f_{di}(\lambda_s, \theta_0) \left[\nabla_\theta \log f_{di} f(\lambda_s, \theta_0)\right]'$$

where

$$\lambda_s = \left[(2\pi s_1/T), \ldots, (2\pi s_n/T)\right]', \quad s_j = 1, 2, \ldots, \quad j = 1,, \ldots, n.$$

Remark 5.4.3. Now let $\xi(x) = U(t, x)$, $x \in R^1$ and $t > 0$ fixed, be a Gaussian field which is stationary in x and has the spectral density of the form (4.3.5). This field is called a solution of the one-dimensional Burgers' equation with weakly dependent random initial condition (see Theorem

4.3.1.). Then the spectral densities of the Gaussian fields $\zeta_{di}(x)$, $x \in Z^1$, $i = 1, 2$ (see condition S with $n = 1$) are of the form

$$f_{d1}(\lambda) = \frac{q}{32}\frac{h^2}{\mu^2 t^2}\sqrt{\frac{\pi}{2\mu t}}\left[\frac{8\mu t}{h^2}\theta\left(\lambda, \frac{h^2}{8\mu t}\right) + 2\frac{\partial^2}{\partial\lambda^2}\theta\left(\lambda, \frac{h^2}{8\mu t}\right)\right],$$

$$f_{d2}(\lambda) = \frac{4q}{h^2}\sin^2\frac{\lambda}{2}\sqrt{\frac{\pi}{2\mu t}}\theta\left(\lambda, \frac{h^2}{8\mu t}\right), \qquad \lambda \in [-\pi, \pi],$$

where $q = 2c\mu^2/\pi$ (see (4.3.5)), and

$$\theta(x, s) = \frac{1}{2\pi}\sum_{k=-\infty}^{\infty} e^{ikx - k^2 s}, \qquad s > 0, \qquad x \in R^1$$

is an elliptic theta function of Jacobi (see Widder (1975)). Note that

$$f_{d2}(\lambda) \sim \lambda^2 \,\text{const} \quad \text{as } \lambda \to 0,$$

but

$$\lim_{\lambda \to 0} f_{d1}(\lambda) > 0$$

(see Leonenko and Woyczynski (1998 b)). In the first case the unknown parameter $\theta = (c, \mu)$ may be estimated based on Theorem 5.4.3. In the second case the rescaled discretized data should be transformed by special technique (see Leonenko and Woyczynski (1998 b)).

Remark 5.4.4. Let $\xi(x) = Y(t, x)$, $x \in R^1$, and $t > 0$ fixed be a Gaussian field which is stationary in x and has the spectral density

$$f(\lambda) = p\,|\lambda|^{1+\alpha} e^{-\mu\lambda^2(t+s)}, \qquad \lambda \in R^1,\ 0 < \alpha < 1, \quad p > 0,$$

(see formula (4.4.3) with $n = 1$). This random field is rescaled solution of the one-dimensional Burgers' equation with strongly dependent random initial condition (see Theorem 4.4.1.). Then

$$f_{d1}(\lambda) = \frac{p}{h^{2+\alpha}}\frac{1}{2\pi}\sum_{m=-\infty}^{\infty} e^{im\lambda} b_m, \qquad \lambda \in [-\pi, \pi],$$

$$b_m = \frac{1}{2\pi}\int_{-\infty}^{\infty} e^{-ims}\,|s|^{1+\alpha} \exp\left\{-\frac{2\mu t\lambda^2}{h^2}\right\} ds,$$

and

$$f_{d2}(\lambda) = \frac{p}{h^{2+\alpha}} 4\sin^2\left(\frac{\lambda}{2}\right)\frac{1}{2\pi}\sum_{m=-\infty}^{\infty} e^{im\lambda} a_m, \qquad \lambda \in [-\pi, \pi],$$

$$a_m = \frac{1}{2\pi}\int_{-\infty}^{\infty} e^{-ims}|s|^{\alpha-1}\exp\left\{-\frac{2\mu t\lambda^2}{h^2}\right\}ds.$$

Note that

$$\lim_{\lambda\to 0} f_{d1}(\lambda) > 0,$$

but

$$f_{d2}(\lambda) \sim \lambda^{1+\alpha}\,\text{const} \quad \text{as } \lambda \to 0$$

(see Leonenko and Woyczynski (1999)). In the first case the unknown parameter $\theta = (p, \alpha, \mu)'$ can be estimated based on Theorem 5.4.3.

Remark 5.4.5. The unknown parameters of multi-dimensional Burgers' equation with random data can be estimated also by using Theorem 5.4.3. (see Leonenko and Woyczynski (1998 a,b, 1999) for details).

2. Estimation of the singularity parameter

We consider the spectral problem of estimation of the singularity parameter $\alpha \in (0, n)$ of random field with singular spectrum. This parameter is called by various author the parameter of self-similarity, the Hurst parameter or the fractional parameter.

T. Let $\xi(x)$, $x \in R^n$, be a homogeneous isotropic mean-square continuous random field with spectral density

$$f(\lambda) = f(\lambda, \theta) = \|\lambda\|^{\theta-n} f_*(\lambda), \quad \lambda \in R^n, \quad 0 < \theta < n,$$

where $f_*(\lambda)$, $\lambda \in R^n$, is bounded and continuous function in a neighborhood of zero, $f_*(0) \neq 0$.

The spectral density then has an integrable pole at the origin with the characteristic effect that the covariance function decays to zero at a very slow rate.

In fact, from Lemma 1.4.4. we obtain that for $n = 1$, $\theta \in (0, 1)$

$$B(|x|) = \frac{|x|^{-\theta} f_*(0)}{c_5(1,\theta)}(1 + o(1)) \quad \text{as} \quad |x| \to \infty,$$

where

$$c_5(1,\theta) = 2\Gamma(\theta)\cos\left(\frac{\theta\pi}{2}\right),$$

and Lemma 1.4.6. we have for $\theta \in (0,(n-1)/2)$, $n \geq 2$:

$$\begin{aligned} B(\|x\|) &= \operatorname{cov}(\xi(0),\xi(x)) \\ &= \frac{\|x\|^{-\theta} f_*(0)}{c_5(n,\theta)}(1+o(1)) \quad \text{as} \quad \|x\| \to \infty, \end{aligned}$$

where $c_5(n,\theta)$ is defined in (1.4.13).

Remark 5.4.6 . Note that assumptions T holds for special densities (1.3.11), (1.3.14), (1.3.18) and assumptions S holds for spectral densities (1.3.9), (1.3.20) (with $k \in (1/2, 3/2)$) and spectral density (1.3.22) with $(n-1)/4 < \nu < n/4$.

Consider the spectral problem of estimation of a parameter $\theta \in \Theta \subset (0,n)$, based on observation $\xi(x)$, $x \in [0,T]^n$, where Θ is compact and the true value θ_0 is in the interior of Θ.

From (1.2.8), (1.2.11) and Theorem and lemma 5.2.1. we obtain:

Theorem 5.4.4. Let $i = 1$ or $i = 2$, and assume that condition T holds. Then the estimate θ_{T_i} is consistent for θ_0 and asymptotically normal:

$$T^{n/2}(\theta_{i_T} - \theta_0) \xrightarrow{D} \mathcal{N}(0, 2\Sigma^{-1}), \qquad \text{as} \quad T \to \infty,$$

where

$$\Sigma = (2\pi)^{-n} \int\limits_{\Pi} \left[\frac{\partial}{\partial\theta} \log\, f_{di}(\lambda,\theta_0)\right]^2 d\lambda,$$

the $f_{di}(\lambda,\theta)$ is defined in condition S and θ_{i_T} is a minimum contrast estimator defined by (5.4.2)–(5.4.3) with $f_{di}(\lambda,\theta)$ in place of $f(\lambda,\theta)$.

Considerable recent attention has been devoted to semi-parametric estimation of the dependent index using method based on information in either frequency or time domains.

It reduced the problem to the regression problem with asymptotically independent errors in frequency domain (see Geweke and Porter–Hudak (1993)) and dependent errors in time domain (see Hall, Koul and Turlach (1997)).

Unfortunately, for processes with singular spectrum it is not true that the normalized periodogram's ordinates at the Fourier frequencies may be treated as asymptotically i.i.d. exponential random variables (see Künsch (1986), Hurvich and Beltrao (1993, 1994), Igloi (1994), Terrin and Hurvich

(1994), Robinson (1995a)). In fact, both the limiting distribution and limiting expectation of the normalized periodogram's ordinates at the Fourier frequencies λ_j depend on j. In view of these anomalies in the behavior of the periodogram at very low frequencies (i.e., frequencies of order $1/T$) it seems natural to contemplate removal from the regression scheme of a block of very low frequencies of the periodogram.

Suppose that $\zeta(x)$, $x \in Z^1$, is a centered stationary Gaussian process with spectral density

$$f(\lambda) = |\lambda|^{\theta-1} f_*(\lambda), \quad \lambda \in [-\pi, \pi],$$

where $f_*(\lambda)$ is a bounded and continuous function at zero, $f_*(0) \neq 0$, and $\theta \in (0,1)$ is unknown parameter (dependent index).

Let $I_T(\lambda_j)$, $\lambda_j = 2\pi j/T$, $j = 1, 2, \ldots$ be a periodogram of observation $\zeta(x)$, $x \in \{1, \ldots, T\}$.

Geweke and Porter–Hudak (1993), Igloi (1994) and Robinson (1995a) proposed to choose the ordinary least square estimates for θ associated with the regression scheme

$$Z_j = b + (\theta - 1)d_j + \xi_j, \qquad j = l+1, \ldots m,$$

where $b = -\gamma$ and

$$\gamma = \exp\left\{ - \int_0^\infty e^{-t} \log t \; dt \right\} \simeq 0,5772\ldots$$

is the Euler constant, thus

$$Z_j = \log I_T(\lambda_j), \quad d_j = \log \lambda_j, \quad \xi_j = \log \frac{I_T(\lambda_j)}{|\lambda_j|^{\theta-1}} + \gamma.$$

The least square estimator of parameter θ has the form

$$\hat{\theta}_T = 1 + Z'V(V'V)^{-1},$$

where

$$Z = (Z_{l+1}, \ldots Z_m)', \quad V = (d_{l+1}, \ldots d_m)'.$$

Utilizing Robinson's (1995) result for the one-dimensional case, we obtain that

$$E\,(\hat{\theta}_T - \theta)^2 = O(1/m), \qquad 2\sqrt{m}(\hat{\theta}_T - \theta_0) \xrightarrow{D} \mathcal{N}(0, \pi^2/6) \quad \text{as} \quad l, m \to \infty$$

in such way that, as $T \to \infty$,

$$\frac{l}{m} + \frac{\sqrt{m}\,\log m}{l} + \frac{\log^2 T}{m} + \frac{m^{3/2}}{T} \to 0.$$

Hall, Koul and Turlach (1997) proposed a regression model in the time domain for semi-parametic estimation of dependence index of stationary time series.

Recent results on the semi-parametric method of estimation of dependence index can be found in Robinson (1994a, b, 1995a, b).

Heuristic approaches to this problems (the popular R/S method and some others) have been discussed by Beran (1994).

3. Concluding remarks

The problems under consideration belong to a large and recently aggressively studied area of statistics of long-memory processes and fields which are characterized by certain singular properties (e.g., vanishing, or unboundedness) of their spectral densities. Beran's (1994) book contains a sufficiently complete bibliography of the subject. Recent results can be found in Robinson (1995a, b), Robinson and Hidalgo (1997), Hosoya (1997), Giraitis, Koul and Surgailis (1997), among the others. The results are differ significantly from the statictical theorem of weakly dependent fields. In particular, the sample mean of long-memory time series does not lose much efficiency compared to the best linear unbiased estimator (Adenstedt (1974), Samarov and Taqqu (1988)). A surprising consequence of long-range dependence is that robust estimation of the unknown mean can be done without losing efficiency (Beran (1994)). Further, the asymptotic distribution of normalized correlogram of a Gaussian processes and fields are non-Gaussian (see section 5.2 and Rosenblatt (1979), Leonenko and Portnova (1993)).

In contrast to sample mean, the sample variance is a bad estimator for unknown variance. It has a large bias and loses much efficiency. So, the scale parameter should be estimated in the frequency domain. Robinson and Hidalgo (1997) have considered the time series linear and non-linear regression with long-memory dependence in both errors and stochastic regressors (spectral singularities are permitted at any frequency). The results differ significantly from the classic work on regression (see Holevo (1973, 1976), Rasulov (1976), Grenander (1981)). The development of statistical methods for data with singular spectrum will certainly be a rewarding task for future research.

Comments

Comments to Chapter 1

1.1. Most of the notations and results presented in section 1.1 are contained in the books of Gikhman and Skorokhod (1974), Grenander (1981), Adler (1981), Vanmarke (1984), Kwapien and Woyczynski (1992).

1.2. A detailed exposition of results of the spectral theory of random fields is contained in books by Hannan (1970), Yadrenko (1983), Yaglom (1987), Kwapien and Woyczynski (1992) and the papers by Yaglom (1952, 1957), Ogura (1990) and others. The above mentioned books and papers also outline the history of the problem. We only note that spectral decomposition (1.2.3) and (1.2.6) were considered in a paper of Cramer (1942) for random processes ($n = 1$), the spectral decomposition of homogeneous and isotropic fields (1.2.21) was considered by Yadrenko (1961) and Yaglom (1961). The spectral representations (1.2.32), (1.2.35), (1.2.39), (1.2.41) of isotropic fields on sphere and isotropic fields on n-dimensional Euclidean space were proposed by Yadrenko (1959, 1971). The partial case of spectral decompositions (1.2.32) and (1.2.35) for three-dimensional isotropic fields was introduced by Obukhov (1947) and Jones (1963). An interesting generalization of a topic on a class of harmonizable fields was proposed by Rao (1991), Swift (1994, 1997). Generalized spectral analysis of a class of nonstationary and fractional fields was derived by Anh and Lunney (1992, 1997).

The discretization problem for continuous parameter random processes and fields was considered, from different viewpoints, by Sinai (1976), Grenander (1981), Grenander and Rosenblatt (1984), Rosenblatt (1985), Stein (1995).

The definition and properties of the indicated special functions are presented in a book Bateman and Erdelyi (1953), Müller (1966), Vilenkin (1968). The relation between spectral functions and group representation theory is studied in the book by Vilenkin (1968).

1.3. For the list of references on the probabilistic theory of processes with singular spectrum, we refer the reader to Cox (1984), Taqqu (1985, 1988), Varwaat (1987), Beran (1992, 1994).

For the applications of FARMA processes to economic data, see, for example, Diebold and Rudebush (1989), Lo (1991), Sowell (1992), Backus and Zin (1993), Robinson (1994 c), Hosking (1996). Some properties of fractional Brownian motion are presented in Taqqu (1978), Muraoka (1992), Dai and Heyde (1996), Ciozek-Georges and Mandelbrot (1996).

Another example of random fields with singular spectrum can be found in Dobrushin (1979), Anh and Lunney (1992, 1997) Beran (1994), Samorodnitsky and Taqqu (1994), Barndorff-Nielsen (1998). An introduction to the role of long memory in the context of critical phenomena in physics is given in Cassandro and Jona-Lasinio (1978).

1.4. A detailed exposition of Tauberian and Abelian Theorems is contained in book by Bingham, Goldie and Teugels (1989).

Theorems of Tauberian type for covariance function of strongly dependent random fields with discrete parameter were considered in the works of Dobrushin and Major (1979) and Major (1981). The exposition is based on works Leonenko and Olenko (1991, 1993) and Olenko (1991, 1996). Theorems 1.4.1.–1.4.4. in the case $L(t) \equiv 1$ are presented in Leonenko and Olenko (1991, 1993). The proofs of these theorems with arbitrary slowly varying function L(t) are given by Olenko (1991, 1993).

The proof of Theorems 1.4.1.–1.4.4. is based on the ideas in the works of Laue (1973, 1987). Bingham (1972) presents an alternative variant of Tauberian and Abelian theorems for Hunkel type transform. These problems are discussed also in books of Mirochin (1981) and Vladimirov, Drozhizhnov and Zav'yalov (1986).

Comments to Chapter 2

2.1. The ideas of construction of special classes of random processes with given one-dimensional distributions and given covariance function were proposed by Sarmanov (1961) (see also, Berman (1984)). The exposition of the present section is based on article of Leonenko (1989).

2.2. Reduction conditions for random processes and fields with strong dependence have been considered in Taqqu (1975, 1979), Dobrushin and Major (1979), Berman (1979, 1984), Maejima (1981, 1982, 1985, 1986a, b), Ivanov and Leonenko (1989). The exposition based on the paper of Leonenko (1989) (see also, Ivanov and Leonenko (1989)).

2.3. A more detailed exposition of this topic is presented in books of Major (1980), Engel (1982), Kwapien and Woyczynski (1992), and in articles of Dobrushin (1979), Fox and Taqqu (1987), Sanchez (1993), Houdre, Perez-Abreu and Üstünel (1994), Doukhan and Leon (1996). Multiple stochastic integrals with respect to a Poisson measure and α-stable measures were examined in Surgailis (1984, 1985), Kwapien and Woyczynski (1992), and others.

2.4. Non-central limit theorems were first derived by Rosenblatt (1961), Ibragimov (1963), Taqqu (1975). The main references on this topic the papers of Taqqu (1979), Dobrushin and Major (1979), Rosenblatt (1979, 1981, 1987) (see also, Gorodetskii (1980), Surgailis (1982), Major (1981, 1982)). Non-central limit theorems for nonlinear functionals of linear sequences were investigated by Davydov (1970), Surgailis (1982), Giraitis and Surgailis (1985), Fox and Taqqu (1985), Avram and Taqqu (1987). Other generalization are discussed in Sanchez de Naranjo (1993), and Ivanov and Leonenko (1989).

The exposition based on the papers of Leonenko (1989) and Leonenko and Olenko (1991, 1993).

The central limit Theorems for non-linear functionals of Gaussian random processes and fields with weak dependence (regular spectrum) were investigated by Sun (1963, 1965), Breuer and Major (1983), Maruyama (1985), Giraitis and Surgailis (1985),Leonenko and Rybasov (1986), Ho and Sun (1987), Chambers and Slud (1989), Ivanov and Leonenko (1989), Leonenko and Parkhomenko (1990), Deriev (1993), Arcones (1994).

Comments to Chapter 3

3.1. The functionals of a geometric nature are of major importance in applications. Monographs by Adler (1981) and Wschebor (1985) are devoted to the geometry of random fields. Characteristics of excess above a level for stationary processes were considered in a books of Cramer and Leadbetter (1967), Leadbetter, Lindgren and Rootzen (1983) and Berman (1992). Many generalizations of these results were obtained by Belayev and his students (see a survey of Belayev (1969)) and Nosko (1982 a, b, 1986, 1990, 1994). For $n = 2$ the first two moments for some geometric functionals are derived in Orsingher (1983, 1985). Functionals of geometric type for strongly dependent random fields were investigated in a book Ivanov and Leonenko (1989).

The ideas of proofs of the Theorems 3.1.3.–3.1.5. are presented in a paper of Berman (1979). The exposition follows Leonenko (1987 a) and Leonenko and Sabirov (1988, 1992). Exactness of normal approximation of geometric functionals was investigated in Leonenko (1988 a). The spherical functionals of geometrical nature considered by Leonenko and El-Bassiouny (1986). Wiener–Itô expansions for functionals related to level crossing counts are presented in Slud (1991, 1994).

3.2. The sojourn time problems for strongly dependent Gaussian processes have been considered by Berman (1979, 1984), Maejima (1981, 1982) and for

Gaussian random fields by Leonenko (1984 a, b, 1987 b). The exposition follows Leonenko (1987, 1989).

3.2. The limiting distributions of spherical measures of excess over a moving level for Gaussian field were obtained by Rybasov (1987 a, b) and for χ-squared field by Leonenko and Sabirov (1989).

3.4. The sojourn time problems for strongly dependent vector Gaussian processes have been considered by Berman (1984), Taqqu (1986) and Maejima (1985, 1986 a, b). We present corrected version of a papers of Leonenko (1990 a, b) and Leonenko and Parkhomenko (1992). Other sojourn problems for random fields considered in Leonenko and Parkhomenko (1991, 1992).

3.5. A detailed exposition of results on geometry of random fields is contained in a book by Adler (1981), Wschebor (1985), Ivanov and Leonenko (1989) and also in the works by Belayev (1969), Nosko (1982 a, b, 1986, 1990, 1994).The exposition of the present section is based on article of Leonenko and Parkhomenko (1990).

3.6. The limiting distributions for local times for stationary Gaussian processes with singular spectrum have been obtained by Berman (1982). The exposition of the present section is based on the articles of Sakhno (1990, 1991, 1992), Leonenko and Sakhno (1993). Asymptotics for occupation densities some classes of strongly dependent vector-valued Gaussian random fields are presented in Sakhno (1992) and Doukhan and Leon (1996).

Comments to Chapter 4

4.1. Equations of mathematical physics with random data were considered by Kampe de Feriet (1955), Ratanov (1984), Ratanov, Shuhov and Suhov (1991), Kozachenko and Endzhyrgly (1995, 1996), Leonenko and Woyczynski (1998c) and others. The history of the Burgers' equation is discussed in the books of Burgers (1974), Whitam (1974) Gurbatov, Malachov and Saichev (1991), Holden, Øksendall, Ubøe and Zhang (1996).

4.2. Hopf–Cole solutions of the Burgers' equation are considered in Burgers (1974), Rosenblatt (1987), Bulinski and Molchanov (1992), Albeverio, Molchanov and Surgailis (1994), among others.

4.3. Bulinski and Molchanov (1991) announced limit theorems for the solutions of Burgers' equation with Gaussian weak dependent random initial condition. Albeverio, Molchanov and Surgailis (1993), Surgailis and Woyczynski (1993, 1994 b), Leonenko and Deriev (1994) discuss large time asymptotics of the solutions of the Burgers' equation with different types of weakly dependent Gaussian and non-Gaussian initial conditions. The exposition follows paper Leonenko and Deriev (1994) (for more general results, see, Deriev and Leonenko (1997)).

4.4. The proof of Theorem 4.4.1. is based on the ideas of Dobrushin (1979). Scaling limits of Burgers' equation with Gaussian strongly dependent initial conditions were investigated by Albeverio, Molchanov and Sugrailis (1994), Surgailis and Woyczynski (1994). The exposition follows Leonenko, Orsingher, Rybasov (1994), Leonenko and Orsingher (1995) and Leonenko, Parkhomenko and Woyczynski (1996).

4.5. Giraitis, Molchanov and Surgailis (1993), Funaki, Surgailis and Woyczynski (1995), Surgailis and Woyczynski (1993) studied non-Gaussian limits distributions of rescaled solution of Burgers' equation with short noise (or Gibbs–Cox) random initial conditions with singular spectrum. The exposition follows Leonenko, Orsingher and Rybasov (1994), Leonenko and Orsingher (1995), Leonenko, Orsingher and Parkhomenko(1995), Leonenko and Li Zhanbing (1994), Leonenko, Li Zhanbing and Rybasov (1995).

4.6. The rate of convergence to normal law of non-linear functionals of Gaussian random fields with long-range dependence was considered by Leonenko (1988) (see also, Ivanov and Leonenko (1989), pp. 64–70). The exposition follows Leonenko, Orsingher and Parkhomenko(1996). The paper of Leonenko, Woyczynski (1998 a) provides the rate of convergence (in the uniform Kolmogorov's distance) of probability distributions of the parabolically rescaled solutions of the multidimensional Burgers' equation with random singular Gaussian initial data to a limit Gaussian random field.

4.7. The exposition follows Albeverio, Molchanov and Surgailis (1994) and Molchanov, Surgailis and Woyczynski (1995).

Comments to Chapter 5

5.1. The regression models with singular errors have been considered by Yajima (1988, 1991), Koul (1992), Koul and Muhkerjee (1993), Dahlhaus (1995), Robinson and Hidalgo (1997). Asymptotic expansion of M-estimators of the location parameter obtains by Koul and Surgailis (1997) (see also, Csörgo and Melniczuk (1995) Ho and Hsing (1996)). The exposition of the section is based on articles of Leonenko and Šilasc-Bensic (1996 a, b, 1997, 1998) (see also, Leonenko and Sharapov (1998)). Note that the theory of linear and non-linear regression of random processes and fields with weak dependence is contained in the books of Yadrenko (1983), Ibragimov and

Rozanov (1970), Grenander (1981), Grenander and Rosenbatt (1984), Ivanov and Leonenko (1989).

5.2. The theory of estimation of unknown covariance function of random field with weak dependence is discussed in the book of Ivanov and Leonenko (1989) and the articles of Buldigin and Zayats (1992), Buldigin (1995), Buldigin and Dychovichny (1995) and Buldigin and Demianenko (1997). For the stationary Gaussian sequence with long-range dependence the limit distribution of the correlogram was obtained by Rosenblatt (1979). The paper of Fox and Taqqu (1985, 1987), Terrin and Taqqu (1990, 1991), Avram and Taqqu (1987), Avram (1988), Giraitis and Surgailis (1991), Doukhan and Leon (1991), Ginovian (1994) devoted to non-central limit theorems for quadratic forms of random processes. The exposition follows to Leonenko and Portnova (1993, 1994), Leonenko and Kepich (1994) and Kepich and Leonenko (1994).

5.3. Efficiency problem of regression coefficients of random fields observed in the sphere was considered by Leonenko, Yadrenko and Il'icheva (1996). The asymptotic distributions of sample mean of Gaussian random fields observed on the sphere have been studied by Il'icheva and Leonenko (1996).

5.4. The problem of estimation of unknown parameter of random processes in frequency domain was investigated by Whittle (1951, 1953), Rice (1979), Walker (1964), Ibragimov (1967), Hannan (1973), Dzhaparidze and Yaglom (1983), Dzhaparidze (1986), Dahlhaus (1989), Heyde and Gay (1989, 1993), Guyon (1982, 1985), Anh and Lunney (1995), Dahlhaus and Wefelmeyer (1996), Leonenko, Sikorskii and Terdik (1998), among others. In particular the problem of estimation of a parameter of strong dependence (singular parameter, parameter of self-similarity, Hurst parameter) was considered by Geweke and Porter-Hadak (1983), Fox and Taqqu (1986), Dahlhaus (1989), Giraitis and Surgailis (1990), Agiakoglou, Newbold and Wohar (1993), Beran and Terrin (1994), Beran (1994), Hurvich and Beltrao (1993, 1994), Igloi (1994), Robinson (1994 a, b, 1995 a, b), Hall Koul and Turlach (1997), Hosoya (1997), Giraitis and Koul (1997). Beran's book (1994) contains a pretty complete bibliography of the statistical inference for random processes with singular spectrum.

The exposition follows Leonenko and Woyczynski (1998 b, c, d, 1999).

Bibliography

[1] Adenstedt, R.K. (1974) On large sample estimation for the mean of a stationary random sequences. *Ann. Statist.*, **2**, pp. 1095–1107.

[2] Adler, R.J. (1981) *Geometry of Random Fields.* Wiley, New York.

[3] Agiakloglou, Ch., Newbold, P. and Wohar, M. (1993) Bias in an estimator of the fractional difference parameter. *J. Time. Ser. Anal.*, **14**, N3, pp. 235–246.

[4] Albeverio, S., Molchanov, S.A. and Surgailis, D. (1994) Stratified structure of the Universe and Burgers' equation: A probabilistic approach. *Prob. Theory Rel. Fields.*, **100**, pp. 457–484.

[5] Anh, V.V., Angulo, J.M. and Ruiz-Medina, M.D. (1999) Possible long-range dependence of fractional random fields. *J. Statist. Planning and Inference*, in press

[6] Anh, V.V. and Heyde C.C.(Eds.) (1999) Special Issue on Long-range dependence. *J. Statist. Planning and Inference*, in press

[7] Anh, V.V. and Lunney, K.E. (1992) Generalized harmonic analysis of a class of non-stationary random fields. *J. Austral. Math. Soc.*, (Ser. A), **53**, N3, pp. 385–401.

[8] Anh, V.V. and Lunney, K.E. (1995) Parameter estimation of random fields with long-range dependence. *Math. Comput. Modelling,* **21**, pp. 66–77.

[9] Anh, V.V. and Lunney, K.E. (1997) Generalized spectral analysis of fractional random fields. *J. Austral. Math. Soc.*, (Ser. A), **62**, pp. 64–83.

[10] Arcones, M.A. (1994) Limit theorems for non-linear functionals of a stationary Gaussian sequence of vectors. *Ann. Probab.*, **24**, N4, pp. 2242–2274.

[11] Avellaneda, M. (1995) Statistical properties of shocks in Burgers turbulence, II :Tail probabilities for velocities,chock-strengths and rare faction intervals. *Comm. Math. Phys.*, **169**, pp.45–59.

[12] Avellaneda, M. and E W. (1995) Statistical properties of shocks in Burgers turbulence. *Comm. Math. Phys.*, **172**, pp. 13–38.

[13] Avellaneda, M. and Majda, A. (1990) Mathematical models with exact renormalization for turbulence transport. *Comm. Math. Phys.*, **131**, pp. 381–429.

[14] Avellaneda, M. and Majda, A. (1991) An integral representation and bounds on the effective diffusivity in passive advection by laminar and turbulence flows. *Comm. Math. Phys.*, **138**, pp. 339–391.

[15] Avellaneda, M. and Majda, A. (1992a) Mathematical models with exact renormalization for turbulent transport, II: fractal interfaces, non-Gaussian statistics and the sweeping effect. *Comm. Math. Phys.*, **146**, pp. 139–204.

[16] Avellaneda, M. and Majda, A. (1992b) Approximation and exact renormalization theories for a model for turbulent transport. *Phys. Fluids*, **A4(1)**, pp. 41–52.

[17] Avram, F. and Taqqu, M.S. (1987) Noncentral limit theorems and Appel polynoms. *Ann. Probab.*, **15**, pp. 767–775.

[18] Backus, D.K. and Zin, S.E. (1993) Long-memory inflation uncertainty: evidence from the term structure of interest rates. *J. Money, Credit and Banking*, **25**, 3, pp. 681–708.

[19] Barndorff-Nielsen, O.E.(1998) Processes of normal inverse Gaussian type. *Finance Stochast.*, **2**, pp.41-68.

[20] Barrasa de la Krus, E. and Kozachenko, Yu.V. (1995) Boundary-value problems for equations of mathematical physics with stricly Orlicz random initial conditions. *Random Oper. Stoch. Equ.*, **3**, pp. 201–220.

[21] Bateman, H. and Erdelyi, A. (1953) *Higher Transcendental Functions.* Vol. II, McGraw-Hill, New York.

[22] Bateman, H. and Erdelyi, A. (1964) *Tables of the Integral Transforms. Transforms of Fourier, Laplace, Mellin.* Vol. I, Nauka, Moscow. (Russian).

[23] Belayev, Yu.K. (1969) New results and generalization of the problem of crossing level, Addition to a Russian Edition of the book Cramer and Leadbetter (1969) *Stationary Random Processes*, Mir, Moscow, pp. 34–378. (Russian).

[24] Beran, J. (1992) Statistical methods for data with long-range dependence. *Statistical Science*, **7**, N4, pp. 404–427.

[25] Beran, J. (1994) *Statistics for Long-Memory Processes.* Chapman and Hall, New York.

[26] Beran, J. and Terrin, N. (1994) Estimation of the long-memory parameter, based on a multivariate central limit theorems. *J. Time. Ser. Anal.*, **15**, N3, pp. 269–278.

[27] Berman, S.M. (1979) High level sojourns for strongly dependent Gaussian processes. *Z. Wahrsch. verw. Gebiete*, **50**, pp. 223–236.

[28] Berman, S.M. (1982) Local times of stochastic processes with positive definite bivariate densities. *Stoch. Proc. Appl.*, **12**, pp. 1–26.

[29] Berman, S.M. (1984) Sojourns of vector Gaussian processes inside and outside spheres. *Z. Wahrsch. verw. Gebiete*, **66**, pp. 529–542.

[30] Berman, S.M. (1992) *Sojourns and Extremes of Stochastic Processes.* Wadsworth and Brooks, California.

[31] Bertini, L., Carcrini, N. and Jona-Lasinio, G. (1994) The stochastic Burgers equation. *Comm. Math. Phys.*, **165**, pp. 211–232.

[32] Bertoin, J. (1998a) The inviscid Burgers equation with Brownian initial velocity. *Comm. Math. Phys.*, **179**.

[33] Bertoin, J. (1998b) Large-deviation estimates in Burgers turbulence with stable noise initial data. *J.Stat.Phys.*, **91**, pp.655–667.

[34] Biler, P., Funaki, T. and Woyczynsky, W.A. (1998) Fractal Burgers equation, to appear

[35] Bingham, N.H. (1972) A Tauberian theorem for integral transforms of Hankel type. *J. London. Math. Soc.*, **5**, N3, pp. 493–503.

[36] Bingham, N.H., Goldie, C.M. and Teugels, T.L. (1987) Regular Variation. Cambridge University Press, Cambridge.

[37] Bolthausen, E. (1982) On the C.L.T. for stationary mixing random fields. *Ann. Probab.*, **10**, pp. 1047–1050.

[38] Borell, C. (1986) Tail probabilities in Gauss space. In: *Vector Space Measures and Applications, Lecture Notes in Mathematics*, **644**, Springer-Verlag, New-York, pp. 73–82.

[39] Bossy, M. and Talay, D. (1996) Convergence rate for the approximation of the limit law of weakly interacting particles: application to the Burgers equation. *Ann. Appl. Probab.*, **6**, pp. 818–861.

[40] Bossy, M. and Talay, D. (1997) A stochastic particle method for McKean-Vlasov and the Burgers equation. *Mathematics and Computation*, **66**, N217, pp. 157–192.

[41] Bradley, R. (1985) On the central limit question under absolute regularity. *Ann. Probab*, **13**, pp. 1314–1325.

[42] Bradley, R. (1992) On the spectral density and asymptotic normality of weakly dependent random fields. *J. Theor.Probab.*, **5**, pp. 355–373.

[43] Breuer, P. and Major, P. (1983) Central limit theorems for nonlinear functionals of Gaussian fields. *J. Multiv. Anal.*, **13**, pp. 425–441.

[44] Brockwell, P.J. and Davis, R.A. (1987) *Time Series: Theory and Methods.* Springer, New York.

[45] Buldigin, V.V. (1995) On properties of empirical periodogram of Gaussian process with square integrable spectral density. *Ukrainian Math. J.*, **47**, pp. 876–889.

[46] Buldigin, V.V. and Demianenko, O.O. (1997) Asymptotic properties of empirical correlogram of Gaussian vector fields. *Dopovidi of National Academy of Sciences of Ukraine*, **2**, pp. 45–49.

[47] Buldigin, V.V. and Dychovichny, A.A. (1995) On some singular integrals and its application in the problem of statistical estimation. *Theor. Probab. Math. Statist.*, **53**, pp. 17–28. (Ukrainian).

[48] Buldigin, V.V. and Zayats, V. (1992) Asymptotic normality of an estimate of the correlation function in different functional space. In: *Probab. Theory and Math. Statist. World Scientific.*, pp. 19–31.

[49] Bulinskii, A.V. (1992) Central limit theorem for the solution of the multidimensional Burgers equation with random data. *Ann. Acad. Scient. Fennicae, Ser. A. I. Mathematica*, **17**, pp. 11–22.

[50] Bulinskii, A.V. and Molchanov, S.A. (1991) Asymptotic Gaussianess of solutions of the Burgers' equation with random initial data. *Theor. Prob. Appl.*, **36**, pp. 217–235.

[51] Burgers J. (1974) *The Nonlinear Diffusion Equation.* Kluwer, Dordrecht.

[52] Carlin, J.B., Dempster, P. and Jones, A.B. (1985) On methods and models for Bayesian time series analysis. *J. Econometrics*, **30**, pp. 67–90.

[53] Carmona, R.A. and Lacroix, J. (1990) *Spectral theory of random Schrödinger operators.* Birkhaüser, Boston.

[54] Carmona, R.A. and Molchanov, S.A. (1991) Stationary parabolic Anderson model and intermittences. *Probab. Theory Relat. Fields*, **102**, pp. 433–453.

[55] Cassandro, M. and Jona-Lasinio, G. (1978) Critical behaviour and probability theory. *Physics*, **27**, pp. 913–941.

[56] Chambers, D. and Slud, E. (1989) Central limit theorems for nonlinear functionals of stationary Gaussian processes. *Probab. Theory and Related Fields*, **80**, pp. 323–346.

[57] Chambers, M.J. (1996) The estimation of continuous parameter long-memory time series models. *Econometric Theory.*, **12**, pp. 374–390.

[58] Chang, C.F. (1996a) A generalized fractional integrated autoregressive moving-average process. *J. Time Series Anal.*, **17**, pp. 111–140.

[59] Chang, C.F. (1996b) Estimating a generalized long memory process. *J. of Economics*, **73**, pp. 237–259.

[60] Chorin, A.J. (1975) *Lecture Notes in Turbulence Theory.* Publish, or Perish. Berkeley, CA.

[61] Ciozek-Georges, R. and Mandelbrot, B.B. (1996) Alternative micropulses and fractional Brownian motion. *Stoch. Proc. Appl.*, **64**, pp. 143–152.

[62] Comte, F. (1996) Simulation and estimation of long memory continuous time models. *J. Time Series Anal.*, **17**, pp. 19–36.

[63] Comte, F. and Renault, E. (1996a) Long memory continuous time models. *Journal Econometrics*, **73**, pp. 19–36, 101–149.

[64] Comte, F. and Renault, E. (1996b) Noncausiality in continuous time models. *Econometric Theory*, **12**, pp. 215–256.

[65] Cox, D.R. (1984) Long-range dependence: a review. In: *Statistics: an appraisal* (H.A. David and H.T. David, eds.), Iowa State Univ., Ames, pp. 55–74.

[66] Cox, D.R. and Townsend, M.W.H. (1948) The use of the correlation in measuring yarn irregularity. *Proc. Roy. Soc. Edinburgh*, Sect. **A**, **63**, pp. 290–311.

[67] Cramer, H. (1942) On harmonic analysis of certain functional space. *Ark. Mat. Astron. Fys.*, **28B**, 7p

[68] Cramer, H. and Leadbetter, M.R. (1967) *Stationary and Related Stochastic Processes.* Wiley. New York.

[69] Csörgo, S. and Mielniczuk, J. (1995) Distant long-range dependent sums and regression estimation. *Stoch. Proc. Appl.*, **59**, pp. 143–155.

[70] Dahlhaus, R. (1989) Efficient parameter estimation for self-similar processes. *Ann. Statist.*, **17**, pp. 1749–1766.

[71] Dahlhaus, R. (1995) Efficient location and regression estimation for long-range dependent regression models. *Ann. Statist.*, **23**, pp. 1029–1047.

[72] Dahlhaus, R. and Wefelmeyer, W. (1996) Asymptotic optimal estimation in misspecified time series models. *Ann. Statist.*, **24**, pp. 952–974.

[73] Dai, W. and Heyde, C.C. (1996) Itô's formula with respect to fractional Brownian motion and its application. *J. Appl. Math. Stoch. Anal.*, **9**, pp. 439–448.

[74] Daley, D.J. and Vesilo, R. (1997) Long range dependence of point processes with queueing examples. *Stoch.Proc.Applic.* **70**, pp. 265–282.

[75] Damerou, F.J. and Mandelbrot, B.B. (1973) Tests of the degree of word clustering in samples of written English. *Linguistics*, **102**, pp. 95–102.

[76] Davydov, Yu. A. (1970) The invariance principle for stationary processes. *Theor. Prob. Appl.*, **15**, pp. 487–498.

[77] Deriev, I.I. (1993) Asymptotic normality for spherical averages of non-linear functionals of Gaussian random fields. *Ukr. Math. J.*, **45**, pp. 472–480. (Russian).

[78] Deriev, I.I. and Leonenko, N.N. (1997) Limit Gaussian behaviour of the solutions of the multidimensional Burgers' equation with weak-dependent initial conditions. *Acta Appl. Math.*, **47**, pp. 1–18.

[79] Diebold, F.X. and Rudebush, G.D. (1989) Long memory and persistence in aggregate output. *J. of Monetary Economics*, **24**, pp. 189–209.

[80] Dix, D.B. (1996) Nonuniqueness and uniqueness in the initial-value problem for Burgers' equation. *SIAM J. Math. Anal.*, **27**, pp. 708–724.

[81] Dobrushin, R.L. (1979) Gaussian and their subordinated self-similar random generalized fields. *Ann. Probab.*, **7**, pp. 1–28.

[82] Dobrushin, R.L. and Major, P. (1979) Non-central limit theorems for non-linear functionals of Gaussian fields. *Z. Wahrsch. verw. Gebiete*, **50**, N1, pp. 1–28, 27–52.

[83] Donoghue, W. J. (1969) *Distribution and Fourier Transform.* Academic Press,New York.

[84] Doukhan, P. (1994) *Mixing.* Springer-Verlag, New York.

[85] Doukhan, P. and León, J.R. (1991) Estimation du spectra d'une processes Gaussein forment dependant. *C.R. Acad. Sci.*, Série I, **313**, 523–526.

[86] Doukhan, P. and León, J.R. (1996) Asymptotics for local time of a strongly dependent vector-valued Gaussian random field. *Acta. Math. Hung.*, **70(4)**, pp. 329–351.

[87] Dzhaparidze, K.O. (1986) *Parameter Estimation and Hypothesis Testing in Spectral Analysis of Stationary Time Series.* Springer, Berlin.

[88] Dzhaparidze, K.O. and Yaglom, A.M. (1983) Spectrum parameter estimation in time series. In: *Developments in Statistics* (P.R. Krishaiah. ed) Vol. **4**, Chapter 1, pp. 1–96, Academic Press, New York.

[89] Engel, D.D. (1982) The Multiple Stochastic Integrals. *Mem. Amer. Math. Soc.*, **38**.

[90] Fan, H. (1995) Large time behaviour of elementary waves of Burgers equation under white noise perturbation. *Comm. Partial Diff. Equations*, **20**, pp. 1699–1723.

[91] Fannjiang, A., Papanicolau, G. (1996) Diffusion in turbulence. *Probab. Theory Relat. Fields*, **105**, 279–334.

[92] Farge, M., Kavlahan. N.K.-R., Perrier, V. and Schneider, K. (1997) Turbulence analysis, modelling and computing using wavelets. In: *Wavelets and Physics (*Hans and den Berg, eds.*)*. Cambridge University Press.

[93] Fox, R. and Taqqu, M.S. (1985) Noncentral limit theorems for quadratic forms in random variables having long-range dependence. *Ann. Probab.*, **70**, pp. 191–212.

[94] Fox, R. and Taqqu, M.S. (1986) Large sample properties of parameter estimates for strongly dependent Gaussian stationary time series. *Ann. Statist.*, **14**, pp. 517–532.

[95] Fox, R. and Taqqu, M.S. (1987a) Central limit theorems for quadratic forms in random variables having long-range dependence. *Probab. Theory and Relat. Fields*, **74**, pp. 213–240.

[96] Fox, R. and Taqqu, M.S. (1987b) Multiple stochastic integrals with dependent integrators. *J. Multivar. Analysis.*, **21**, pp. 105–127.

[97] Funaki, T., Surgailis, D. and Woyczynski, W.A. (1995) Gibbs–Cox random fields and Burgers turbulence. *Ann. Appl. Probab.*, **5**, pp. 461–492.

[98] Gay, R. and Heyde, C.C. (1990) On class of random field models which allows long range dependence. *Biometrica*, **77**, pp. 401–403.

[99] Geman, D., Horowitz, T. (1980) Occupation densities. *Ann. Probab.*, **8**, pp. 1–67.

[100] Geweke, T. and Porter-Hudak (1983) The estimation and application of long memory time series model. *J. Time Series Anal.*, **4**, N4, pp. 221–238.

[101] Gikhman, I.I. and Skorohod, A.V. (1974) *Theory of Random Processes.* Vol. **1**, Nauka, Moscow. (Russian). English transl.: Springer-Verlag.

[102] Ginovian, M.S. (1994) On Toeplitz type quadratic functionals of stationary Gaussian processes. *Probab. Theory Relat. Fields*, **100**, pp. 395–406.

[103] Giraitis, L. and Koul, H. (1997) Estimation of the dependence parameter in linear regression with long-range-dependente errors. *Stoch. Proc.Applic.* 71, pp. 207–224.

[104] Giraitis, L., Molchanov, S.A. and Surgailis, D. (1993) Long memory shot noises and limit theorems with applications to Burgers equation. In: *New Directions in Time Series Analysis,* Part II (D. Brillinger et al., eds.), pp. 153–176, Springer, Berlin.

[105] Giraitis, L. and Surgailis, D. (1985) Central limit theorems and other limit theorems for functionals of Gaussian process. *Z. Wahrsch. verw. Gebiete*, **70**, pp. 191–212.

[106] Giraitis, L. and Surgailis, D. (1991) A central limit theorem for quadratic forms in strongly dependent linear variables and its application to asymptotic normality of Whittle's estimates. *Prob. Theory Rel. Fields*, **86**, pp. 87–104.

[107] Gorodetskii, V.V. (1977) On convergence to semistable Gaussian processes. *Theor. Probab. Appl.*, **22**, pp. 513–522.

[108] Gorodetskii, V.V. (1980) Invariant principle for functions of jointly stationary Gaussian variables. *Zap. Nauchn. Sem. LOMI*, **50**, pp. 32–44. (Russian).

[109] Gradstein, I.S. and Ryzhin, I.M. (1962) *Tables of Integrals, Sums, Series and Products.* Fizmatgiz, Moscow. (Russian).

[110] Graf, H.P. Hampel, F.R. and Tacier, J. (1984) The problem of unsuspected serial correlations. *Robust and Nonlinear Time Series Analysis, Lecture Notes in Statist.*, **26**, Springer, New York.

[111] Granger, C.W.J. (1966) The typical spectral shape of an economic variable. *Econometrica*, **34**, pp. 150–161.

[112] Granger, C.W.J. (1980) Long memory relationship and the aggregation of dynamic models. *J. Econometrics*, **14**, pp. 227–238.

[113] Granger, C.W.J. (1981) Some properties of time series data and their use in econometric model specification. *J. of Econometrics*, **16**, pp. 121–130.

[114] Granger, C.W.J. and Jeyeux (1980) An introduction to long-range time series models and fractional differencing. *J. Time Ser. Anal.*, **1**, pp. 15–30.

[115] Gray, H.L., Zhang, N. and Woodward, W.A. (1989) On generalized fractional processes. *J. Time Ser. Anal.*, **10**, pp. 233–257.

[116] Grenander, U. (1981) *Abstract Inference.* Wiley, New York.

[117] Grenander, U. and Rosenblatt, M. (1984) Statistical Analysis of Stationary Time Series. Chelsea Publ. Company, New York.

[118] Gripenberg, G. and Norros, I. (1996) On the prediction of fractional Brownian motion. *J. Appl. Probab.*, **33**, pp. 400–410.

[119] Gurbatov, S.N., Malakhov, A. and Saichev, A. (1991) *Non-linear Waves and Turbulence in Nondispersive Media: Waves, Rays and Particles.* Manchester University Press, Manchester.

[120] Gurbatov, S.N., Simdyankin, S.I., Aurell, E., Frisch, U. and Toth, G. (1998) On the decay of Burgers turbulence. *F. Fluid Mech.*. (To appear).

[121] Guyon, X. (1982) Parameter estimation for a stationary process on a d-dimentional lattice. *Biometrica*, **69**, pp. 95–105.

[122] Guyon, X. (1995) *Random Fields on a Network. Modeling, Statistics and Applications.*, Springer, Berlin.

[123] Haensler, E. (1984) An exact rate of convergence in the functional central limit theorem for special martingale difference array. *Z. Wahrsch. verw. Gebiete*, **65**, pp. 523–534.

[124] Hall, P., Koul, H.L. and Turlach, A. (1997) Note on convergence rates of semiparametric estimators of dependence index. *Ann. Statist.*, **25**, pp. 1725–1739.

[125] Handa, K. (1995) On a stochastic PDE related to Burgers' equation with noise. In: *Non-linear stochastic PDE's*; (T. Funaki and W.A. Woyczyndki, eds.). The IMA volumes in *Mathematics and its Applications*, vol. **77**, Springer, New-York, pp. 147–156.

[126] Hannan, E. (1965) Group representation and applied probability. *J. Appl. Prob.*, **2**, pp. 1–68.

[127] Hannan, E. (1970) *Multiple Time Series.* Wiley, New York.

[128] Hannan, E. (1973) The asymptotic theory of linear time series models. *J. Appl. Prob.*, **10**, pp. 130–145.

[129] Hassler, E.J. (1994) Misspecification of long memory in seasonal time series. *J. Time Ser. Anal.*, **15**, pp. 19–30.

[130] Heyde, C.C. and Gay, R. (1989) On asymptotic quasi-likelihood, *Stoch. Process Appl.*, **31**, pp. 223–236.

[131] Heyde, C.C. and Gay, R. (1992) Thoughts on the modelling and identification of random processes and fields subject to possible long-range dependdence. In: *Probability Theory*, de Gruyter and Co., Berlin, pp. 75–81.

[132] Heyde, C.C. and Gay, R. (1993) Smoothed periodogram asymptotics and estimation for processes and fields with possible long-range dependence. *Stoch. Process. Appl.*, **45**, pp. 169–182.

[133] Ho, H.-C. and Sun, T.–C. (1987) A central limit theorem for non-instantaneous fielters of a stationary Gaussian process. *J. Multiv. Anal.*, **22**, pp. 144–155.

[134] Ho, H.-C. and Sun, T.–C. (1990) Limit distributions of non-linear vector functions of stationary Gaussian processes. *Ann. Probab.*, **18**, pp. 1159–1173.

[135] Hodges, S. and Carverhill, A. (1993) Quasi-mean reversion in an efficient stock market; the characterization of economic equilibria which support Black–Schols option pricing. *Econ. J.*, **102**, pp. 395–405.

[136] Holden, M., Øksendall, B., Ubøe, J. and Zhang, T.–S. (1996) *Stochastic Partial Differential Equations. A Modeling, White Noise Functional Approach.* Birkhaüser, Boston.

[137] Holevo, A.S. (1973) On the general problem of mean estimation. *J. Multiv. Anal.*, **3**, pp. 262–275.

[138] Holevo, A.S. (1976) On the asymptotic efficient regression estimates in the case of degenerate spectrum. *Theor. Probab. Appl.*, **21**, pp. 324–333.

[139] Holst, L., Quine, M.P. and Robinson, J. (1995) A general stochastic model for nucleation and linear growth. *Ann. Appl. Probab.*, **23**.

[140] Hosking, J.R.M. (1981) Fractional differencing. *Biometrica*, **68**, pp. 165–176.

[141] Hosking, J.R.M. (1996) Asymptotic distributions of the sample mean, autocovariance and autocorrelations of long-memory time series. *J. of Econometrics*, **73**, pp. 261–284.

[142] Hosoya, Y. (1997) A limit theory for long-range dependence and statistical inference on related models. *Ann. Statist.*, **25**, pp. 105–137.

[143] Houdre, C., Pérez-Abreu V. and Üstünel, A.S. (1994) Multiple Wiener-Itô integrals: an introductional survey. In: *Chaos Expansions, Multiple Wiener-Itô Integrals and Their Applications (Houdre, C. and P'erez-Abreu, V., eds.), CRS Press, Boca Raton*, pp. 1–27.

[144] Hu, Y. and Woyczynski, W.A. (1994) An extremal rearrangement properties of statistical solutions of Burgers' equation. *Ann. Appl. Probl.*, **4**, pp. 838–858.

[145] Hu, Y. and Woyczynski, W.A. (1995a) Shock density in Burgers turbulence, in *Non-linear stochastic PDE's*; *Hydrodynamic Limit and Burgers Turbulence* (T. Funaki and W.A. Woyczyndki, eds.) IMA, Volume **77**, pp. 167–192, Springer-Verlag, Berlin.

[146] Hu, Y. and Woyczynski, W.A. (1995b) Limit behaviour of quadratic forms of moving averages and statistical solutions of the Brugers equation. *J. Multiv. Anal.*, **52**, pp. 15–44.

[147] Huang, S. and Cambanis, S. (1978) Stochastic and multiple Wienet integrals for Gaussian processes. *Ann. Probab.*, **6**, pp. 585–614.

[148] Hurst H.E. (1951) Long-term storage capacity of reservoirs. *Transactions of the American Society of Civil Engineerings*, **116**, pp. 770–779.

[149] Hurvich, C.M. and Beltrao, K.I. (1993) Asymptotics for the low-frequency ordinates of the periodogram of a long-memory time series. *J. of Time Series Anal.*, vol. **14**, No. 5, pp. 455–472.

[150] Hurvich, C.M. and Ray, B.K. (1995) Estimation of the memory parameter for nonstationary or noninvertible fractionally integrated processes. *J. of Time Series Anal.*, vol. **16**, N1, pp. 17–42.

[151] Huslett, J. and Raftery, A.E. (1989) Space-time modelling with long memory dependence: Assessing Ireland's wind power resource (with discussion). *J. Roy. Statist. Soc.,* Ser. **C**, **38**, pp. 1–21.

[152] Ibragimov, I.A. (1963) On estimation of the spectral function of stationary Gaussian process. *Theor. Probab. Appl.*, **8**, pp. 391–430. (Russian).

[153] Ibragimov, I.A. (1967) On maximum likelihood estimation of parameters of the spectral density of stationary time series. *Theor. Probab. Applic.*, **12**, pp. 115–119.

[154] Ibragimov, I.A. and Rozanov, Yu.A. (1970) *Gaussian Random Processes.* Nauka, Moscow. (Russian).

[155] Iglói, E. (1994) On periodogram based least squares estimation of the long memory parameter of FARMA process. *Publ. Math., Debrecen.*, **44**, pp. 367–380.

[156] Iglói, E. and Terdik, G. (1997) Bilinear stochastic systems with long range dependence in continuous time. In: *Stoch. Diff. and Diff. Equ.* (Csiszar and Michaletsky, eds.). *Progress in Systems and Control Theory*, vol. **23**, Birkhaüser, Boston, pp. 299–309.

[157] Il'icheva, L.M. an d Leonenko, N.N. (1996) Asymptotic properties of sample mean of some class of a non-Gaussian random fields. *Theor. Probab. Math. Stat.*, **54**, pp. 38–46.

[158] Imkeller, P. (1994) On exact tails for limiting distributions of U-statistics in the second Gaussian chaos. In: *Chaos Expansions, Multiple Wiener-Itô Integrals and Their Applications* (Houdre, C. and P´ erez-Abreu, V., eds.), CRS Press, Boca Raton, pp. 179–193.

[159] Inoue, A. (1993) On the equations of stationary processes with divergent diffusion coefficients. *J. Fac. Sci. Univ. Tokyo*, Sect. **IA**, Math. **40**, pp. 307–336.

[160] Itô, K. (1951) Multiple Wiener integral. *J. Math. Soc. Japan.*, **3**, pp. 157–164.

[161] Itô, K. (1953) Complex multiple Wiener integral. *Japan J. Math.*, **22**, pp. 63–86.

[162] Ivanov, A.V. and Leonenko, N.N. (1989) *Statistical Analysis of Random Fields.* Kluwer, Dordrecht.

[163] Janicki, A.W. ans Woyczynsky, W.A. (1997) Hanudorff dimension of regular points in stochastic Burgers flows with Lévy α-stable initial data. *J. Stat. Phys.*, **86**, N1/2, pp. 277–299.

[164] Jeffreys, H. (1939) *Theory of Probability.* Clarendon Press, Oxford.

[165] Jones, H. (1963) Stochastic processes on the sphere. *Ann. Math. Statist.*, **34**, pp. 213–218.

[166] Kampe de Feriet, J. (1955) Random solutions of the partial differential equations. In:*Proc. 3rd Berkeley Sump. Math. Stat. Probab.*, vol. III, University of California Press, Barkeley, Ca., pp. 199–208.

[167] Kardar, M., Parisi, G. and Zhang, Y.C. (1986) Dynamical scaling of growing interfaces, *Phys. Rev. Lett.*, **56**, pp. 889–892.

[168] Karhunen, K. (1947) Über linear Methoden in der Wahrscheinlichkeitsrechung. *Ann. Acad. Scient. Fennicae*, **37**, pp. 1–79.

[169] Kendall, M.G. and Moran, P.A.P. (1963) Geometrical probability. *Statistical Monographys and Courses*, N**5**, Griffin, London.

[170] Kelbert, M.Ya. and Suhov, Yu.M. (1995) The branching random work and systems of reaction–diffusion (Kolmogorov–Petrovskii–Piskunov) equations. *Comm. Math. Phys.*, **167**, pp. 607–634.

[171] Kelbert, M.Ya. and Sazonov, I.A. (1996) Pulses and other Wave Processes. In: *Fluids,* Kluwer Academic Publishers, Dordrecht, Boston, London.

[172] Kepich, A.Yu. and Leonenko, N.N. (1994) On the limit distribution of correlogram of random field with long-range dependence and unknown mean. *Random. Oper. and Stoch. Equ.*, **2**, N3, pp. 203–210.

[173] Kifer, Yu. (1997) The Burgers equation with a random force and a general model for directed polymers in random enviroments. *Probab. Theory Relat. Fields* **108**, pp. 29–65.

[174] Kiivery, H.T. (1992) Fitting spatial correlation models: approximating a likelihood approximation. *Austral. J. Statist.*, **34(3)**, pp. 497–512.

[175] Klyatskin, V.I., Woyczynski, W.A. and Gurarie, D. (1996) Diffusing passive trancers in random incompressible flows: statistical tomography aspects. *J. Stat. Phys.*, **84**, pp. 797–836.

[176] Kofman, L., Pogosyan, D., Shandarin, S. and Melott, A (1992) Coherent structures in the Universe and the adhesion model. *Astrophysical J.*, **393**, pp. 437–449.

[177] Kolmogorov, A.N. (1940) The Wiener helix and other interesting curves in Hilbert space. *Doklady Acad. of Sciences of USSR*, **26**, pp. 115–118. (Russian).

[178] Kolmogorov, A.N. (1941) Local structure of turbulence in fluid for very large Reynolds numbers. Transl. in *Turbulence* (S.K. Freidlander and L.Topper, eds.), Interscience Publishers, New York, pp. 151–155.

[179] Kotelenez, P. (1995a) A class of quasilinear stochastic partial differential equation of McKean-Vlasov type with mass conservation. *Prob. Theory Rel. Fields.*, **102**, pp. 159–188.

[180] Kotelenez, P. (1995b) A stochastic Navier-Stokes equation for the vorticity of a two-dimensional fluid. *Ann. Appl. Probab.*, **5**, pp. 1126–1160.

[181] Koul, H.L. (1992) M-estimators in linear models with long range dependent errors. *Statist. and Probab. Lett.*, **14**, pp. 153–164.

[182] Koul, H.L. and Hsing, T. (1996) On the asymptotic expansion of the empirical process of long memory moving avareges. *Ann. Statist.*, **24**, pp. 992–1024.

[183] Koul, H.L. and Mukherjee, K. (1993) Asymptotics of R-, MD- and LAD-estomators in linear regression models with long-range dependent errors. *Probab. Theory and Related Fields*, **95**, pp. 185–201.

[184] Koul, H.L. and Surgailis, D. (1997) Asymptotic expansion of M-estimators with long memory errors. *Ann. Statist.*, **25**, pp. 818–850.

[185] Kozachenko, Yu.V. and Endzhyrgly, M.V. (1995, 1996) Justification of application of the Fourier method to boundary-value problems subject to random initial conditions. I,II. *Theory Probab. and Math. Statist.*, **51**, pp. 53, 65–75, 81–92.

[186] Kuksin, S.B. (1996) Growth and oscillation of solution of non-linear Schrödinger equation. *Comm. Math. Phys.*, **178**, pp. 265–280.

[187] Künsch, H.R. (1986) Discrimination between monotonic trends and long-range dependence. *J. Appl. Probab.*, **23**, pp. 1025–1030.

[188] Künsch, H.R., Beran, J., Hampel, F. (1993) Contrasts under long-range correlations. *Ann. Stat.*, **21**, N2, pp. 943–964.

[189] Kwapien S., and Woyczynski, W.A. (1992) *Random Series and Stochastic Integrals: Single and Multiple.* Birkhaüser, Boston.

[190] Lamperti, J.W. (1962) Semi-stable stochasic processes. *Trans. Amer. Math. Soc.*, **104**, pp. 62–78.

[191] Laue, G. (1973) Zur theorie der charakteristischen Funktionen nichnegativer zufallsgroben. *Trans. 6th Prague Conf. Inform. Thepry*, Prague, pp. 529–546.

[192] Laue, G. (1987) Tauberian and Abelian theorems for characteristic functions. *Teor. Veroyatn. Mat. Stat.*, **37**, pp. 78–92. (Russian). English transl. in *Theory Probab. Math. Statist.*, **37**.

[193] Lawrence, A.J. and Kategoda, N.T. (1977) Stochastic modelling of riverflow time series (with discussion). *J. Roy. Statist. Soc.*, Ser. **A**, **140**, pp. 1–47.

[194] Leadbetter, M.R., Lindgren, G. and Rootzen, H. (1983) *Extremal and Related Properties of Random Sequences and Processes.* Springer-Verlag, Berlin.

[195] Leonenko, N.N. (1984a) On measures of the excess over a level for a Gaussian isotropic random fields. *Theory Probab. Math. Stat.*, **31**, pp. 64–82.

[196] Leonenko, N.N. (1984b) Reduction conditions for measures of the excess above moving level for strongly dependent homogeneous isotropic Gaussian fields. *Theory Probab. Appl.*, **29**, pp. 611–612.

[197] Leonenko, N.N. (1987a) Limit theorems for characterisrics of the excess above a level for a Gaussian random field. *Math. Notes*, **41**, pp. 608–618. (Russian).

[198] Leonenko, N.N. (1987b) Limit theorems for functionals of geometric type of homogeneous isotropic random fields. In: *Theory Probab. Math. Stat.* (V. Sazonov, Yu. Prohorov, eds.), vol. **2**, VNU SP, Netherlands, pp. 173–202.

[199] Leonenko, N.N. (1988a) On the exactness of normal approximation for a functionals of Gaussian random fields. *Math. Notes*, **43**, pp. 283–299. (Russian).

[200] Leonenko, N.N. (1988b) Sharpness of normal approximation of functionals of strongly correlated Gaussian random fields. *Math. Notes.*, **43**, pp. 283–298.

[201] Leonenko, N.N. (1989) Reduction conditions for functional of geometric type of homogeneous isotropic random fields of Gamma-correlation. I,II. *Ukrain. Math. J.*, **41**, N1, pp. 49–55, N3, pp. 335–340.

[202] Leonenko, N.N. (1990a) Sojourns of multidimensional Gaussian random fields with dependent components. *Comput. Math. Appl.*, **19**, pp. 101–108.

[203] Leonenko, N.N. (1990b) Limit theorems for measures of sojourn in domains of Gaussian vector fields. *Ukr. Math. J.*, **42**, pp. 625–634. (Russian).

[204] Leonenko, N.N. and Deriev, I.I. (1994) Limit theorems for solutions of multidimentional Burgers equation with weak dependent random initial conditions. *Theor. Prob. and Math. Statist.*, **51**, pp. 103–115.

[205] Leonenko, N.N. and El-Bassiony, A.H. (1986) Limit theorems for some characteristics of excess above level for a Gaussian field with strong dependence. *Theor. Probab. Math. Statist.*, **35**, pp. 54–60. (Russian).

[206] Leonenko, N.N. and Kepich, A.Yu. (1994) The non-Gaussian limit distribution of the correlogram of a random field. *Probab. Theory and Math. Statist.*, **51**, pp. 117–128.

[207] Leonenko, N.N. and Li Zhanbing (1994) Non-Gaussian limit distributions of solutions of the Burgers' equation with strongly dependent random initial conditions. *Random Oper. Stoch. Equ.*, **2**, pp. 95–102.

[208] Leonenko, N.N., Li Zhanbing and Rybasov, K.V. (1995) Non-Gaussian limitting distributions of solutions of multidimensional Burgers equation with random data. *Ukrainian Math. J.*, **47**, pp. 330–336.

[209] Leonenko, N.N., Olenko, A.Ya. (1991) Tauberian theorems for correlation function of homogeneous isptropic random fields. *Ukrainian Math. J.*, **43**, 12, pp. 1652–1664.

[210] Leonenko, N.N. and Olenko, A.Ya. (1991) Tauberian and Abelian theorems for correlation function of a homogeneous isotropic random field. *Ukr. Math. J.*, **43**, pp. 1652–1664.

[211] Leonenko, N.N. and Olenko, A.Ya. (1993) Tauberian theorems for correlation function and limit theorems for spherical averages of random fields. *Random Oper. and Stoch. Equ.*, **1**, N1, pp. 57–67.

[212] Leonenko, N.N. and Orsingher, E. (1995) Limit theorems for solutions of Burgers equation with Guassian and non-Gaussian initial data. *Theory Prob. and Applic.*, **40**, pp. 387–403.

[213] Leonenko, N.N., Orsingher, E. and Parkhomenko, V.N. (1995) Scaling limits of solutions of burgers equation with singular non-Gaussian data. *Random Oper. Stoch. Equ.*, **3**, pp. 101–112.

[214] Leonenko, N.N., Orsingher, E. and Parkhomenko, V.N. (1996) On the rate of convergence to the normal law for solutions of the Burgers equation with singular initial data. *J. Stat. Phys.*, **82**, pp. 915–929. (Errata, ibib. **84**, pp. 1389–1390.)

[215] Leonenko, N.N., Orsingher, E. and Rybasov, K.V. (1994) Limit distributions of solutions of multidimentional Burgers equation with random initial data. I,II. *Ukrain. Math. J.*, **46**, pp. 870–877, 1003–1110.

[216] Leonenko, N.N. and Parkhomenko, V.N. (1990a) Central limit theorem for nonlinear transformations of vector Gaussian fields. *Ukr. Math. J.*, **42**, pp. 1057–1063. (Russian).

[217] Leonenko, N.N. and Parkhomenko, V.N. (1990b) Asymptotic normality of a functional of geometric character of Gaussian random field. *Theor. Probab. Math. Statist.*, **41**, pp. 72–82. (Russian).

[218] Leonenko, N.N. and Parkhomenko, V.N. (1991) Comparision theorems for measures of the sojourn of a Gaussian vector field in a 'small' sphere. *Theor. Probab. Math. Statist.*, **45**, pp. 71–76. (Russian).

[219] Leonenko, N.N. and Parkhomenko, V.N. (1992) Limit distributions of spherical sojourn measures for vector Gaussian random field. *Theor. Probab. Math. Statist.*, **47**, pp. 83–90. (Ukrainian).

[220] Leonenko, N.N. Parkhomenko, V.N. and Woyczynski, W.A. (1996) Spectral properties at the scaling limit solutions of the Burgers' equation with singular data. *Random Oper. Stoch. Equ.*, **4**, pp. 229–238.

[221] Leonenko, N.N. Portnova, A.Yu. (1993) On limit distribition of correlogram of stationary Gaussian processes with slowly decay of correlation. *Ukr. Math. J.*, **45**, N12, pp. 1635–1641.

[222] Leonenko, N.N. Portnova, A.Yu. (1994) Convergence of the correlation of a Gaussian field to a non-Gaussian distribution. *Theory Probab. and Math. Statist.*, **49**, pp. 97–102.

[223] Leonenko, N.N. and Rybasov, K.V. (1986) Conditions for the convergence of spherical averages of local functionals of Gaussian fields to a Wiener process. *Theor. Probab. Math. Stat.*, **34**, pp. 85–93. (Russian).

[224] Leonenko, N.N. and Sabirov, Sh.O. (1988) On the normal approximation of functional of strongly correlated random field. *Theory of Random Processes*, **6**, pp. 55–60. (Russian).

[225] Leonenko, N.N. and Sabirov, Sh.O. (1989) Spherical measures of excess of level for a class of random fields. *Kibernetika*, **2**, pp. 108–111. (Russian).

[226] Leonenko, N.N. and Sabirov, Sh.O. (1992) Limit theorems for characteristic of intersection of lower level for Gaussian random fields. *Theor. Probab. Math. Statist.*, **47**, pp. 88–89. (Ukrainian).

[227] Leonenko, N.N. and Sakhno, L.M. (1993) On the limit distribution of the local time of a homogeneous isotropic χ^2 random field. *Theory Probab. and Math. Statist.*, **48**, pp. 111–124. (Ukrainian).

[228] Leonenko, N.N. and Sharapov, M.M. (1998) Reduction theorems and their application in case of statistical estimation of regression coefficients in long-memory time series. *Dopovidi of National Academy of Sciences of Ukraine*, **2**.

[229] Leonenko, N.N., Sikorskii, A.Yu. and Terdik, G. (1998) On spectral and bispectral estimator of the parameter of non-Gaussian data. *Random Oper. Stoch. Equ.*, **6**, N2, pp. 153–176.(Errata, ibib. **7**, N 1)

[230] Leonenko, N.N. and Šilac-Benšic, M. (1996a) Asymptotic properties of the LSE in a regression model with long-memory Gaussian and non-Gaussian stationary errors. *Random Oper. and Stoch. Equ.*, **4**, N1, pp. 17–32.

[231] Leonenko, N.N. and Šilac-Benšic, M. (1996b) On estimation of regression coefficients in the case of long-memory noise. *Theory of Stoch. Processes*, **18**, pp. 3–4.

[232] Leonenko, N.N. and Šilac-Benšic, M. (1997) On the asymptotic distributions of the least square estimations in a regression model with singular errors. *Dopovidi of National Academy of Sciences of Ukraine*, **7**, pp. 26–31.

[233] Leonenko, N.N. and Benšic, M. (1998) On estimation of regression coefficients of long memory random fields observed on the arrays. *Random Oper. Stoch. Equ.*, **5**, N3, pp. 237–252.

[234] Leonenko, N.N. and Woyczynski, W.A. (1998a) Exact probabilic asymptotics for singular n-D Burgers' random fields: Gaussian approximation. *Stoch. Proc. Appl.*, 76, pp. 141–165.

[235] Leonenko, N.N. and Woyczynski, W.A. (1998b) Parameter identification for stochastic Burgeres' flows via parabolic rescaling. (To appear).

[236] Leonenko, N.N. and Woyczynski, W.A. (1998c) Statistical inference for long-memory random fields arising in Burgers' turbulence. In: *Proc. 2nd Scandinavian-Ukrainian Conference in Mathematical Statistics.* (To appear).

[237] Leonenko, N.N. and Woyczynski, W.A. (1998d) Parameter estimation of random fields arising in three-dimensional Burgers' turbulence. In:*Proc. of the Conference: Functional Analysis-V, Dubrovnik.* (To appear).

[238] Leonenko N.N. and Woyczynski, W.A. (1998e) Scaling limits of solutions of the heat equation for singular non-Gaussian data. *J. Stat. Phys.*, *91* ,N 1/2.

[239] Leonenko N.N. and Woyczynski, W.A. (1999) Parameter identification for singular random fields arising in Burgers' turbulence. *J. of Statistical Planning and Inference.* (To appear).

[240] Leonenko N.N., Yadrenko, M.I. and Il'icheva, L.M. (1996) Efficient estimation of regression coefficients of a homogeneous isotropic random field observed on a sphere. *Dopovidi of National Academy of Sciences of Ukraine*, **3**, pp. 21–25.

[241] Lin, S.J. (1995) Stochastic analysis of fractional Brownian motion. *Stochastics and Stochastics Reports*, **55**, pp. 121–140.

[242] Liptser, R. and Shiryaev, A. (1986) *Theory of Martingales.* Nauka, Moscow. (Russian).

[243] Liu, S.–D. and Liu, S.–K. (1992) KdV-Burgers equation modelling of turbulence. *Sci. China*, Ser.**A, 35**, pp. 576–586.

[244] Lo, W.A (1991) Long-term memory in stock market prices. *Econometrica*, **59**, pp. 1279–1313.

[245] Maejima, M. (1981) Some sojourn time problems for strongly dependent Gaussian processes. *Z. Wahrsch. verw. Gebiete*, **57**, pp. 1–14.

[246] Maejima, M. (1982) A limit theorem for sojourn time for strongly dependent Gaussian processes. *Z. Wahrsch. verw. Gebiete*, **60**, pp. 359–380.

[247] Maejima, M. (1985) Sojourns of multi-dimensional Gaussian processes with dependent components. *Yokohama Math. J.*, **33**, pp. 121–130.

[248] Maejima, M. (1986a) Some sojourn time problems for 2-dimensional Gaussian processes. *J. Multiv. Anal.*, **18**, pp. 52–69.

[249] Maejima, M. (1986b) Sojourns of multi-dimensional Gaussian processes. In:*Dependence in Probability and statistics* (E. Eberein and M.S. Taqqu eds.), Birkhaüser, Boston, pp. 165–192.

[250] Major, P. (1981a) Limit theorems for non-linear functionals of Gaussian sequences. *Z. Wahrsch. verw. Gebiete*, **57**, pp. 129–158.

[251] Major, P. (1981b) Multiple Wiener–Itô Integrals. *Lecture Notes in Math.*, **849**. Berlin-New York: Springer-Verlag.

[252] Major, P. (1982) On renormalizing Gaussian fields. *Z. Wahrsch. verw. Gebiete*, **59**, pp. 515–533.

[253] Malyshev, V.A. and Minlos, R.A. (1985) Gibbsian Random Fields. Nauka, Moscow. (Russian).

[254] Mandelbrot, B.B. (1973) Le problème de la réalité des cycles lents et le syndrome de Joseph. *Economie Appliquée*, **26**, 349–365.

[255] Mandelbrot, B.B. (1982) *The Fractal Geometry of Nature.* W.H. Freeman and Co., San Francisco.

[256] Mandelbrot, B.B. and van Ness (1958) Fractional Brownian motions, fractional noises and applications. *SIAM Rev.*, **10**, pp. 422–437.

[257] Mandelbrot, B.B. and Wallis, J.R. (1968) Noah, Joseph and operational hydrology. *Water Resources Research*, **4**, pp. 903–918.

[258] Maruyama, G. (1985) Wiener functionals and ptobability limit theorems I: the central limit theorems. Osaka. *J. Math.*, **22**, pp. 697–732.

[259] McKean, H.M. (1974) Wiener's theory of nonlinear noise. In: *Stoch. Diff. Equ., Proc.* SIAM-AMS. **6**, pp. 191–289.

[260] Merser, T. (1909) Functions of positive and negative type and their connection with the theory of integral equation. *Trans. London Phil. Soc.*, (A), **29**, pp. 415–446.

[261] Michel, R. and Pfanzagl, J. (1971) The accuracy of the normal approximation for minimum contrast estimates. *Z. Wahrsch. verw. Gebiete*, **18**, pp. 73–84.

[262] Miroshin, R.N. (1981) *Intersection of curves by Gaussian processes.* Leningrad State Univ. Press, Leningrad. (Russian).

[263] Molchan, G. (1969) Gaussian processes with spectra which are asymptotically equivalent to a power of λ. *Theory Probab. Applic.*, **14**, pp. 530–532.

[264] Molchanov, S.A. (1997) Wiener-Kolmogorov conception of the stochastic organization of nature. *Proc. of Sumposia in Applied Mathematics, AMS*, **52**, pp. 1–36.

[265] Molchanov, S.A., Surgailis, D. and Woyczynski, W.A, (1995) Hyperbolic asymptotics in Burgers turbulence. *Commun. Math. Phys.*, **168**, pp. 209–226.

[266] Molchanov, S.A., Surgailis, D. and Woyczynski, W.A. (1997) The large-scale structure of the Universe and quasi-Voronoi tesselation of shock fronts in forced Burgers' turbulence in R^d. *Ann. App. Prob.*, **7**, N1, pp. 220–223.

[267] Møller, J. (1994) *Lectures on random Voronoi tesselations.* Springer's lecture notes in statistics. Berlin.

[268] Monin, A.S. and Yaglom, A.M. (1975) *Statistical fluid Mechanics: Mechanics of Turbulence*, I,II. The MIT Press, Cambridge, Massachusetts.

[269] Monin, A.S. and Yaglom, A.M. (1980) *Statistical fluid Mechanics.* The MIT Press, Cambridge, Massachusetts.

[270] Müller, C. (1966) Spherical Harmonics. *Lecture Notes in Statist.*, **17**, Springer-Verlag, Berlin.

[271] Mumford, D. (1983) Tata Lectures on Theta-Functions. *Progress in Mathematics*, vol. **28, 43**, Birkhaüser, New York.

[272] Muraoka, H. (1992) A fractional Brownian motion and the canonical representations of its coefficients processes. *Soochow J. Math.*, **18**, pp. 361–377.

[273] Noble, J.M. (1997) Evolution equation with Gaussian potential. *Nonlinear Analysis Theory Methods and Applications*, **28**, N1, pp. 103–135.

[274] Norros, I., Valkeila, E. and Virtamo, J. (1997) *An elementary approach to a Girsanov formula and other analytical results on fractional Brownian motion.* Preprint, Univ. of Helsinki.

[275] Nosko, V.P. (1970) On bursts of Gaussian random fields. *Vestnik MGU, Ser. Mat., Mech.*, **4**, pp. 18–22. (Russian).

[276] Nosko, V.P. (1982a) On horizon of a random field of cohes on a plane. The mean number of the horizon corners. *Theor. Probab. Appl.*, **24**, pp. 259–265. (Russian).

[277] Nosko, V.P. (1982b) Weak convergence of the horizon of a random field of cones in an expanding strip. *Theor. Probab. Appl.*, **24**, pp. 693–706. (Russian).

[278] Nosko, V.P. (1984) Asymptotic distribution of qurvature of gomogeneous Gaussian field in a points of high local maximums. *Theor. Probab. Appl.*, **29**, pp. 782–787. (Russian).

[279] Nosko, V.P. (1986) On distribution of length of arc of outlier on the high level of stationary Gaussian process. *Theor. Probab. Appl.*, **31**, pp. 592–594.

[280] Nualart, D., Mayer-Wolf, E. and Pérez-Abreu V. (1992) Large deviations for multiple Wiener–Itô processes. In: *Sem. Probab. XXVI, Lecture Notes in Mathematics*, **1526**, Springer-Verlag, New-York, pp. 11–31.

[281] Nualart, D., Üstünel, A.S. and Zakai, M. (1988) On the moment of a multiple Wiener-Itô integral and the space induced by the polynomial of the integral. *Stochastics*, **25**, pp. 232–340.

[282] Nualart, D. and Zakai, M. (1989) Generalized Brownian functionals and the solution to a stochastic partial differential equation. *J. Func. Anal.*, **84**, pp. 279–296.

[283] Obukhov, A.M. (1947) Statistical homogeneous fields on sphere. *Uspekhi Matemat. Nauk.*, **2**, pp. 196–198 (Russian).

[284] Ogura, H. (1990) Representation of the random fields on a sphere. *Mem. Fac. Eng.*, Kyoto Univ., **52**, pp. 81–105.

[285] Olenko, A.Ya. (1991) Some problems of the correlation and spectral theory of random fields. *Candidate of Sciences Thesis*, Kiev University, Kiev, pp. 1–96. (Russian).

[286] Olenko, A.Ya. (1996) Tauberian and Abelian theorems for random fields with strong dependence. *Ukr. Math. J.*, **48**, pp. 368–382. (Ukrainian).

[287] Orsingher, E. (1983) Some results on geometry of Gaussian random fields. *Rev. Roumanie Math. Pures and Appl.*, **28**, pp. 493–511.

[288] Orsingher, E. (1985) Evaluting the area of Gaussian isotropic fields on the sphere. *Rev. Roumanie Math. Pures and Appl.*, **30**, pp. 111–129.

[289] Pearson, K. (1902) On the mathematical theory of errors of judgement, with special reference to the personal equation. *Philos. Trans. Roy. Soc.*, Ser. **A**, **198**, pp. 235–299.

[290] Pérez-Abreu V. and Tudor, C. (1994) Large deviations for a class of chaos expansions. *J. Theor. Probab.* (To appear).

[291] Petrov, V.V. (1995) *Limit Theorems of Probability Theory. Sequences of Independent Random Variables.* Claredon Press, Oxford.

[292] Pinsky, M.A.(1993) Fourier inversion for multidimensional characterictic functions.*J. Theor. Probab,* **6**, 187–193.

[293] Piterbarg, L. and Rozovskii, B. (1996) Maximum likelihood estimators in the equations of physical oceanography. In: *Stochastic Modelling in Physical Oceanography (R. Adler, P.Müller and B.Rozovskii, eds.) Progress in Probability*, vol. **39**, Birkhaüser, Boston.

[294] Pitman, E.J. (1968) On the behaviour of the characteristic function of a probability distribution in the neighbourhood of the origin. *J. Austral. Math. Soc.*, **8**, N3, pp. 423–443.

[295] Plikusas, A. (1981) Some properties of the multiple Itô integral. *Lithuanian Math. J.*, **21**, pp. 184–191. (Russian).

[296] Porter-Hudak, S. (1990) An application of the seasonal fractionally differenced model to the monetary aggregates. *J. Amer. Statist. Assoc.*, **85**, pp. 338–344.

[297] Priestley, M.B. (1981) *Spectral Analysis and Time Series.* Academic Press, New York.

[298] Rao, M.M. (1991) Sampling for harmonizable isotropic random fields. *J. Combinatorics, Information and Sciences*, **16**, pp. 207–220.

[299] Rasulov, N.P. (1976) On asymptotically efficient estimates of regression coefficients under spectral density of noise degeneration. *Theor. Probab. Appl.*, **21**, pp. 316–324.

[300] Ratanov, N.E. (1984) Stabilization of solution of hyperbolic second-order equation. In: *Uspekhi Matemat. Nauk.*, **39**, pp. 151–152 (Russian).

[301] Ratanov, N.E., Shuhov, A.G. and Suhov, Yu.M. (1991) Stabilization of the parabolic equation. *Acta Appl. Math.*, **22**, pp. 411–443.

[302] Rice, J. (1979) On the estimation of the parameters of power spectrum. *J. Multiv. Anal.*, **9**, pp. 378–392.

[303] Robinson, P.M. (1994a) Semiparametric analysis of long-memory time series. *Ann. Statist.*, **22**, pp. 515–539.

[304] Robinson, P.M. (1994b) Rates of convergence and optimal spectral bandwidth for long range dependence. *Probab. Theory Related Fields*, **99**, pp. 443–473.

[305] Robinson, P.M. (1994d) Time series with strong dependence. In: *Advances in ecenometrics: the 6th world congress*, vol. **1**, (C.A. Sims, eds.), Cambridge Univ. Press, Cambridge, pp. 47–95.

[306] Robinson, P.M. (1995a) Log-periodogram regression of time series with long-range dependence. *Ann. Statist.*, **23**, pp. 1048–1072.

[307] Robinson, P.M. (1995b) Gaussian semiparametric estimation of long-range dependence. *Ann. Statist.*, **23**, pp. 1630–1661.

[308] Robinson, P.M. (1997) Large-sample inference for non-parametric regression with dependent errors. *Ann. Statist.*, **25**, pp. 2054–2083.

[309] Robinson, P.M. and Hidalgo, F.J. (1997) Time series regression with long-range dependence. *Ann. Statist.*, **25**, pp. 77–104.

[310] Rogers, L.C.G. (1997) Arbitrage with fractional Brownian motion. *Mathematical Finance*, **7**, pp. 95–105.

[311] Rosenblatt, M. (1961) Independence and dependence. In: *Proc. 4th Sympos. Math. Statist. Probab.*, University of California, pp. 431–443.

[312] Rosenblatt, M. (1968) Remark on the Burgers equation. *J. Math. Phys.*, **9**, pp. 1129–1136.

[313] Rosenblatt, M. (1979) Some limit theorems for partial sums of quadratic form in stationary Gaussian variables. *Z. Wahrsch. verw. Gebiete*, **49**, pp. 125–132.

[314] Rosenblatt, M. (1981) Limit theorems for Fourier transforms of functionals of Gaussian sequences. *Z. Wahrsch. verw. Gebiete*, **55**, pp. 123–132.

[315] Rosenblatt, M. (1985) *Stationary Sequences and Random Fields.* Birkhaüser, Boston.

[316] Rosenblatt, M. (1987a) Remarks on limit theorems for nonlinear functionals of Gaussian sequences. *Probab. Theory and Related Fields*, **75**, pp. 1–10.

[317] Rosenblatt, M. (1987b) Scale renormalization and random solutions of Burgers equation. *J. Appl. Prob.*, **24**, pp. 328–338.

[318] Rossberg, H.J., Jesiak, B. and Siegel, G. (1985) *Analytic Methods of Probability Theory.* Academic Press, Berlin.

[319] Rozanov, Yu.A. (1967) *Stationary Random Processes.* Holden-Day.

[320] Ryan, R (1998a) The statistics of Burgers turbulence initialized with fractional Brownian-noise data. *Comm. Math. Phys.* **191**, pp.71–86.

[321] Ryan, R (1998b) Large-deviation analysis of Burgers turbulence with white-noise initial data. *Comm.Pure Appl.Math.*, **51**, pp.47–75.

[322] Rybasov, K.V. (1987a) Limit theorems for spherical functionals for homogeneous isotropic Gaussian random fields. *Candidate of Sciences Thesis*, Kiev University, Kiev. (Russian).

[323] Rybasov, K.V. (1987b) On asymptotic normality of a functional of homogeneous isotropic Gaussian random fields. *Theor. Probab. Math. Statist.*, **37**, pp. 111–117.

[324] Rykov, E.W., Yu, G. and Sinai, Ya.G. (1996) Generalization variational principles, global weak solutions and behaviour with random initial data for systems of conservation laws arising in adhesion particle dunamics. *Comm. Math. Phys.*, **177**, pp. 349–380.

[325] Saichev, A.I. and Woyczynski, W.A. (1996a) Density fields in Burgers and KdV-Burgers turbulence. *SIAM J. Appl. Math.*, **56**, pp. 1008–1038.

[326] Saichev, A.I. and Woyczynski, W.A. (1996b) Model description of passive tracer density fields in the framework of Burgers' and other related model equation. In: *Nonlinear stochastic PDE's: Burgers turbulence and Hydrodinamic Limits.*, IMA vol.**77**, Springer, pp. 167–192.

[327] Saichev, A.I. and Woyczynski, W.A. (1997a) Evolution of Burgers' turbulence in the presence of external forces. *J. Fluid Mech.*, **331**, pp. 313–343.

[328] Saichev, A.I. and Woyczynski, W.A. (1997b) Distributions in the Physical and Engineering Sciences. *Volume 1: Distributional and Fractal Calculus, Integral Transforms and Wavelets*, Birkhaüser, Boston.

[329] Sakhno, L.M. (1990) Local times of homogeneous isotropic chi-square random fields. *Ukr. Math. J.*, **42**, N10, pp. 1412–1415. (Russian).

[330] Sakhno, L.M. (1991) Local times for a class of random fields. *Theory Probab. and Math. Statist.*, **44**, pp. 118–125. (Russian).

[331] Sakhno, L.M. (1992) Limit theorems for local times of homogeneous isotropic random fields. *Candidate of Sciences Thesis*, Kiev University, 101p. (Russian).

[332] Samarov, A. and Taqqu, M.S. (1988) On the efficiency of the sample mean in long memory noise. *J. Time. Ser. Anal.*, **9**, pp. 191–200.

[333] Samorodnitsky, G and Taqqu, M.S. (1994) *Stable Non- Gaussian Random Processes.* Chapman and Hall, New York.

[334] Sánchez de Naranjo, M.V. (1993) Non-central limit theorems for nonlinear functionals of K Gaussian random fields. *J. Multivar. Anal.*, **44**, pp. 227–255.

[335] Santaló, L.A. (1976) *Integral Geometry and Geometric Probability.* Addison-Wesley Reading, Mass.

[336] Sarmanov, O.V. (1961) Investigation of stationary Markov processes by the method of eigen functions expansion. *Trudy Mat. Inst. Steklov*, **60**, pp. 238–261. (Russian). English translation: *Selected Translations in Math. Statist. Probab.*, **4**, (1963) pp. 245–269.

[337] Schneider, K. and Farge, M. (1997) Wavelet forcing for numerical simulation of two-dimensional turbulence. *C.R. Acad. Sci. Paris*, Série II, t.**325**, 263–270.

[338] Schneider, K., Kavlahan. N.K.-R. and Farge, M. (1997) Comparision of an adaptive wavelet method and nonlinearly filtered pseudo-spectral methods for the two-dimensional Navier–Stokes equations. *Theoretical and Computational Fluid Dunamics*, **9(314)**, pp. 191–206.

[339] Shandarin, S.F. and Zeldovich, Ya.B. (1989) Turbulence, intermittency, structures in a self-gravitating medium: the large scale structure of the Universe. *Rev. Modern. Phys.*, **61**, pp. 185–220.

[340] Shigekava, I. (1980) Derivatives of Wiener functionals and absolute continuity of induced measure. *J. Math. Kyoto Univ.*, **20**, pp. 263–289.

[341] Sinai, Ya.G. (1976) Self-similar probability distributions. *Theory Prob. and Applic.*, **21**, pp. 64–80.

[342] Sinai, Ya. G. (1991) Two results concerning asymptotic behaviour of solutions of the Burgers equation with force. *J. Stat. Phys.*, **64**, pp. 1–12.

[343] Sinai, Ya. G. (1992) Statistics of shocks in solutions of inviscid Burgers equation. *Comm. Math. Phys.*, **148**, pp. 601–621.

[344] Slud, E. (1991) Multiple Wiener–Itô integral expansion for level-crossing-count functionals. *Probab. Theory and Related Fields*, **87**, pp. 349–34.

[345] Slud, E. (1994) Multiple Wiener–Itô representation of the number of curve-crossing by a differentiable Gaussian process, with applications. *Ann. Probab.*, **22**, pp. 1355–1380.

[346] Smith, H.F. (1938) An empirical law describing heterogeneity in the yields of agricultural crops. *J. Agricultural Science*, **28**, pp. 1–23.

[347] Sowell, F.B. (1992) Modelling long-run behaviour with the fractional ARIMA model. *J. of Monetary Economics*, **29**, pp. 277–302.

[348] Stein, M.L. (1995) Fixed-domain asymptotics for spatial periodograms. *JASA*, **90**, 432, pp. 1277–1288.

[349] Stroock, D.W. and Waradhan, S.R.S. (1979) Multidimensional diffusion processes. Springer, Berlin.

[350] Student (1927) Errors of routine analysis. *Biometrica*, **19**, pp. 151–164.

[351] Sun, T.C. (1963) A central limit theorem for non-linear functions of a normal stationary process. *J. Math. Mech.*, **12**, pp. 945–978.

[352] Sun, T.C. (1965) Some further results on central limit theorems for non-linear functions of a normal stationary process. *J. Math. Mech.*, **14**, pp. 71–85.

[353] Surgailis, D. (1984) On multiple Poisson stochastic integrals and associated Markov semigroups. *Probab. and Math. Stat.*, **3**, pp. 217–239.

[354] Surgailis, D. (1985) On the multiple stable integral. *Z. Wahrsch. verw. Gebiete*, **70**, pp. 621–632.

[355] Surgailis, D. (1996) Intermediate asymptotics of statistical solutions of Burgers' equation. In: *Stochastic Modelling in Physical Oceanography* (R. Adler et al., eds.). *Progress in Probability*, vol. **39**, pp. 37–145, Birkhaüser, Boston.

[356] Surgailis, D. (1997) Asymptotics of solutions of Burgers equation with random pieswise constant data. In: *Stochastic Models in Geosystems* (S.A.Molchanov and W.A.Woyczynski, ads.).pp.427-441,Springer, New York.

[357] Surgailis, D. and Woyczynski, W.A. (1993) Long range prediction and scaling limit for statistical solutions of the Burgers' equation. In: *Nonlinear Waves and Weak Turbulence, with Applications in Oceanography and Condensed Matter Physics* (Fitzmaurice, N. et al., eds.), 313–338, Birkhaüser, Boston.

[358] Surgailis, D. and Woyczynski, W.A. (1994a) Burger's equation with non-local shot noise data. *J. Appl. Prob.*, **31A**, pp. 351–362.

[359] Surgailis, D. and Woyczynski, W.A. (1994b) Scaling limits of solutions of the Burgers' equation with singular Gaussian initial data. In: *Chaos Expansions. Multiple Wiener–Itô Integrals and Their Applications* (Perez-Abreu, V. et al., eds.), pp. 145–161, CRC Press.

[360] Swift, R.J. (1994) The structure of harmonizable isotropic random fields. *Stoch. Anal. Appl.*, **12**, pp. 583–616.

[361] Swift, R.J. (1997) Locally time-varying harmonizable spatially isotropic random fields. *Indian J.* pure and *Appl. Math.*, **28(3)**, pp. 295–310.

[362] Tanigushi, M. (1991) Higher Order Asymptotic Theory for Time Series Analysis. *Lecture Notes in Statistics*, **68**, Springer, Berlin.

[363] Taqqu, M.S. (1975) Weak converegence to fractional Brownian motion and to the Rosenblatt process. *Z. Wahrsch. verw. Gebiete*, **31**, pp. 287–302.

[364] Taqqu, M.S. (1977) Law of the iterated logarithm for sums of non-linear functions of Gaussian variables that exhibit a long-range dependence. *Z. Wahrsch. verw. Gebiete*, **40**, pp. 203–238.

[365] Taqqu, M.S. (1978) A representation for self-similar processes. *Stoch. Proc. Appl.*, **17**, pp. 55–64.

[366] Taqqu, M.S. (1979) Converegence of iterated processes of arbitraty Hermite range. *Z. Wahrsch. verw. Gebiete*, **50**, pp. 53–83.

[367] Taqqu, M.S. (1985) A bibliographic guide to self-similar processes and long-range dependence. In: *Dependence in Probability and Statistics* (E. Eberlein and M.S. Taqqu, eds.), Birkhaüser, Boston, pp. 137–165.

[368] Taqqu, M.S. (1986) Sojourn in an elliptical domain. *Stoch. Proc. Appl.*, **21**, pp. 319–326.

[369] Taqqu, M.S. (1988) Self-similar processes. In: *Encyclopedia of Statistical Sciences*, **8**, Wiley, New York, pp. 352–357.

[370] Terdik, G. (1992) Stationary in fourth order and the marginal bispectrum for bilinear models with Gaussian residuals. *Stoch. Proc. Appl.*, **42**, pp. 315–327.

[371] Terdik, G. (1995) On problem of identification for stochastic bilinear systems. *SAMS*, **17**, pp. 85–102.

[372] Terdik, G. and Meaux, L. (1991) The exact bispectra for bilinear realizable processes with Hermite degree 2. *Adv. Appl. Prob.*, **23**, pp. 798–808.

[373] Terdik, G. and Subba Rao T. (1989) On Wiener–Itô representation and the best linear predictors for bilinear time series. *J. Appl. Probab.*, **26**, pp. 274–286.

[374] Terrin, N. and Hurvich, C.M. (1994) An asymototic Wiener-Itô representation for the low frequency ordinates of the periodogram of a long memory time series. *Stoch. Proc. Appl.*, **54**, pp. 297–307.

[375] Terrin, N. and Taqqu, M.S. (1990) A non-central limit theorem for quadratic forms of Gaussian stationary sequences. *J. Theor. Probab.*, **3**, pp. 449–475.

[376] Terrin, N. and Taqqu M.S. (1991) Convergence in distribution of sums of bivariate Appel polynomials with long range dependence. *Probab. Theory and Related Fields*, **90**, pp. 57–81.

[377] Titchmarch, E.C. (1937) *Introduction to the Theory of Fourier Integrals.* Clarendon Press, Oxford.

[378] Truman, A. and Zhao, H.Z. (1996) On stochastic diffusion and stochastic Burgers' equations. *J. Math. Phys.*, **37**, pp. 283–307.

[379] Truman, A. and Zhao, H.Z. (1998) Stochastic Burgers'equation and their semi-classical expansions. *Comm. Math. Phys.*, **194**, pp. 231–248.

[380] Tutsumi, T., Kida, S. (1972) Statistical mechanics of the Burgers model of turbulence. *J. Fluid. Mech.*, **55**, pp. 659–675.

[381] Ubøe, J. and Zhang, T.-S. (1995) A stability property of the stochastic heat equation. *Stoch. Process Appl.*, **60**, pp. 247–260.

[382] Vanmarke, E. (1984) *Random Fields: Analysis and Synthesis.* MIT Press, Cambridge.

[383] Varberg, D.E. (1966) Convergence of quadratic forms of independent random variables. *Ann. Math. Statist.*, **37**, pp. 567–576.

[384] Vergassola, M., Dubrulle, B.D., Frisch, V. and Noullez, A. (1994) Burgers equation, devils staircases and mass distribution for the large-scale structure *Ast. and Astrophys.*, **289**, pp. 325–356.

[385] Verwaat, W. (1987) Properties of general self-similar processes. *Bull. Inst. Internat. Statist.*, **52**, pp. 199–216.

[386] Viano, M.C., Deniau, Cl. and Oppenheim, G. (1994) Continuous-time fractional ARMA process. *Statist. and Probab. Lett.*, **21**, pp. 323–326.

[387] Viano, M.C., Deniau, Cl. and Oppenheim, G. (1995) Long-range dependence and mixing for discrete time fractional processes. *J. Time Ser. Anal.*, **16**, pp. 323–338.

[388] Vilenkin, N.Ya. (1968) *Special Functions and Theory of Group Representation. Amer. Math. Soc.*, Providence.

[389] Vishik, M.J. and Fursikov, A.V. (1988) *Mathematical Problems of Statistical Hydromechanics.* Kluwer, Dordrecht.

[390] Vladimirov, V.S., Drozhzhinov, Yu.N. and Zav'yalov, B.I. (1986) *Multidimensional Tauberian Theorems for Generalized Functions.* Nauka, Moscow. (Russian).

[391] Walker, A. (1964) Asymptotic properties of least-square estimates of parameters of the spectrum of a stationary nondeterministic time series. *J. Austral. Math. Soc.*, **35**, pp. 363–384.

[392] Wehr, J. and J. Xin (1996) White noise perturbation of the viscous chock fronts of the Burgers equation. *Comm. Math. Phys.*, **181**, pp. 183–203.

[393] Weinberg, D.H. and Gunn, T.E. (1990) Large scale structure and the adhesion approximation. *Monthly Not. Royal Astron. Soc.*, **247**, pp. 260–286.

[394] Whittle, P. (1951) *Hypothesis Testing in Time Series Analysis.* Almquist and Wickell, Uppsala.

[395] Whittle, P. (1953) Estimation and information in stationary time series. *Ark. Math.*, **2**, pp. 423–434.

[396] Whittle, P. (1956) On the variation of yield variance with plot size. *Biometrica*, **43**, pp. 337–343.

[397] Whittle, P. (1962) Topographic correlation, power-law covariance functions and diffusion. *Biometrica*, **49**, pp. 304–314.

[398] Wicksell, S.D. (1933) On correlation functions of type III. *Biometrica*, **25**, pp. 121–133.

[399] Widder, D.V. (1975) The Heat Equation. Academic Press, New York.

[400] Willinger, W., Taqqu, M.S., Leland, W.E. and Wilson, D.V. (1994) Self-similarity in high-speed packet traffic: analysis and modellinf of ethernet traffic measurements. *Statist. Sci.*, **10**, pp. 67–85.

[401] Wiener, N. (1941) The homogeneous haos. *Amer. J. Math.*, **60**, pp. 897–936.

[402] Witham, G.B. (1974) *Linear and Nonlinear Waves.* Wiley, New York.

[403] Wong, R. (1986) Errors bounds for asymptotic expansion of Hankel transforms. *SIAM J. Math. Anal.*, **7**, pp. 799–808.

[404] Woyczynski, W.A. (1993) Stochastic Burgers flows, in *Nonlinear Waves and Weak Turbulence* (W. Fitzmaurice et al., eds.), pp. 279–311. Birkhaüser, Boston.

[405] Wschebor, M. (1985) Surface Aléatoires. *Lecture Notes in Math.*, **1147**, Springer, New York.

[406] Yadrenko, M.I. (1959) On isotropic random fields on Markovian type on sphere. *Dopovidi of National Academy of Sciences of Ukraine*, **12**, pp. 1105–1107. (Ukrainian).

[407] Yadrenko, M.I. (1961) Some problems of the theory of random fields. *Candidate of Sciences Thesis*, Math. Inst., Kiev, pp. 1–98. (Russian).

[408] Yadrenko, M.I. (1971) Statistical problems for isotropic random fields. *Proc. VIII Summer Math. School, Math. Instit.*, Kiev, pp. 237–282. (Russian).

[409] Yadrenko, M.I. (1983) *The Spectral Theory of Random Fields.* Optimization Software Inc., New York.

[410] Yaglom, A.M. (1952) An introduction to the theory of stationary random functions. *Uspekhi Matemat. Nauk.*, **7**:5(51), pp. 1–168 (Russian).

[411] Yaglom, A.M. (1957) Some classes of random fields in n-dimensional space related to stationary random processes. *Teor. Veroyatnosty i Primenen.*, **2**, pp. 292–338. (Russian). English transl. in *Theor. Probab. Appl.*, **2**, pp. 273–320.

[412] Yaglom, A.M. (1961) Second-order homogeneous random fields. *Proc. IV Berkeley Symp. on Math. Stat. and Probab.*, vol. **2**.

[413] Yaglom, A.M. (1987) *Correlation Theory of Stationary and Related Random Functions.* I,II. Springer-Series in Statistics, Berlin.

[414] Yajima, Y. (1988) On estimation of a regression model with long-memory stationary errors. *Ann. Stat.*, **16**, N2, pp. 791–807.

[415] Yajima, Y. (1991) Asymptotic properties of the LSE in a regression model with long-memory stationary errors. *Ann. Stat.*, **19**, N1, pp. 158–177.

[416] Yong, C.H. (1974) *Asymptotic behaviour of trigonometric series.* Chinese Univ. Hong Kong.

[417] Zeldovich, Ya.B. (1970) Gravitational instability: an approximate theory for large density perturbation. *Astron. Astrophys.*, bf 5, pp. 84–89.

[418] Zeldovich, Ya.B., Molchanov, S.A., Ruzmaikin, A.A. and Sokolov, D.D. (1988) Intermittency, diffusion and generalization in a nonstationary random medium. *Math. Phys. Rev.*, **7**, pp. 3–110.

[419] Zeldovich, Ya.B. and Novikov, I.D. (1975) *The structure and evolution of the Universe.* Moscow, Nauka. (Russian).

[420] Zorich, V.A. (1984) *Mathematical Analysis*, vol.II, Nauka, Moscow. (Russian).

[421] Zygmund, A. (1977) *Trigonometric Series.* Cambridge Univ. Press.

Index

Other *Mathematics and Its Applications* titles of interest:

P.M. Alberti and A. Uhlmann: *Stochasticity and Partial Order. Doubly Stochastic Maps and Unitary Mixing.* 1982, 128 pp. ISBN 90-277-1350-2

A.V. Skorohod: *Random Linear Operators.* 1983, 216 pp. ISBN 90-277-1669-2

I.M. Stancu-Minasian: *Stochastic Programming with Multiple Objective Functions.* 1985, 352 pp. ISBN 90-277-1714-1

L. Arnold and P. Kotelenez (eds.): *Stochastic Space-Time Models and Limit Theorems.* 1985, 280 pp. ISBN 90-277-2038-X

Y. Ben-Haim: *The Assay of Spatially Random Material.* 1985, 336 pp. ISBN 90-277-2066-5

A. Pazman: *Foundations of Optimum Experimental Design.* 1986, 248 pp. ISBN 90-277-1865-2

P. Kree and C. Soize: *Mathematics of Random Phenomena. Random Vibrations of Mechanical Structures.* 1986, 456 pp. ISBN 90-277-2355-9

Y. Sakamoto, M. Ishiguro and G. Kitagawa: *Akaike Information Criterion Statistics.* 1986, 312 pp. ISBN 90-277-2253-6

G.J. Szekely: *Paradoxes in Probability Theory and Mathematical Statistics.* 1987, 264 pp. ISBN 90-277-1899-7

O.I. Aven, E.G. Coffman (Jr.) and Y.A. Kogan: *Stochastic Analysis of Computer Storage.* 1987, 264 pp. ISBN 90-277-2515-2

N.N. Vakhania, V.I. Tarieladze and S.A. Chobanyan: *Probability Distributions on Banach Spaces.* 1987, 512 pp. ISBN 90-277-2496-2

A.V. Skorohod: *Stochastic Equations for Complex Systems.* 1987, 196 pp. ISBN 90-277-2408-3

S. Albeverio, Ph. Blanchard, M. Hazewinkel and L. Streit (eds.): *Stochastic Processes in Physics and Engineering.* 1988, 430 pp. ISBN 90-277-2659-0

A. Liemant, K. Matthes and A. Wakolbinger: *Equilibrium Distributions of Branching Processes.* 1988, 240 pp. ISBN 90-277-2774-0

G. Adomian: *Nonlinear Stochastic Systems Theory and Applications to Physics.* 1988, 244 pp. ISBN 90-277-2525-X

J. Stoyanov, O. Mirazchiiski, Z. Ignatov and M. Tanushev: *Exercise Manual in Probability Theory.* 1988, 368 pp. ISBN 90-277-2687-6

E.A. Nadaraya: *Nonparametric Estimation of Probability Densities and Regression Curves.* 1988, 224 pp. ISBN 90-277-2757-0

H. Akaike and T. Nakagawa: *Statistical Analysis and Control of Dynamic Systems.* 1998, 224 pp. ISBN 90-277-2786-4

A.V. Ivanov and N.N. Leonenko: *Statistical Analysis of Random Fields.* 1989, 256 pp. ISBN 90-277-2800-3

Other *Mathematics and Its Applications* titles of interest:

V. Paulauskas and A. Rackauskas: *Approximation Theory in the Central Limit Theorem. Exact Results in Banach Spaces.* 1989, 176 pp. ISBN 90-277-2825-9

R.Sh. Liptser and A.N. Shiryayev: *Theory of Martingales.* 1989, 808 pp. ISBN 0-7923-0395-4

S.M. Ermakov, V.V. Nekrutkin and A.S. Sipin: *Random Processes for Classical Equations of Mathematical Physics.* 1989, 304 pp. ISBN 0-7923-0036-X

G. Constantin and I. Istratescu: *Elements of Probabilistic Analysis and Applications.* 1989, 488 pp. ISBN 90-277-2838-0

S. Albeverio, Ph. Blanchard and D. Testard (eds.): *Stochastics, Algebra and Analysis in Classical and Quantum Dynamics.* 1990, 264 pp. ISBN 0-7923-0637-6

Ya.I. Belopolskaya and Yu.L. Dalecky: *Stochastic Equations and Differential Geometry.* 1990, 288 pp. ISBN 90-277-2807-0

A.V. Gheorghe: *Decision Processes in Dynamic Probabilistic Systems.* 1990, 372 pp. ISBN 0-7923-0544-2

V.L. Girko: *Theory of Random Determinants.* 1990, 702 pp. ISBN 0-7923-0233-8

S. Albeverio, PH. Blanchard and L. Streit: *Stochastic Processes and their Applications in Mathematics and Physics.* 1990, 416 pp. ISBN 0-9023-0894-8

B.L. Rozovskii: *Stochastic Evolution Systems. Linear Theory and Applications to Non-linear Filtering.* 1990, 330 pp. ISBN 0-7923-0037-8

A.D. Wentzell: *Limit Theorems on Large Deviations for Markov Stochastic Process.* 1990, 192 pp. ISBN 0-7923-0143-9

K. Sobczyk: *Stochastic Differential Equations. Applications in Physics, Engineering and Mechanics.* 1991, 410 pp. ISBN 0-7923-0339-3

G. Dallaglio, S. Kotz and G. Salinetti: *Distributions with Given Marginals.* 1991, 300 pp. ISBN 0-7923-1156-6

A.V. Skorohod: *Random Processes with Independent Increments.* 1991, 280 pp. ISBN 0-7923-0340-7

L. Saulis and V.A. Statulevicius: *Limit Theorems for Large Deviations.* 1991, 232 pp. ISBN 0-7923-1475-1

A.N. Shiryaev (ed.): *Selected Works of A.N. Kolmogorov, Vol. 2: Probability Theory and Mathematical Statistics.* 1992, 598 pp. ISBN 90-277-2795-X

Yu.I. Neimark and P.S. Landa: *Stochastic and Chaotic Oscillations.* 1992, 502 pp. ISBN 0-7923-1530-8

Y. Sakamoto: *Categorical Data Analysis by AIC.* 1992, 260 pp. ISBN 0-7923-1429-8

Lin Zhengyan and Lu Zhuarong: *Strong Limit Theorems.* 1992, 200 pp. ISBN 0-7923-1798-0

Other *Mathematics and Its Applications* titles of interest:

J. Galambos and I. Katai (eds.): *Probability Theory and Applications.* 1992, 350 pp. ISBN 0-7923-1922-2

N. Bellomo, Z. Brzezniak and L.M. de Socio: *Nonlinear Stochastic Evolution Problems in Applied Sciences.* 1992, 220 pp. ISBN 0-7923-2042-5

A.K. Gupta and T. Varga: *Elliptically Contoured Models in Statistics.* 1993, 328 pp. ISBN 0-7923-2115-4

B.E. Brodsky and B.S. Darkhovsky: *Nonparametric Methods in Change-Point Problems.* 1993, 210 pp. ISBN 0-7923-2122-7

V.G. Voinov and M.S. Nikulin: *Unbiased Estimators and Their Applications. Volume 1: Univariate Case.* 1993, 522 pp. ISBN 0-7923-2382-3

V.S. Koroljuk and Yu.V. Borovskich: *Theory of* U-*Statistics.* 1993, 552 pp. ISBN 0-7923-2608-3

A.P. Godbole and S.G. Papastavridis (eds.): *Runs and Patterns in Probability: Selected Papers.* 1994, 358 pp. ISBN 0-7923-2834-5

Yu. Kutoyants: *Identification of Dynamical Systems with Small Noise.* 1994, 298 pp. ISBN 0-7923-3053-6

M.A. Lifshits: *Gaussian Random Functions.* 1995, 346 pp. ISBN 0-7923-3385-3

M.M. Rao: *Stochastic Processes: General Theory.* 1995, 635 pp. ISBN 0-7923-3725-5

Yu.A. Rozanov: *Probability Theory, Random Processes and Mathematical Statistics.* 1995, 267 pp. ISBN 0-7923-3764-6

L. Zhengyan and L. Chuanrong: *Limit Theory for Mixing Dependent Random Variables.* 1996, 352 pp. ISBN 0-7923-4219-4

A.V. Ivanov: *Asymptotic Theory of Nonlinear Regression.* 1997, 333 pp. ISBN 0-7923-4335-2

E.M.J. Bertin†, I. Cuculescu and R. Theodorescu: *Unimodality of Probability Measures.* 1997, 264 pp. ISBN 0-7923-4318-2

A. Swishchuk: *Random Evolutions and Their Applications.* 1997, 215 pp. ISBN 0-7923-4533-9

V. Kalashnikov: *Geometric Sums: Bounds for Rare Events with Applications.* Risk Analysis, Reliability, Queueing. 1997, 283 pp. ISBN 0-7923-4616-5

V. Buldygin and S. Solntsev: *Asymptotic Behaviour of Linearly Transformed Sums of Random Variables.* 1997, 516 pp. ISBN 0-7923-4632-7

Yu.A. Rozanov: *Random Fields and Stochastic Partial Differential Equations.* 1998, 238 pp. ISBN 0-7923-4984-9

N. Leonenko: *Limit Theorems for Random Fields with Singular Spectrum.* 1999, 404 pp. ISBN 0-7923-5635-7

Other *Mathematics and Its Applications* titles of interest:

V.S. Korolyuk and V.V. Korolyuk: *Stochastic Models of Systems*. 1999, 200 pp.
ISBN 0-7923-5606-3

GPSR Compliance

The European Union's (EU) General Product Safety Regulation (GPSR) is a set of rules that requires consumer products to be safe and our obligations to ensure this.

If you have any concerns about our products, you can contact us on ProductSafety@springernature.com

In case Publisher is established outside the EU, the EU authorized representative is:

Springer Nature Customer Service Center GmbH
Europaplatz 3
69115 Heidelberg, Germany

Batch number: 10584518

Printed by Printforce, the Netherlands